ADVANCES IN DESERT AND ARID LAND TECHNOLOGY AND DEVELOPMENT

VOLUME 1

ADVANCES IN DESERT AND ARID LAND TECHNOLOGY AND DEVELOPMENT

VOLUME 1

Papers presented at the International Conference on the Applications of Science and Technology for Desert Development. The American University in Cairo, Egypt, September 9-15, 1978.

Editors:

ADLI BISHAY
The American University in Cairo

WILLIAM G. McGINNIES
The University of Arizona

1979

harwood academic publishers
chur • london • new york

Published under license by:

harwood academic publishers gmbh
postrasse 22
7000 Chur, Schweiz

Editorial Office for the United Kingdom:
Chansitor House
37/38, Chancery Lane
London WC2A 1EL

Editorial Office for the United States of America:
Post Office Box 786
Cooper Station
New York, New York 10003

Library of Congress Cataloging in Publication Data
International Conference on the Applications of Science
and Technology for Desert Development, Cairo, 1978.
Applications of science and technology for desert
development.

(Advances in desert and arid land technology and
development ; v. 1 ISSN 0142-5889)
1. Desert resources development-Congresses.
2. Desertification-Congresses.
I. Bishay, Adli. II. McGinnies, William Grovenor, 1899-
III. Title. IV. Series.
GB612.157 1978 333.7'3 79-5165
ISBN3-7186-0002-1

Library of Congress catalog card number 79-5165, ISSN: 0142-5889.
ISBN: 3-7186 0002-1. Printed in the United States of America.

TABLE OF CONTENTS

Technology Problems and Desert Communities

Biosaline Research

INTRODUCTION TO THE SERIES

The general purpose of the Advances in Desert and Arid Land Technology series will be to bring an interdisciplinary approach to the problems of desert technology and development. The series will include original work and review articles covering science, technology, engineering, agriculture, architecture, sociology, management and economics of desert and arid land utilization and development. Both invited and unsolicited papers of high merit will be selected for inclusion in the series.

W.G. McGinnies
Tucson

Adli Bishay
Cairo

PREFACE

It is appropriate to the aims of this series that the first volume contains the proceedings of the Cairo International Workshop on the "Application of Science and Technology for Desert Development" held at The American University in Cairo, Egypt, 9-15 September, 1978. The Workshop was sponsored by The American University in Cairo and supported by the U.S. National Science Foundation.

The Workshop, with an international attendance, was oriented particularly toward Egyptian problems, but also included papers relating to other desert regions. The specific goal of the Workshop was to provide guidelines for the establishment of the Desert Development Demonstration and Training Program at The American University in Cairo, but the scope of the papers was much wider and hence it is believed they will have great value for other similar programs and for other areas of desert environment. The topics discussed and the speakers invited provide authoritative coverage of many of the problems needing solution for the improvement of human life in arid environments throughout the world.

The forty-five invited papers covered seven general topics: (1) Egyptian Deserts; (2) Desert Development and Combating Desertification; (3) Water Resources and Irrigation; (4) Desert Plants and Environment; (5) Energy and Mineral Resources; (6) Technology Problems and Desert Communities, and (7) Biosaline Research. The three Plenary lectures by Mustafa K. Tolba, Harold E. Dregne and Mohamed El Kassas dealt with Desertification, Technological Limitations, and Education and Training for Desert Development. The sessions on Biosaline Research were organized with the help of NSF personnel and Professor A. San Pietro.

The success of the Workshop as indicated by the publication of this volume and the initiation of the "Desert Development Demonstration and Training Project" at The American University in Cairo has only been made possible through the cooperation, encouragement and hard work of many individuals and institutions in Egypt and abroad.

President Anwar El Sadat kindly accepted to be the Honorary Chairman of the Workshop and the Minister of Development and New Communities, Eng. H. Kafrawy, was delegated to give the President's inaugural speech. Minister Kafrawy also announced at that time the donation by the Government of Egypt of $50,000 towards the first building of the Desert Project to be constructed on the 500 acres of desert land allocated by the Ministry of Land Reclamation to The American University in Cairo.

The U.S. National Science Foundation not only financially supported the Workshop, but also offered its help and encouragement to secure its success. Mr. Selim Selcuk and Mr. Gilbert Devey are to be especially thanked for their interest and assistance.

The support and encouragement of President Richard Pedersen and senior administrators of The American University in Cairo, and the hard work of our colleagues in the Department of Materials Engineering and Physical Sciences contributed immeasurably to the success of the Workshop.

Special thanks are due to members of the Organizing Committee: Professors Mohamed El Kassas; W.G. Matlock; Rushdi Said; Anthony San Pietro; Abdu A. Shata, Mustafa K. Tolba, Salah Arafa, Fadel Assabghy, Mahmoud Farag, Gabriel Habashy, Farkhonda Hassan, and Hosny Omar. The discussions of the Workshop Consulting Committee resulted in improving the draft recommendations for the Desert Development Demonstration and Training Program which formed an integral and highly effective component of the Workshop activities. This committee included: Eng. El Ahmady Abdel Raouf; Professors Donald H. Baepler; Harold E. Dregne; Saad Eddin Ibrahim; Mohammed El Kassas; Osama El Kholy; William G. McGinnies; Rushdi Said; Gerald W. Thomas, and Mustafa K. Tolba.

Thanks are also due to Aziz Kirollos, who typed a large part of the manuscript of this volume, to Sameh Butros who was responsible for implementing many of the activities of the Workshop and to Nevine Seif, who was a valuable assistant in all stages preceding and following the publication of this volume. The Assistant Editor of the Series, Mary Dungan Megalli, has applied her considerable editorial skills to the preparation of the Workshop papers and the final preparation of this volume.

To my graduate and undergraduate students whose hard work and great enthusiasm contributed effectively to the success of the Workshop, this volume is dedicated.

Adli Bishay
Cairo, May 1979

DESERT DEVELOPMENT DEMONSTRATION AND TRAINING AT THE AMERICAN UNIVERSITY IN CAIRO

As a result of the recommendations of the Cairo International Workshop on "Applications of Science and Technology for Desert Development," organized by The American University in Cairo and held September 9-15, 1978, the "Desert Development Demonstration and Training Project" was officially initiated in January, 1979 by the President of AUC, after the approval of its Board of Trustees.

The concept of this new AUC activity is based on the establishment of a demonstration project in a newly reclaimed desert area, with various approaches and alternatives to development of integrated desert communities. The ultimate goal is to establish different patterns of desert development as alternatives for what could be done in larger scale programs for development of other desert areas by the Egyptian Government or the private sector. The project closely links application of existing knowledge, research and educational aspects and will serve as a training ground for students and visiting scientists from Egypt and other countries.

The demonstration and training project will be primarily concerned with arid lands technologies appropriate for Egypt's climate, physical, human and capital resources, its environmental concerns and its socioeconomic and tradition patterns.

The program emphasis will be on arid lands agriculture where water supply is the limiting factor. Optimum land and water resources management is a major goal. Also, the uses of underground water as a source for partial irrigation, and problems of salinization of such sources will be emphasized.

The program will also be directed at the establishment of small scale agriculturally based industries.

The program will include the production of solar, wind and biomass energy for rural communities and the most appropriate designs of energy efficient low cost housing for hot, arid environments.

Raw materials resources will be studied, particularly to determine the suitability of indigenous materials for construction and other development activities.

Desert communities, based on alternative patterns of development, particularly those concerning agricultural and water use practices, will be established in an integrated rural development context. Social science research will go hand in hand with technical research to identify and insure a productive, viable life style during the development of these communities.

In order to aid the University in implementing this project, the Government of Egypt, through the Ministry of Land Reclamation, has (1) allocated 500 feddans in the "Al Intelaq" area for the establishment of research in desert development, and (2) set aside approximately 7000 feddans from the "Al Intelaq" area for a period of two years during which The American University in Cairo would conduct a feasibility study for establishing an integrated community in this area, with the provision that at the end of this period agreement will be reached between the Ministry of Land Reclamation and The American University in Cairo for the implementation of an integrated, self sufficient desert community based upon alternative agricultural and industrial development activities as recommended by the feasibility study.

While the 500 feddan area will be the center for testing, training and applying research and alternative approaches for desert development, it will also be the center for administering the major project of creating an economically viable and integrated desert community on the 7000 feddan area. In order to encourage immediate implementation of this project, the Ministry of Development and New Communities has granted the University $50,000 (£E 35,000) towards the construction of the project's first building facility.

In all the areas of activity planned for the project, the primary concern of The American University in Cairo and cooperating agencies is to ensure that an interdisciplinary approach is effected in real terms. This means that every activity will be under the immediate supervision, monitoring, study and evaluation of a group of persons representing all disciplines involved in the activity (Physical and Social Sciences, Agriculture, Engineering, Management, Economics, etc.) and will be subject to constant examination and discussion by this group. It is the expressed commitment of The American University in Cairo and the group of local and foreign experts and interested persons involved in this project to see that the interdisciplinary group works efficiently and cooperatively in connection with each activity and with a view of the overall program.

In the field of Agriculture The American University in Cairo, which does not have a faculty of Agriculture, is establishing cooperation with national universities and research institutions as well as international institutions, with the goal of providing research, training and demonstration of arid lands agriculture in an interdisciplinary program not presently available in any local institution.

Adli Bishay
Director

WHAT COULD BE DONE TO COMBAT DESERTIFICATION

MOSTAFA KAMAL TOLBA
Executive Director, United Nations Environment Program

ABSTRACT

The United Nations Conference on Desertification (Nairobi 29 August-9 September 1977) put on the agenda of world concerns the problem of desertification and adopted a Plan of Action to Combat Desertification. The immediate goal is to arrest, and, where possible, to reclaim desertified lands. Anti-desertification measures are primarily national actions and should be conceived as integral components of national plans for development.

A national plan of action to combat desertification should encompass the following principal activities: Monitoring and assessment, land-use policy and planning, environmentally sound management of land and water, enacting appropriate legislation, application of existing scientific and technological knowledge, training of scientific and technical manpower, socio-cultural studies as bases for mobilization of popular participation, appropriation of investment resources.

Actions for combatting desertification and reclaiming desertified land are closely linked with land use activities and development of resources in desert land and in semi-arid and sub-humid areas. Within this spectrum of habitat types, ecological conditions represent fragile balances that are tenuously maintained under natural conditions and can be easily upset under irrational land-use systems. This means that plans for development should be based on: Maintaining optimal sustainable productivity and not short-term, destructive productivity; application of ecologically sound technologies that will not upset the fragile ecological balance; integrated actions that would resolve conflicts among competing users; development of non-agricultural land uses; development of ecologically and socio-culturally sound human abodes; development of innovative modalities for administration and social services.

Ecologically sound technologies need to cope with various requirements, including: Management of water resources (development, conveyance, conservation, use), management of range resources, rainfed agriculture (dryland farming), irrigated agriculture (river basins, oases), units of human settlements, means of communication (among scattered societies), provision of educational, medical and other services.

National machineries need to be evolved for: Development of indigenous scientific capabilities and processes for application of science and technology and adaptation of technologies to fit prevalent ecological conditions, education and training, extension for meaningful public participation.

International cooperation in combatting desertification offers to individual nations: Exchange of information, consultative groups (donor system), additional resources for financing plans of action (additionality and automaticity), special account in the United Nations, and examples of what can be done at national and regional levels.

INTRODUCTION

It gives me special pleasure to address this gathering of distinguished scientists who came together to discuss an important and indeed urgent subject: desert development. With a long history of experience behind each of you, you came to assemble here today to discuss the crucial issue of the application of science and technology to the development of desert resources. I would like however, at the outset, to stress that we have to differentiate between two distinct though closely related activities: that of combatting desertification and that of developing resources of desert lands. I intend to concentrate more on the first aspect.

As you know, the United Nations Conference on Desertification which concluded its work in Nairobi (Kenya) a year ago today, and the two-year preparatory process that preceded it, put the problem of desertification on the agenda of world concern. The world conference conceived desertification as a process of ecological degradation in arid, semi-arid and sub-humid lands by which the productivity of land is lost or substantially diminished: grazing lands cease to produce palatable pasture, dryland agriculture fails, irrigated fields are abandoned due to salinization, waterlogging and other forms of soil deterioration.

It is true that the terrestrial ecosystems in semi-arid and sub-humid regions maintain a tenuous balance that is often upset by spells of drought, raids of locust swarms and other natural vicissitudes. But under natural conditions, and through rational and ecologically sound management based on the proper application of scientific knowledge and appropriate technology, dryland ecosystems regain their balanced equilibrium, and their inherent resilience eventually undo disturbances caused by natural factors. Under excessive pressures of land use such as overgrazing, farming marginal lands, excessive irrigation and impeded drainage, damage is caused beyond the resilience of the ecosystem and the result is desertification.

There is no convincing evidence that desertification, and in particular its acceleration in recent decades, is caused by large-scale climatic change. We have records of rainfall at Alexandria since 1847 and they show the year-to-year fluctuations that are typical of arid land rainfall but do not indicate a trend towards climatic dessication. Even for earlier times we note that Claudius Ptolemaeus kept weather records of Alexandria for a few years during the second century and the number of rainy days per year was much the same as nowadays.

Desertification is, therefore, considered as essentially due to interactions between social pressures and fragile ecosystems. In other words, it is essentially due to actions of man wrestling to secure a living under ecological conditions with limited potential for natural productivity. Lacking other alternatives, he is often forced to resort to land-use practices that are destructive.

The urgency and magnitude of the problem of desertification may be illustrated by the following estimates:

- On the southern fringes of the African Sahara 650,000 square kilometres (65 million hectares) of once productive land have become desert during the last 50 years;
- Throughout the world, about 60,000 square kilometres (6 million hectares) of productive and fertile lands are lost annually;
- Areas of greater size are partly damaged as their productivity is reduced through soil erosion, waterlogging and other forms of soil degradation;
- Some 600-700 million people (14% of the world's population) live in threatened drylands, of these about 60 million are immediately affected by desertification.

Though its direct influence affects only one third of the land surface and one seventh of the world population, desertification has its indirect influence on a global scale. This influence is manifested in:

- Reduction of the food-producing capacity of the world, causing depletion of world food reserves;
- Dust storms that bring to the global air-mass circulation added loads of particulates with yet to be assessed impact on global climate;
- Loss of genetic resources of plants and animals that are parts of the global heritage;
- Unplanned mass migrations with socio-political repercussions on adjoining areas of the same country or neighbouring countries.

Egypt

One question that comes to mind: is desertification a problem that Egypt has encountered in the past or encounters at present? Let me before answering this question paraphrase the definition of desertification broadly as the retrogressive change of economically productive land into non-productive land. In answering the question I shall dwell briefly on three examples:

1. The first example is the Mareotis land that extends west of Alexandria to the Libyan border. Strabo (66-24 B.C.) describes the country west of Alexandria as studded with towns and cities, and he enumerates 13 major towns including Ammonia, the seat of the Government of the Prefect. Socrates (4th Century A.D.) writes: "Mareotis is a district of Alexandria containing very many villages and an abundant population with numerous splendid churches ...". The first record of the decline of Mareotis is in a manuscript dated A.D. 950 by an anonymous Arab author (National Library, Paris), "Leaving Tarrana and following the road to Barka one comes to Menas which consists of three abandoned towns situated in the midst of a sandy desert. The buildings are still standing. There are many constructed palaces surrounded by walls; some of these buildings are occupied by monks ...". Maqrizi (1365-1441 A.D.) wrote: Mareotis was a district dependent on Alexandria. At one time houses and gardens were dotted over the whole

country westward up to the very frontiers of Barka." Weedon (The Cairo Sci. Journal, 1912) summarizes this 2000 year history when he says "One of the most striking anomalies of modern Egypt consists in the fact that the whole district of Mariout, which in ancient times was famous for its fertility should at the present day be for the most part barren and waste." The area was kept productive till the 11th century through careful methods of soil and water conservation and ecologically sound systems of land use.

In recent history, attention was given to Mareotis early in this century. The Egyptian Ministry of Agriculture carried out its first field experiments with barley in 1917-1919. A horticultural experimental station was established in Borg el Arab (45 Km west of Alexandria). This was instrumental in the widespread cultivation of olives and figs and in establishing the early steps towards regeneration. Since 1954 there have been large-scale programs for reclaiming these once productive lands.

2. The second example is the Western Desert of Egypt, often referred to as the *Sea of Sands*, because of the presence of extensive bodies of sand including long chains of sand dunes. Some of these chains extend in a north-south direction for several hundreds of kilometers. The moving dunes sometimes overwhelm cultivated and inhabited oases, representing the simple form of desertification. Drifts of these sands move southeastwardly towards the Nile Valley, veneering the fringes of the farmlands with sand deposits. But the annual river flood used to cover these lands with silt-laden water and eventually deposited layers of alluvial silt over the veneer of sand. The mixture of silt and sand added to the fertile soil. Since the Nile floods have been controlled, the sandy intrusions over the western fringes of the Nile Valley appear to have increased. Detailed analysis of aerial photographs and satellite imagery over the past several years reveal the extent of this increase. Similar hazards of sand encroachment may affect the newly reclaimed lands in the Tahrir Province.

It is this aspect of sand encroachment over productive land that gave rise to the oversimplified notion of desert creep. We must not forget that this is only one, and may be much less important than other causative agents of desertification.

3. The third, and probably the most serious example, is desertification through waterlogging and salinization. This is an ecological hazard faced by almost all large-scale projects of irrigated lands all over the world, and Egypt is no exception. This was dramatized here in Egypt by the experience of land reclamation of the West Nubariya schemes. The calcereous silty soil responded to irrigation and the early vigor of biomass productivity seemed most promising. But it did not take many years to notice that the ground water-table was rising at an alarming rate of 2-3 metres annually. It eventually reached critical levels and the consequences of waterlogging became evident. Extensive studies, supported by national and international resources, were initiated and showed that irrigation and drainage systems are inseparably entwined, and that failure of the drainage system is another cause of desertification. West Nubariya is not the only example. Several of the case studies presented to the United Nations Conference on Desertification (Iraq, Pakistan, USSR) and several studies presented to the International Conference on Environmental Consequences of Irrigation of Arid Lands that was held in Alexandria in February 1977, tell the same tale.

The Sahel Drought

I would like to turn now to world concern over desertification. As you know, a spell of years with below-average rainfall (1968-1973) befell the Sahel region of Africa. It eventually extended eastward to Ethiopia and Somalia. During this episode of drought, several processes and events became manifest. Lake Chad shrunk to one third of its usual size. Failure of the Niger and Senegal rivers deprived much of the irrigated croplands in Niger, Mali, Upper Volta, Senegal and and Mauritania of needed water. Annual pastures in the northern territories were lost. Death of shrubs and trees was widespread. Famished and depauperate livestock were driven southward, leaving behind land stripped of its plant cover, soil scorched under burning sun and dry air, and patches of newly formed desert. Further south droughtstricken farmers were harassed by repeated crop failure. By 1973 the agricultural base of Mauritania, Upper Volta, Mali, Niger and Chad (already amongst the least developed countries of the world) collapsed, and that of Senegal and Gambia was severely damaged. It was estimated that between 100,000 and 250,000 people died. Two million nomadic pastoralists lost about 50% of their animals, and in certain instances losses exceeded 90%. The harvests of 1968-72 were less than half of the usual crop for some 15 million farmers, and in worse situations there was no crop to harvest. This breakdown of the life-supporting systems was followed by mass exodus towards towns and less affected areas further south not prepared to receive such an influx and the resultant socio-economic disruptions.

Similar catastrophic droughts have occurred in the past, and will recur in the future. They are an inherent feature of rainfall in semi-arid and sub-humid territories and a principal cause of the fragility of their terrestrial ecosystems. The Sahel drought dramatized for the world the hazards of climatic aberrations that have been recurrent all through history. The world's first response was an international relief effort that mitigated the impact of what could otherwise be a holocaust of ecological events. Its second response was the United Nations General Assembly resolution in December 1974 (Resolution 29/3337):

- To initiate concerted international action to combat the spread of desert conditions, and
- To convene a United Nations Conference on Desertification which would produce an effective, comprehensive and coordinated *Plan of Action to Combat Desertification*.

The Plan of Action, which was produced by the United Nations Conference on Desertification and endorsed by the United Nations General Assembly, had as its immediate goal the arrest of desertification and, where possible, reclamation of desertified land. But as action to combat desertification is implemented primarily at a national level and should be conceived as an integral component of national plans for development, the Plan of Action underlines the importance of establishing and developing national capabilities for combat-desertification.

The principal pivot of the Plan of Action is the application of existing knowledge in science and technology in the development of ecologically and socially appropriate programs aiming at halting further desertification, reclaiming desertified lands, and proper utilization and development of resources of areas prone to desertification as a means of combating desertification.

NATIONAL PLAN OF ACTION

The Plan of Action to Combat Desertification comprises recommendations for action at national, regional and international levels. Countries facing hazards of desertification are expected to make use of these recommendations in formulating their own national plans for the control of desertification menace. Egypt is an example of such countries. A comprehensive national plan to combat desertification should comprise two basic elements: (1) monitoring and assessment, and (2) land-use policy and planning.

Monitoring and Assessment

Monitoring and assessment activities aim at providing basic information needed for formulating national land-use policies and for drawing plans for the three components I mentioned earlier: combating desertification, reclaiming desertified lands, and developing resources of areas prone to desertification. Such information can be collected through monitoring and surveying of atmosphere, hydrological elements, soil conditions, plant cover, animal life, human conditions, land use etc. Three points need to be stressed in this respect.

(a) Individual components of the various monitoring activities are incapable of leading to assessment of the magnitudes of desertification problems or of the prospects and potentials of the available land. Such assessment requires the integration of information collected about all individual components. Recent development in methodologies of mathematical modeling and systems analysis makes it possible to assess ecological situations objectively and comprehensively.

(b) Monitoring and survey activities require integration of different methodologies and technologies: field observation, airborne surveys, satellite imagery, and so on. Archeological surveys may also unearth some ancient experiences and practices in using and developing resources of land in the areas under study.

(c) Surveys of natural resources often extend beyond desertified and desertification-prone areas, and cover areas of extremely arid deserts where underground water, oil or mineral resources may be present.

Land-Use Policy and Planning

The resource data so obtained provide, together with social and economic objectives, sound bases for national land-use policy that provides guidance to land-use planning. The identification of policy options is the concern of the techno-scientific machineries of government, whereas the formulation of policies is a political process vested with the political authority. This suggests the need for feedback mechanisms within the national structures that allow for information-decision flow and exchange.

A national land-use plan may take the form of thematic map or maps of land-use and its future prospects. Land-use plans covering territories that are prone to desertification, desertifided, or natural deserts, should aim at: (i) halting further degradation of productive land; (ii) reclaiming desertified areas; (iii) maintaining optimal sustained productivity of land; (iv) resolving conflicts among competing users; (v) developing non-agricultural land uses (oil, mineral, recreation, urban settlement, etc.); (vi) providing for basic human needs including environmentally sound and socio-culturally acceptable habitats for local population. At local (state, province, district) levels, authorities should formulate respective land-use plans within the framework of a national plan.

REQUIREMENTS FOR IMPLEMENTATION

The implementation of national plans of action requires a number of national activities that provide the prerequisites for success. Four principal activities may be underlined.

(a) *Training and Education.* The prevalent disciplinary structures of educational systems produce discipline-wise trained personnel. But the inter-disciplinary scope of environmental problems, desertification being a prominent one, requires training of scientists, planners and managers in a way that mixes a wide range of educational processes combining disciplinary and multidisciplinary aspects.

One of the recommendations of the *Plan of Action to Combat Desertification* underlines the need for a broad base for training land-use planners: "Though planners' primary expertise should be in the theory and technique of assessment, and planning and management of human and natural resources, they should have access to expertise in law, economics, demography geology, soil sciences, agricultural sciences, hydrology, hydrogeology, climatogology, remote sensing, statistics and cartography." This may be a long list, but the problems faced extend over a wide front of interactive issues.

A second objective of education and training aims at ensuring public participation and meaningful and effective contributions to programs of combating desertification. This is a two-prong objective: (i) vocational training and extension activities that facilitate the application of science, transfer of technology and introduction of innovative means of management; (ii) sensitizing societies concerned to accept new concepts that are congenial to rational management and discard those that are incompatible with it.

(b) *Legislation*. Implementation of national land-use plans also requires surveys of existing legislative and administrative structures concerned with water resources and water rights, systems of land tenure, limits of land use patterns, environmental impact assessments, and protection of flora and fauna.

As gaps in existing legislation are recognized, new legislation may be introduced or existing laws reformed or developed. Such legislation must be socially acceptable, and requires effective machinery for implementation and for ensuring observance.

(c) *Socio-Economic Aspects*. The national plan of action to combat desertification should be conceived as an integral part of the national plan for development. Both are concerned with the improvement of the quality of life of the people, and their implementation depends on the enthusiastic participation of the people. Public participation and contribution should be ensured at two stages: (i) Planning processes, at a time when such participation can constructively influence choice among alternatives; and (ii) Implementation of programs. Programs for preventing ecological degradations that lead to desertification and for sustaining productivity of agro-ecosystems in reclaimed and developed dryland should be supported by economic and demographic policies which address maintenance of adequate rural labour force, sedentarization of nomads where appropriate, resettlement of migrants, provision for alternative opportunities for employment, monitoring human conditions with special attention to health and habitat conditions, and insurance mechanisms that are compatible with socio-economic needs.

(d) *Application of Available Scientific and Technological Knowledge.* Almost every recommendation of the UN Plan of Action to Combat Desertification refers to the need for developing scientific capabilities and stresses the need for applying available scientific knowledge in development and management of resources in desertified areas and areas prone to desertification.

Scientific capabilities are needed for a variety of activities including: monitoring and assessment, land use policy and planning, management of resources development. Such capabilities should cover natural sciences, social sciences and application of science and technology.

Scientific machineries in most of the countries affected by desertification are government departments or institutions. The sectorial structure of government bodies does not lend itself easily to the required integration of sciences, hence the need for coordinating mechanisms.

Technology transfer, whether it is the application of scientific findings of national institutes of research or the adaptation of imported technology, requires mechanisms for pilot studies. Though there is need to make optimum use of indigenous experiences and knowledge, there is equal need to ensure social acceptability of appropriate technologies that are new.

THE INTEGRATED APPROACH

Actions for combating desertification should be guided by concepts that by necessity differ from those conceived for less fragile ecosystems. These concepts comprise the following:

(a) *Optimum Sustainable Yield.* Maintenance of optimum sustainable level of land productivity cannot be achieved unless based upon long-term use. This is in contrast with the concept of maximum productivity which is destructive when it surpasses the carrying capacity of the ecosystem. Overuse almost always causes desertification. But to resist the temptation of maximizing yield and revenue and to accept a sustainable lower level of productivity not only requires changes in the economic bases of land-use but may also require changes in land tenure systems.

(b) *Application of Existing Scientific Knowledge.* There is ample evidence to show that failures to maintain balanced livelihood systems in dry lands are associated with failures to apply available scientific knowledge. Accordingly, priority should be given to removal of obstacles that hinder the application of existing knowledge and to the adaptation of this knowledge to local physical and social situations. UNESCO has carried out a preliminary study and the United Nations University is now carrying out a world-wide survey on such obstacles.

(c) *Ecologically Sound Land Use.* According to ecological concepts, land-use is a human activity manipulating complexes of interdependent elements (plants, animals, soil, water, energy) and processes (material cycling and energy flow) that form ecosystems. Changes in any one element or process have repercussions on all others. Ecologically sound land-use practices manipulate the ecosystem processes and do not destroy their interactions, and conform with climatic patterns; i.e., seasonal extremes and non-predictable year-to-year variations.

Ecologically sound land use in pastural systems means grazing practices that allow vegetation to recuperate. In rainfed farming it means land zoning in which farming should not extend beyond certain climatic and soil limits and should always combine means for conservation of soil and prevention of its erosion. Ecologically sound design of large-scale agricultural projects in irrigated systems means taking fully into account the structure, including hydraulic properties, of the soil at surface and subsurface levels, characteristics of irrigation water, efficiency of the irrigation network, and efficiency of the drainage network. Maintenance of both networks is of special importance in preventing the main cause of desertification in these systems, i.e. waterlogging leading also to building up of destructive salinity and alkalinity.

(d) *Integration Among Users.* Within the framework of an ecologically sound national plan for land use, various forms and types of land use could be integrated and stratified on bases of scientific assessment of ecosystem capacity and resilience. Competing claims on land in dryland areas may include:

(i) Dryland agriculture which comprises pastoralism, rainfed cropping and irrigated agriculture. There are conflicting interests among users within this group. The accelerated desertification of recent decades is in part attributed to the breakdown of traditional practices of harmonized relationships between users. This breakdown has been caused by pressures of population growth, aspiration towards higher incomes, haphazard introduction of technological innovations, and incorporation of dryland agriculture into the commercial marketing system.

(ii) Arid lands hold a considerable proportion of the world mineral and oil resources. Revenues from these resources constitute substantial parts of the national economy of many dry land countries.

(iii) It is estimated that 20-30% of the populations of dry lands live in urban centres. Historically, dryland cities have developed as commercial, and administrative centers, as ports, mining centers and so on. Growth of dryland cities in the developed world (e.g. Southewest USA) is related, apart from the development of mining, to the amenities of sun and mild winters that attract new residents and tourists.

Within the framework of national land use planning, all these competing claims may find ample space in drylands. The national reports presented by Iraq, Algeria and Iran to the U.N. Conference on Desertification, and the recent Egyptian policy of establishing new urban centers in the eastern and western deserts, offer concrete examples in this respect.

(e) *Ecologically Sound Human Settlements*. Arid climates (dry and often hot) impose certain stresses that should be reduced by sound design of housing schemes. Plans for establishing new towns in deserts should include environmental impact assessments covering estimates of future demand for, and available resources of, water and energy, requirements for waste disposal, layout and orientation of housing units, architectural design of houses, and perimeter reserves and shelterbelts.

(f) *Modalities of Social Services*. Populations in desertification-prone areas will remain sparse, especially those involved in various forms of land use other than irrigated agriculture. But these populations need to have access to medical, educational, recreational and other services that are usually accessible in more densely populated areas.

The Australian experience in extensive use of advanced technological means for providing schooling via radio and television, medical care via effective communication systems and air-borne doctors, etc. could be adapted to provide for conditions prevalent in other territories of arid lands.

Means for relief services during years of drought should be conceived as part of a system for insurance against drought hazards, and should include stocks of reserve materials and a machinery that can bring relief materials to populations in stress.

The Syrian national report to the United Nations Conference on Desertification describes an elaborate and extensive network of pastoralist cooperatives that have managed to promote ecologically sound systems of management of dryland pastures and to establish considerable storage facilities for reserve feed materials. This system, receiving support from FAO and IBRD, is being introduced in Jordan.

NATIONAL MACHINERIES

It is evident from all these considerations that activities to combat desertification are varied and transcend concerns of various departments of government. For this reason, national machinery for combating desertification is best conceived as comprising a number of components, each with appropriate structure as befits its role, but all coordinated to function within the framework of the national plan of action.

The world conference felt the need for a national body to act as *policy and planning authority*. It recommended that this body should be at a high level in the government structure. It did not feel that this "authority" need be concerned solely with desertification problems but may be a national authority for land-use policy and planning. Such an authority could carry the responsibility of approving national land use policies, and a master plan of land use.

Development of policies and plans for strengthening indigenous capabilities for combating desertification in the fields of scientific research, application of available scientific and technological knowledge and adaptation of technologies to fit prevalent ecological conditions should be one of the principal objectives of such a national authority.

There may be need for a technical group to provide substantive support to the authority. This group could be a *national commission* for desertification control or desert development. This commission could include high-ranking representatives of appropriate government departments and agencies and community leaders. It would ensure coordination of implementation of the national plan, carry out data processing and synthesis andprovide the much needed channels of communication and feedback links between national institutions for research and data collection on the one hand and the policy-making authority on the other.

INTERNATIONAL COOPERATION

So much, then, for national efforts. Desertification, though manifest most forcibly at the local and national levels, often extends beyond political boundaries and its combat requires regional cooperation. The United Nations Conference on Desertification approved a number of recommendations in the Plan of Action calling for the world-wide cooperation especially in fields of exchange of information and research programs. It also reviewed various forms of regional cooperative action. Six studies on the feasibility of regional projects related to combating desertification were presented to the Conference.

FINANCIAL RESOURCES

Though action to combat desertification will mainly depend on national resources, there is yet an urgent and pressing need for international financial support if these national efforts are to bear fruit within a reasonable period of time. The Conference and the United Nations General Assembly identified three possibilities for mobilizing international financial support.

(a) Consultative Group. The General Assembly authorized the Executive Director of the United Nations Environment Program to convene a Consultative group to assist in mobilizing financial resources for implemention of the Plan of Action. The group comprises representatives of the United Nations organizations, donor countries, multi-lateral financial agencies, and developing countries having substantial interest in combating desertification.

At its first session last May the Group adopted a policy statement describing its functions and working modalities and considered the six regional projects that were earlier presented to the United Nations Conference on Desertification. The Group proposed specific refinements of the projects and suggested that the refined versions be transmitted to members for consideration of financial support.

It is hoped that the Group will function as a forum for mobilizing resources from donor countries and financial institutions and for directing these resources for national and international programs for combating desertification. The United Nations Environment Program has established a Desertification Unit to serve the Group and to liaise between the Group and countries requiring support for their national programs to combat desertification.

(b) Additional Resources. The U.N. General Assembly noted the fact that the annual cost of preventive measures that would halt continued loss of world productive land through various forms of desertification is estimated at $400 million. This does not include national and regional costs of infrastructure and administrative and support machineries. These are the funds required for achieving and maintaining zero desert growth.

It also noted that to turn the tide of deterioration and to reclaim desertified lands would require considerably greater expenditure. A program for recovering lands lost during the last 25 years would cost $20-25 billions.

The Community of Nations at the U.N. Conference on Desertification and the General Assembly felt that the implementation of the Plan of Action at a scale commensurate with the magnitude and urgency of desertification hazards, will require substantial resources not at present available for this purpose. These additional resources should have a reliable flow over a considerable period of time (20-25 years). Reliability of flow suggests automaticity. In other words, a world-wide program to combat desertification was clearly felt to require international resources that are *additional* to the present or prospective resources for development assistance and that have an *automatic* flow.

At the request of the General Assembly, a group of highly qualified specialists in international finance was invited by the United Nations Environment Program to prepare a study on the subject of additionality and automaticity of resources for combating desertification. Their report will be discussed by the General Assembly at its session which will begin ten days from now.

The study enumerates a number of sources for financing international programs of development including combat of desertification and development of dry land resources. These likely sources include: (a) Increased Official Development Assistance (ODA). This was considered by the United Nations General Assembly to reach 0.7% of the GNP of developed (donor) countries during the present decade. By 1975 it fell down to 0.33%. It is stressed that OAD resources should be increased to meet the target. (b) Loans with concessional terms to finance anti-desertification projects. (c) A share of revenues from exploitation of international commons (ocean sources), taxes on polluters of marine environment, rents for telecommunications wavelengths and satellite parking space, international taxation on trade flows, and taxes on military expenditure and arms transfer.

(c) United Nations Special Account. In addition, the General Assembly endorsed in principle the creation of a special account within the United Nations for implementing the Plan of Action to Combat Desertification. The establishment and operation of this account will also be considered by the General Assembly at its forthcoming session.

In conclusion, I wish to say that loss of productive land by desertification is an especially serious example of environmental degradation that faces our world. It is made particularly acute by the growing need for more food resources to meet the basic requirements of the ever increasing populations of the world. Today there are already hundreds of millions who are underfed. Growing populations require more and more villages, towns and urban agglomerates, representing an added pressure on food producing lands. The experience of the Egyptian towns and cities spilling out over cultivated lands is a worldwide phenomenon.

Desertification is a truly global problem as it relates directly to the satisfaction of basic human needs and to the survival of millions yet to be born. The world has become aware of this problem. Admittedly, there are serious gaps in our understanding of the complex systems that determine climate and of the complex and very varied soil-water-biota relationships in arid lands. But the main bulk of scientific knowledge and technological means necessary for combating desertification and developing the resources of arid lands are available. Several countries have succeeded in halting this menace. Experience during the intermediary years tell us that success in halting the menace, and any other environmental menace, comes through the coordinated and integrated actions of applications of science and technology, reform of socio-political institutions including land tenure systems, mobilization of popular participation, and sound and efficient management systems. You have the experience and you are fully equipped to tackle the set of basic questions involved in the problem you are addressing and come up with concrete practical recommendations. I wish you every success in your difficult task.

TECHNOLOGICAL LIMITATIONS TO ARID ZONE DEVELOPMENT

H.E. DREGNE
Horn Professor and Director, International Center for Arid and Semi-arid Land Studies, Texas Tech University, Lubbock, Texas, U.S.A.

ABSTRACT

Technology, when properly fitted to local conditions, can be a powerful force for improving human welfare. There is little likelihood that demands for a better life by a constantly increasing population can be met unless the technological component of development is expanded and improved. The basic technology to bring about that better life is, in the main, available now.

Two major technological limitations to arid land development can be identified. One is the shortage of applied research on the adaptability of new technologies to local conditions. The other is - paradoxically - the rapidly increasing availability of automated analytical techniques and their use in decision making. The first limitation handicaps development efforts by delaying the effective introduction of new technologies. The second, which substitutes machines for men, may well have catastrophic effects in the long term wherever biological systems are involved.

Minimizing the adverse effect on development of a lack of local testing can be done by supporting applied research and by bringing about information exchanges through formal-informal networks. Coping with the fascination that complex equipments has will become more and more difficult as automation proceeds. Only a determined effort can reverse the trend toward decision making by machines.

A central thesis of the United Nations conference on Desertification held in Nairobi, Kenya in 1977 was that the technology needed to combat desertification is already available. The consensus was that further delay in coping with the massive problem of land degradation could not be justified by claims that more research was needed before action could be taken. In all likelihood, that same belief in the adequacy of the present scientific and technological base to serve as a beginning for the development process will be supported and accepted at the 1979 United Nations Conference on Science and Technology for Development.

Insofar as knowledge of the basic principles underlying specific technologies is concerned, the proposition that we may not know everything we would like to know but we do know enough to be able to expect a reasonable degree of success in applying the technologies seems to be generally correct. Further delays in facing up to the problems of development are not warranted. But the adequacy of understanding of basic principles does not mean to me that technology transfer can be accomplished without additional adaptive research.

As we meet to consider ways of bringing about successful arid zone development, it is appropriate to consider two problems that I see facing us now and in the future. My first concern is with the continuing need for adaptive research. My second concern is with what I consider to be an alarming reliance on automated devices in the search for answers to questions.

ADAPTIVE RESEARCH

In my opinion, there are few situations where technology transfer can be done without at least some adaptation to local field conditions. It may be that certain industrial processes can be transferred completely from one country to another in a so-called turnkey installation. I doubt that such an operation is possible for any technology subject to environmental influences. The design of homes in El Paso, Texas is going to be different - if they are to be acceptable - from the design of homes in Luxor, even if the national environment of the two places is practically identical. But the difference in the final product should not obscure the fact that home design is - or should be - based on physical principles, functional requirements, and social needs. Calculation of heat storage and release in a house is fairly simple and straightforward, given a few basic facts, and determination of the functions buildings and streets are to serve and of the social needs that the potential occupants perceive does not appear to be an inordinately difficult task. Yet, once construction has been completed, changes inevitably are made and complaints go up about the poor planning that was done.

DREGNE'S LAW

Since I am not an architect, I am willing to believe that some of the blame for post-construction problems can be attributed to faulty planning by the designers. However, I have seen so much of it in laboratory construction, irrigation projects, cropping schemes, and plant introduction programs that I have formulated Dregne's Law: "If a technology looks as though it is immediately transferable, look again."

To cope with the impact of Dregne's Law I propose that the logical procedure to follow in taking an innovative approach to building design (or any other development process) is to set goals, analyze previous experiences, evaluate materials and designs in light of the goals, build prototypes with which to experiment, make changes as experience dictates, and then proceed with full-scale development. It seems to me it is completely unrealistic to expect that any new development, no matter how thoroughly it has been researched, will not need to be modified in some respect after it has been put into use. That is particularly true for biological systems but it applies, I suspect, nearly everywhere. And if it is true, we need to accept it and plan for it.

Afforestation Research

Afforestation is a prime example of a practice that requires adaptive research even though much is known about it. It is a practice that is accepted as a valuable means of stabilizing soils, providing firewood, and reducing duststorms. Every country in the world probably has had some experience with shrub and tree establishment. Yet, despite this experience, further adaptive experimentation is required whenever there is a change in the plant source or in climatic or edaphic conditions.

Granting that at least some adaptation will have to be made in any new or introduced technique does not mean that no effort should be made to learn from other peoples' experience. It is unconscionable to waste time, personnel, and money to learn what is already described in accessible literature sources. Unfortunately, even diligent searches sometimes fail to uncover the news that someone at another institution is working on the same problem or has already solved it. And sometimes, to our great embarrassment, that other person is working in the building next to us.

The value of imaginative adaptive research is illustrated, for me, by experiences with afforestation of sand dunes in Iran. The first efforts to use cheap bare-rooted tree stock for planting on sand dunes were unsuccessful; practically all of the transplants died within six months. Container-grown stock, on the other hand, was transplanted successfully. However, containerized stock was expensive and, therefore, in short supply. Only a small fraction of the sand dunes needing stabilization could be treated. After several years of disappointingly slow progress, studies were again undertaken with bare-rooted stock and a successful method of transplanting was developed for some species. The rate of annual afforestation subsequently jumped from 50,000 hectares to over 250,000 hectares, using the new techniques, and there was a corresponding drop in unit production costs. All this would never have happened if the scientists had quit experimenting after they had found a successful - but expensive - method.

Research in Arid Zones

I see a great need for high quality adaptive research in arid zone development. Building construction, including material selection and construction techniques as well as design, is one area needing attention. Among others are irrigation efficiency and water conservation, solar energy development, afforestation techniques, plant introduction methods, legume utilization for nitrogen additions to soils and disposal of organic and inorganic wastes. The principles upon which to act are known; what is needed is to find specific local adaptations that are effective in the field.

If there is to be high quality adaptive research, there obviously must be high quality scientists to carry out the studies. Dedicated researchers are essential and there will never be enough of them even though the inner satisfaction of contributing to the betterment of mankind is a very special reward for solving real world problems.

PROBLEM SOLVING BY AUTOMATION

My second concern about technology is related to the degree to which scientists are relying increasingly upon machines to relieve them of the need to understand their research and to draw conclusions from it. Since thinking is hard work, I am not optimistic that the current trend is going to change significantly in the near future, if it ever does.

Laboratory Analyses

With the veritable explosion in sophisticated laboratory equipment during the past twenty years, analyses of all kind have become easy to make if one has access to the necessary machinery. The consequence of all this mechanical aid is that researchers now call for many more analyses than they did before, and for analyses they hardly even thought of when a buret and a stirring rod were the major pieces of equipment on hand. Progress in equipment is welcome, not only because it makes the analyst's job easier but because it frequently allows a much higher degree of accuracy and the determination of substances that previously were very difficult to measure. Atomic absorption, nuclear magnetic resonance, electron microscopy, gas-liquid chromatography, and activation analysis are analytical techniques that bring joy to the life of a research chemist. What seems to me to be unwelcome is the increasing tendency that I discern for scientists to have analyses made simply because techniques and equipment are available to run them. With the advent of the atomic absorption spectrophotometer as a routine piece of laboratory equipment, I see more and more lengthy lists of data of the uranium, lead, iron, and molybdenum - to name a few - content of soils. Although I am convinced that there probably is not a man alive who could say what the significance was of the data. It has become easier to extend the list of "necessary" analyses than to understand the limitations placed on the data by soil and plant phenomena sampling methods, equipment electronics, interference phenomena, equipment operating characteristics, and similar factors. Then, to top it off, all too frequently there are no accepted criteria by which to establish the significance of the data even if the analytical techniques were perfect. We only delude ourselves when we say such data are useful. And equally bad, the time and effort required to obtain the data have been diverted from research that could have been productive.

Discrimination in deciding upon the kind and number of analyses to make has always been necessary, in the bad old days as well as in the good new days. The decisions are harder to make now, when we have a large number of selections available. For most of us, there is a strong temptation to collect data for the sake of collecting data. More is better. Few things are more effective in keeping administrators from asking embarrassing questions about research productivity than are lists of data and complicated graphs, tables, and statistical analyses. Learned (though irrelevant) discussions of the statistical significance of third, fifth, or twentieth order interactions are guaranteed to impress administrators. This has always been so; it is part of the game. Computer analyses of data and hundreds of printout sheets every hour on the hour have only made matters worse. It takes a truly knowledgable scientist to know when the point of diminishing returns from such efforts has been reached. And the task is becoming increasingly difficult as our equipment gets better and better.

Data Analysis

A second point where automation is substituted for objective thinking is in data analysis. When machines are churning out numbers day after day, a researcher begins to realize that he simply cannot assimilate all that data. He then either becomes panicked and quits trying or he turns to his friendly computer to extricate him from the mess he is in. The helpful computer accepts all the data -- good or bad -- fed into it and is quite willing to spew out statistical analyses in accordance with directions from canned programs. Again, tall piles of printout sheets accumulate that may never be looked at except for the pages carrying the final statistical analyses. The analyses look impressive and are easily obtained. Unfortunately, they may be of no value. Whether or not they are useful depends upon interpretations placed upon them by experienced scientists who are thoroughly familiar with the data and how they were obtained.

The waste of paper, machinery, and money that the careless use of computer analyses represents is bad enough; what is worse is the waste of brainpower when statistical analyses are substituted for critical informed judgment developed from years of careful observation. Statistical analyses not subjected to a critical judgment may be worse than no analyses at all because they mislead the uninformed user.

Role of Machines

In summary, I believe that scientists and technologists should be the masters in the laboratory and in the field. Machines are tools that can only be used effectively if their potentials and limitations are known and if they are employed to respond to real needs rather than simply used because they are there and are expected to be kept busy. That is no easy task when equipment is justified by the number of hours it operates. And it is becoming increasingly difficult for investigators to understand the fundamentals of their disciplines as well as the rudiments of electronics and statistics.

CONCLUSIONS

A vastly stepped-up pace of scientific research throughout the world has generated immense amounts of information on almost everything imaginable. Yet there remains one area -- adaptive research -- in which the transferability of tecnnology frequently is limited in scope. In environmentally affected disciplines, technology transfer is almost always dependent upon adaptation to local conditions. Failure to recognize that has led to much wasteful -- and even harmful -- development activity.

The stepped-up rate of scientific research has also led to increasing reliance on automated equipment to collect and analyze data. One result, all too commonly, is an abdication by the researcher of responsibility for selectively choosing the analyses to be made and for interpreting the results of computer-generated statistical tests. Critical informed judgment is more necessary now than ever before, if we are to make effective use of the almost unlimited opportunities for fast, accurate, and detailed analyses.

EDUCATION AND TRAINING FOR DESERT DEVELOPMENT

M. KASSAS
Department of Botany, Cairo University

ABSTRACT

Education and training programs involving available facilities and machineries aim at providing trained personnel needed for national development and management of resources, including desert development. Trained personnel needed for desert development will be required to manage the hazards of desert environment and the resources available in desert areas. These hazards are either natural (recurrent drought, locusts, etc.) or man-made (ecological deterioration, desertification). National programs for monitoring and assessment of desert hazards and resource potentials and for management and development of resources must be multi-disciplinary and may require innovative restructuring of the training machineries, especially at the tertiary level (universities and institutes of higher education). The success of national plans and their effective implementation depend on the conscious acceptance and positive participation of the community, adding special responsibilities to educational institutions. Information, education and training functions should be integrated and conceived within the framework of a national plan. Regional coordination in certain aspects of training could be effective. International centers for special training may play a very useful role.

INTRODUCTION

Some hundred countries of the world, including the USA and Egypt, face challenges related to deserts. Two classes of problems may be noted. The first, now known as desertification, is a process of environmental degradation caused or triggered by man's action, and by which productivity of land is reduced or lost. Desertification results in reduction of food-producing potential and combat of desertification aims at conserving the productive potential of land. The second relates to development of desert resources and to transforming the desert lands -- which worldwide amount to some 84,350,000 square kilometers, equivalent to about 36% of the earth's land surface, and that in Egypt amount to some 970,000 square kilometers, equivalent to 97% of its land surface. These two interrelated sets of problems are at present put forcibly on the agenda for action of all arid-land countries including Egypt, and thus figure in their plans for national development, plans for scientific studies including research and technology and plans for education and training.

The main objective of systems of education, including training, should be to improve human and institutional competence so that they can cope effectively and rationally with society's requirements of economic and social development. For education to be an effective instrument for social and economic progress, it must provide man with knowledge and a thorough understanding of the environment in which he lives, and with the means for bridging the gap between this knowledge and its application. The extensive studies submitted to the United Nations Conference on Desertification (Nairobi, August 29 to September 9, 1977) showed the availability of a considerable volume of scientific knowledge on desert processes and technological means for combating desertification and developing desert lands, but a good deal of these were yet to be applied in many of the less developed countries. To bridge the gap between knowledge and its application, education should act through four main systems.

Socio-political systems may foster resource abuse. Certain land tenure systems may lead to soil abuse (e.g., ownership of water and not land in desert oases). Investment-allocation policies may lead to degradation of land resources because of shortage of capital. This calls for education of policy makers and decision makers.

Socio-cultural systems may also foster resource abuse. The status of cattle ownership in the social matrix often leads to overgrazing and range degradation. Traditions and deep-rooted social beliefs and customs may prevent a society from utilizing available resources, thinning animal herds when appropriate, etc. This calls for education of the masses with a view of evolving sociocultural concepts and modes of behaviour compatible with environmental requirements.

Underdeveloped technical and scientific infrastructures may fail to assimilate available knowledge. This calls for education of concerned people ranging from farmers (to enable them to utilize scientific advice) to research workers in knowledge-producing centers.

Demographic patterns in arid lands are usually those of sparsely populated territories. Systems of conventional education need to develop and apply innovations in the concept and structure of schools that will befit this demographic pattern.

Clawson[1] estimates that the arable soils, natural grazing lands and water resources in the Middle East region include: 20 million acres of irrigable land, 24 million acres of rainfed arable land, and extensive areas of pasture. These areas are physically and economically capable of greatly increased agricultural output. At present these resources are only partly developed. He enumerates obstacles to land-use development that are mostly social and institutional: unstable and inefficient administrative systems; lack of technical knowledge at the level of the farmer in the field; entrenched customs especially in livestock grazing; obsolete land ownership patterns; shortage of investment capital, and lack of transportation and marketing facilities.

EDUCATION AND TRAINING

Education and training programs aim at providing trained personnel needed for national development and management of resources, and desert development is a part of this. The prevalent disciplinary structures of educational and training systems provide for personnel that are trained discipline-wise. But the multi-disciplinary nature of environmental problems, including those related to desertification control, require innovative approaches to education and training. Recommendation 2 of the *UN Plan of Action to Combat Desertification* underlines the need for a broad base for training land-use planners: "Though the planners' primary expertise should be in theory and technique of assessment, planning and management of human and natural resources, they should have access to expertise in law, economics, demography, sociology, urban planning, ecology, geography, geology, soil science, agricultural sciences, hydrology, hydrogeology, climatology, remote sensing, statistics and cartography. Where such training is not locally available, arrangement should be made to train planners elsewhere until the required training is established locally." This list of scientific disciplines may see long, but the integration of these various disciplines would produce a synthetic body of knowledge that is especially required and that is new. The topics that need to be addressed may not be completely new, but they are acquiring a new place in the national agenda for action.

Let us remind ourselves of a situation in recent history. The USA faced in the 1930's the hazards of drought (years with below average rainfall). As Clark[2] describes it, "The terrible duststorms, or *black blizzards*...swept across the Texas and Oklahoma panhandles and adjacent areas in the spring of 1934, blowing away tons of topsoil and reducing much of the land to drifted sand...overgrazing destroyed the native grasses on millions of acres...The Dust Bowl demonstrated the tragic consequences of attempts to cultivate crops in an area of scant rainfall and high wind." The life-supporting systems in semi-arid and sub-humid territories of the USA were menaced by near-collapse. This was not very different from the events of the African Sudano-Sahelian belt in 1968-1973.

Within the following two decades the USA managed to evolve systems and machineries that enabled the same life-support systems to face comparable droughts without destruction. In these two decades, new sciences were evolved, and included: applied plant ecology, range management, soil conservation and watershed management. These new bodies of knowledge were all synthetic in nature and were needed to resolve problems related to ecological deteriorations associated with drought hazards. A.G. Tansley,[3] one of the principal fathers of plant ecology, after listing the various topics of science that are included in plant ecology, concludes that it is "a means of approach to a large part of detailed botanical study...it is a synthesis of the special knowledge obtained by the study of particular departments of botany in relation to the life of plants and their natural habitat."

During the late 1960's and early 1970's the world became acutely aware of environmental issues. A new science evolved, environmental science. This is again a novel synthesis of bodies of scientific knowledge. This is not just old wine in new bottles, this is a novel blend that meets a new taste.

The first moral that I would underline derives from the concept that the principal objective of an educational system ought to be the improvement of human and institutional competence in dealing with emerging social and economic needs.

Whenever emerging problems require new sciences or new synetheses of bodies of knowledge, the educational system should create these new sciences and the educational processes that serve them. But the evolution of a new science is also dependent on the development of its husbanding congregation of adherents from peerage to students. The world has recognized the threats of desertification and the needs for combating it, recovering territories lost for it during recent decades, and developing the resources of arid lands in general. This needs a new science; let us for convenience call it desertology, and its adherents desertologists.

But national programs in arid lands (including deserts) comprise a variety of interrelated activities that require various aspects of this new science and hence various types of its trained adherents. These activities may be grouped into four principal categories: (1) Monitoring and assessment; (2) Planning; (3) Management of resource development, and (4) Scientific research and technology. These require various categories of trained scientists and technicians. Their training needs to be multi-disciplinary and may require innovative restructuring of the machineries for training, especially at the level of higher education (universities, institutes).

Let me consider, as an example, the situation in Egypt. The Egyptians lived within this great oasis of the Nile Valley for some 7000 years. They established and developed one of the earliest agricultural civilizations of the world, and maintained its fertility and the high productivity of their lands throughout this long history. This unparalleled achievement was closely linked with development of the necessary knowledge of land and crop husbandry and related sciences of climatology, calendar-astronomy, hydrology, etc. Whenever needed, they established agricultural schools and extension systems for transfer of knowledge and exchange of experiences. Later, agricultural colleges and research institutes were developed and provided Egyptian agriculture with needed experts and expertise. This agricultural scientific institution enabled Egypt during the last 100 years to: (a) maintain a high level of land productivity; (b) maintain quality superiority of its cotton; (c) introduce and develop new crops, especially fruits and vegetables; (d) participate in the world-wide green revolution of intensive agriculture, and (e) control the Nile waters and store them. This notable success should be very well appreciated, and credit should be given to the national educational machinery comprising the several faculties and high colleges of agricultural sciences.

But there were equally notable shortcomings in two areas: land-use planning and reclamation of arid lands. Egypt invested a substantial part of its resources during the last 30 years in expanding its agricultural lands. One major project, the High Dam, was designed to provide irrigation for one million acres, and there were other projects. Today the farmed lands are not more, and probably less, than 30 years ago. Why? Because cities, towns, human and industrial settlements encroached on extensive areas of agricultural fields. This is an aspect of failure of national land-use planning. It is true that during the last 2-3 years plans were set for the establishment of new human and industrial settlements in the deserts outside the farmlands of the Nile Valley. This, though belated, is most welcome. But its belatedness indicates that there is a shortage in the synthetic science of land-use planning and its adherents, and this should be a suggestion to the educational machinery.

Projects for reclaiming extensive areas of the Egyptian deserts figured in the national development plans during the last 25 years. Projects included the Tahrir Province, or southwest extension of the Delta; the West Nubariya projects northwest of the Delta; the New Valley, or agricultural development of the oases of the Western Desert; Kom Ombo, Esna, etc., projects aimed at extending irrigated farmlands on both sides of the Nile Valley in Upper Egypt. To these we may add prospected projects in the south of the oases and in Sinai. To make a long story short, let me say that every one of these projects faced very serious problems that nearly undermined their worth.

In his memoirs published in the *Ahram* newspaper (August 17, 1978), Sayed Marei, a senior agriculturalist and statesman, refers to the Tahrir Province as follows: "The idea of relcaiming the desert is important for the future of Egypt... But the problem was in implementation...Millions [of pounds] were evaporated in the desert sun or swallowed by the desert sand..." Here again is an indication of shortage in the synthetic science of desertology, the experts and the expertise. Failures in these new projects were due to the fact that they were addressed by the experts and expertise developed and successfully applied in the farmlands of the Nile Valley. The new lands represent, however, new ecological systems that need to be studied and the reactions and interactions of their component elements and factors must be understood before a rational system of management can be evolved.

The 15 case studies, from 12 countries, and the 39 country papers presented to the UN Conference on Desertification make it abundantly clear that problems in combating desertification and reclaiming arid lands are related to incomplete understanding of the nature of the ecological interactions within the desert habitat. Problems such as those faced in the West Nubariya area of Egypt are encountered in the Colorado Valley in the USA, the newly irrigated lands of the steppe territories in the USSR, the old and new irrigated lands in Iraq, Pakistan, Syria, etc.

Paradoxically, the main bulk of the scientific disciplinary knowledge is available. But what is needed is the synthesis of these bodies of knowledge, the snythesis inside the brains of trained personnel -- since problems are solved by people, not by books. Here is the basic role of education and training, and this is the message to the educational machineries in Egypt and elsewhere, in all countries menaced by the hazards of desertification and where the prospects of developing the resources of arid land are tempting.

The people needed include a range of trained personnel to: (a) do the basic surveys, monitoring and assessment of potentials and hazards; (b) do the planning of land-use and designing of projects; (c) carry out programs of research and pilot experiments; (e) provide the extension services. Although the training of this spectrum of personnel must include a body of basic knowledge common to all, the requirements of their training will vary.

The educational machineries (schools, colleges, universities, training institutes, research centers, etc.) need to address these requirements, not only with respect to desertification and arid lands but also with respect to several other aspects of the educational needs of our changing world. These machineries responded during the 19th and 20th centuries to needs of dissecting knowledge into disciplines of specialization. For example, natural history became botany, zoology and geology, and each was later anatomized into serveral discrete sciences. Chemistry ramified into branches that are now almost alien to each other. These machineries also responded to the needs for hybrid interdisciplinary scineces such as biochemistry, biophysics, etc. In all these instances new institutional units were established in the form of new university departments and colleges, etc., to meet the need. Similarly, the present need for new syntheses of scientific knowledge and for integrated training require new organization restructuring. This should allow the larger body -- the university for example -- to function as an integral whole with oeprational versatility.

This may not be easy. Let me quote one of many likely examples. The Texas Technical College (now University), under its then newly appointed president, Prof. Grover E. Murray, responded in 1966 to the emerging need for training and research in arid lands by creating its International Center for Arid and Semi-arid Land Studies (ICASALS). This was, in effect, a mechanism for mobilizing the resources of the whole university. In the words of one of ICASALS' early documents, "All schools -- agriculture, arts and sciences, business administration, education, engineering, home economics, law, graduate -- will contribute their special capabilities. Thus, ICASALS rests on a most compatible foundation -- a solid academic structure where full advantage can be taken of the resources, accumulated knowledge and research facilities which only a university can offer."

This operational versatility, that draws on the broad capabilities of a university, will allow for a variety of training options that provide: (a) the broad basis that must be a common denominator in training a variety of scientific personnel, and (b) the circumscribed training for each category. The nature and structure of the mechanism will depend on the nature and structure of the university (or institute of higher education) concerned, and there cannot be a single model for that. We may quote relevant parts of the text of Recommendation 20 of the *UN Plan of Action to Combat Desertification*: "The implementation of national programmes for monitoring, studying and combating desertification, and for rational management of resources in dry lands subject to desertification, requires various categories of trained scientists. Their training must take into full consideration the complex and locally varying nature of the process of desertification and the need for integrative measures to combat it, to reclaim lost terrain, and to improve the quality of life for the inhabitants." In goes on to call for national action to: "(a) Provide for the inclusion, evaluation and modification of information on desertification and arid and semi-arid land processes in environmental programs and curricula of education at all levels, including schools and universities; (b) Provide courses and degree programs in universities and institutes of higher studies, including post-graduate training in the multi-disciplinary studies, to support national programs for combating desertification and reclaiming desert lands; (c) Provide special scientific and pedagogic courses related to the ecology of deserts and arid and semi-arid lands in teacher training programs."

EDUCATION FOR COMMUNITY PARTICIPATION

Let me further quote Recommendation 20: "The success of national plans and their efficient implementation depend on the conscious acceptance and positive participation of the societies concerned. It is desirable that national education programmes be available, and that these be based on a collaborative process involving the users of the land, local and national governments, resource management specialists and educators. In the same manner, educational opportunities should be available at all levels of society -- from policy maker and resources manager to student or local citizen. Furthermore, these educational programmes should be linked to national institutions for desertification reserach and management, so as to improve their relevance, facilitate the actual application of new ideas and research results, and provide training opportunities for students. These activities call for a variety of well coordinated mass-media programmes." It goes on to call for national action to "(d) Incorporate in the objectives of out-of-school educational programmes the mobilization of popular support for national programmes to combat desertification and reclaim arid lands. These educational programmes need to take prevalent socio-cultural factors into consideration with a view to fostering attitudes compatible with necessary changes, or gradually changing attitudes opposed to change, and to use languages and means of demonstration that are intelligible to the masses, and means of communication accessible to remote areas; (e) Promote mass-media programmes that comprise continuous and mutually supportive activities, including radio, television, press, cinema, pamphlets, posters, etc; (g) Establish demonstration units where improved or new methodologies of land-use are practiced. These units could organize demonstration and training programmes in which men and women engaged in farming, agriculture or pastoralism could participate for appropriate periods of time; (g) Provide information on desertification processes to high-level decision-makers and civil servants."

There is little to add to this comprehensive list of educational activities. But two points need to be underlined.

First, educational processes need to take part in efforts aiming at (a) removing socio-cultural obstacles that may impede implementation of national programs to combat desertification and to reclaim desert land; and (b) mobilizing positive and effective public participation in, and support of, these programs. Second, educational processes need to take part in efforts aiming at (a) removing socio-political obstacles that may lead to policies and plans that are not based on the best scientific information; and (b) improving organizational structures related to policy making (and decision making) that foster links and feed-back channels between knowledge-making units and policy making bodies. To this we may add improving institutional competence, administrative efficiency and managerial skills -- all need the bracing of educational processes.

Let me conclude by quoting W.H. Walker:[4] "An agrarian revolution is taking place in many parts of the world. The arid zones will become increasingly important in this revolution. If education does not become a major force in bringing about the necessary changes, the battle for *freedom from hunger* in the race with population will be lost."

REFERENCES

1. M. Clawson (1971). Social and institutional obstacles to agriculture of arid lands in the Middle East. In, Food, Fiber and the Arid Lands, W.G. McGinnies, Ed. University of Arizona Press, Tucson, pp. 3-15.

2. I.G. Clark (1963). Historical framework. In, Aridity and Man, C. Hodge and P.C. Duisberg, Eds. AAAS, Washington, D.C., pp. 87-116.

3. A.G. Tansley (1946). Introduction to Plant Ecology. George Allen and Unwin, London.

4. W.H. Walker (1970). Education for resource utilization. In, Arid Lands in Transition, H.E. Dregne, Ed. AAAS, Washington, D.C., pp. 399-409.

POSSIBILITIES OF LAND RECLAMATION IN DESERT AREAS: REVIEW OF PRESENT AND FUTURE STATUS

OSMAN BADRAN
President, Arab Authority for Agriculture and Development, Cairo

S.A. KANDEEL, M.H. LAKANY, M.T. OMRAN, A.B. EL SAYED, M.L. EL OSTA, Alexandria University

ABSTRACT

The impoverishment of arid and semi-arid ecosystems through the impact of man's activities has paved the way to desertification. The overall Egyptian plan to combat desertification through land reclamation of desert areas adjacent to the Nile walley is critically reviewed, including (1) Applications of technology for desert reclamation in light of local and international experience; (2) Critical analysis of current research programs and local experience during the last decades, such as that in species and provenance trials and utilization in newly reclaimed desert areas. Future strategy with respect to possibilities of land reclamation in desert areas, including amelioration of deteriorated soils, improving of drainage and irrigation systems, and development of new urban centers in desert areas, is discussed.

INTRODUCTION

In Egypt, 97% of the total area is desert. The majority of the bioclimate of the country is classified, according to the Desertification Map of the world, as Hyperarid, which means that the ratio between precipitation and evaoptranspiration (P/ETP) is less than 0.03. This is determined less by the climate index than by the absence of vegetation except for ephemerals and shrubs in river beds.

The rest of the country along the Mediterranean coast is classified as arid and semi-arid with P/ETP 0.20 and P/ETP 0.50 respectively. Thus the problem of desertification in Egypt is an acute and pressing one. There is good evidence that most of the country was once green and productive, but due to climatic changes and human pressure, among other factors, deserts now cover vast areas of Egypt.

Most, if not all, factors causing desertification are operating in Egypt. High temperatures accompanied by low and sometimes scanty rainfall lead to high evaporation and accumulation of salts in irrigated and rainfed lands. Sand dunes endanger villages, roads, and in some areas, cause losses of urban areas. Soil fertility and crop production are threatened by sand storms and the amount of sand deposited monthly in certain areas is estimated at 75 tons per square mile.

In spite of all this, Egypt has been able to withstand and tolerate these drastic conditions for a long time, whereas many other civilizations have succumbed.

HISTORY OF LAND RECLAMATION IN EGYPT

Land reclamation in Egypt started on a large scale in the last 20 years and increased with the building of the High Dam.

In the last 20 years experience has been gained in the fields of land reclamation of sandy, calcareous, and saline soils. Before 1952 the process of land reclamation was very slow, about 2500 feddans per year, most of it done by rich land owners. After 1952 the areas increased greatly and exceeded 140,000 feddans in some years, as in 1962. The land reclamation plan is based on reclamation of 2.5 million feddans in the future. The total reclaimed area from 1954 till now is about 912,000 feddans; 759,300 feddans in the delta valley and the rest (152,700 feddans) from deserts.

At the beginning, the choice of the areas to be reclaimed and the building of canals and the drainage system was done by the Ministry of Irrigation. Then the choice of the areas to be reclaimed was done by the Soils Department of the Ministry of Agriculture.

Land reclamation processes are a result of cooperation between different organizations to establish new permanent settlements and to provide facilities for new settlers such as roads, electric power, housing, etc.

CURRENT ACTIVITIES

During the last 30 years, Egypt has carried out extensive activities for reclaiming deserts bordering the Delta and along the North Coast. The main objective of these activities is to increase the area of cultivable land and to combat desertification.

The activities include: (a) Surveying water resources, including surface runoff and ground water, and planning of their use; (b) Establishing some pilot farms; (c) Carrying out irrigation and drainage projects; (d) Zoning pasture land and establishing some animal production farms; (e) Planning of new rural communities on scientific bases; (f) Stabilizing sand dunes, especially on the North coast, and (g) Ameliorating the deteriorated soils of the valley and Delta.

PROGRAMS OF LAND RECLAMATION IN EGYPT

The programs of land reclamation are concentrated on the desert areas mainly in Sinai, the western north coast, oases of the Western Desert, Qattara depression, Fayoum depression and the south of the New Valley.

Sinai: Land reclamation activities in Sinai began in the late 1940's. During this period the following was accomplished:

1. Conservation of surface runoff, building of Rawafah dam and fixation of 400 feddans of sand dunes
2. Study of land resources for agriculture development and settlements in 0.25 million feddans in the Suez Canal area
3. Study of the natural resources for agricultural expansion in the northeast of Sinai in one million feddans
4. Development of projects including irrigation areas in the north and those using underground water in the south

The current plan for land reclamation in Sinai from 1967 to 2000 includes the following:

1. The use of Nile water to reclaim the war-destructed areas; at present 2000 feddans will be expanded to 30,000 and will reach 0.25 million feddans by the year 2000
2. The use of surface runoff and wells to reclaim 10,000 feddans in western Sinai; this area will be increased to 50,000 by the year 2000
3. In the middle of Sinai an area of about 2 million feddans will be cultivated, depending on underground water and

other natural sources. A large part of this area will be used for pasture.

Northwestern Coast: In this region the following projects for land reclamation were carried out:

1. Afforestation and fixation of 200 feddans of sand dunes at El Kaser, El Garawla and Baghosh.
2. Improving drainage systems
3. Reclamation of 45,358 feddans and building of 18 new villages.

Western Desert Oases: In this area a study of the underground water was done to explore the possibility of water use and consequently an area of 1500 feddans was reclaimed and used in addition to the afforestation of sand dunes.

TECHNOLOGIES AND PRACTICES IN DESERT RECLAMATION

New technologies for water and land uses should be used in the newly reclaimed areas. This would increase the efficiency of water use and minimize the damage of excessive or improper water application by surface irrigation, beside increasing soil productivity through the use of the most efficient planting methods. Those technologies should be generally inexpensive, easy to use, acceptable environmentally, and based on locally available resources. The diverse effects of certain technologies must be carefully considered.

Water: The proper application of water to soil is through irrigation methods producing the highest yields with the minimum amount of water loss and without soil deterioration.

In Egypt the flooding and furrow methods of irrigation used by landowners in the Delta have disadvantages but do not require a high level of technology.

Surface irrigation methods result in low to medium water use efficiency. It accelerates salinization and waterlogging in areas where leaching and drainage systems are not properly designed, as in the reclaimed areas of Mariout. Excess water will raise the water table, as has happened in West Nubaria. The flooding system is not recommended for sandy and light textured soils in the newly reclaimed areas.

Great water loss through deep percolation also leads to salt accumulation. In the furrow irrigation system, the uniformity of water application decreases as the furrow becomes longer, with high amounts of water percolation at the beginning of the furrow. New technologies in using water resources should be applied by using improved irrigation methods and control of salinity and waterlogging.

IRRIGATION TECHNIQUES

Sprinkler Irrigation: In this method, irrigation pipes cover all parts of the field. There are different types of sprinklers -- rotating, fixed and perforated pipes. Also, different types of conveyers are used, permanent or fixed, semi-permanent or semi-portable, and fully portable. In the latter, lightweight aluminium tubing or plastic tubing can be used. This type of irrigation is well adapted to the reclaimed lands of high permeability. The rate of water application should not be greater than the permeability rate of the soil. This can be regulated by using specific pressure nozzle sizes and spacing of nozzles. This sytem can be used also for applying fertilizers through water. It is very useful in hot arid areas to reduce and control leaf temperature. This method reduces water loss by seepage and percolation. Water application is uniform and water efficiency is more than 75%. At the same time, this method is costly, and is not suitable in windy or exposed areas without the presence of shelter belts or windbreak systems.

Drip Irrigation: This type of irrigation can be used mainly for perennial crops (as fruit trees and vines) due to its high cost. Irrigation pipes are located on or under the soil with outlets located near the roots of the plants at fixed intervals. Small amounts of water are applied to form wet areas which extend to the plant roots. This can be adjusted to replace water continuously as it is used by the plant. The water delivery is adjusted according to the soil absorbing capacity, seasonable crop requirements and climatic conditions. The aim of this method is to keep soil moisture at the optimum level for plant growth.

This method improves the efficiency of water use and gives good control over the replacement and amount of water. Evaporation is reduced by this method. It is suitable for sandy soils and in arid areas where too much water is lost through evaporation. Some difficulties are involved in this method: the emitters can become clogged, resulting in lack of uniformity in water application; technical advice is required when saline waters are used, and the method must be complemented with another irrigation method, usually sprinkler, to leach excess salinity in the root zone.

Drainage Water: Drainage water can be used in areas where there is not enough fresh water for irrigation and in areas where soils are highly permeable, and salt tolerant plants are grown. In areas where drainage water is of high salinity it can be mixed with fresh water before its use. Under these conditions the following points should be taken into account: (a) the increase in yield due to increasing water supply should be more than the decrease due to increasing water salinity; (b) the leaching requirements when using saline drainage water should be added when computing water requirements of the crop.

Irrigation with drainage water has been used in the Delta region. For instance, in 1976 about 2300 million m^3 were used. Another 2300 million m^3 went to Karun Lake and were reused in irrigation. This gives a total of 4.6 millards.

Underground Water: Underground water is the main source of irrigation in the extension of reclaimed areas in the Western Desert and New Valley. Studies showed that the recharge of this reservoir is about 6000 million m^3 with annual recharge of 815 million m^3. The depth of the water layer varies from 200-2000 meters. Assuming that a layer of 100 meters will be consumed within 200 years, the annual amount that could be used equals 3000 million m^3 besides the annual recharge. This amount is enough to irrigated about 763,000 feddans with 5000 m^3 per feddan annually, which equals the water requirements. An equilibrium between recharge and pumping must be established to avoid indiscriminate over-exploitation which may cause problems in the future.

SOIL

In the reclaimed areas, soil conservation, salinity and waterlogging control and drainage are necessary to maintain soil productivity apart from using fertilizers, and sand dune stabilization.

In newly reclaimed areas control of land degradation by wind and erosion is very essential. This means establishment of windbreaks and shelterbelts of trees and shrubs, to control wind velocity and provide shading. The government can provide farmers with suitable tree seedlings at nominal price with the needed technical assistance to plant them. In areas on the western coast and New Valley that are affected by sand dune encroachment, stabilization by planting with drought resistant trees is an efficient method. Fencing with dried plant materials can provide the region with protection till the tree roots extend to enough depth. This can be accompanied by the use of petrochemicals to stabilize the surface for one season. Later on, trees will do the stabilization for decades.

Control of Salinity and Waterlogging: Soil salinity and water logging in newly reclaimed areas as in Mariout and Western Nubaria is a result of over irrigation which raised the water table and caused salts to accumulate in the soil profile. Seepage control and drainage systems are very essential to prevent waterlogging and also to prevent or reduce salt accumulation in the root zone. The quality of water should be taken into account in some areas, to prevent soil salinity, to choose the crops that can be grown and the irrigation method. Surface drainage is used in some reclaimed areas, but by the end of 1985, subsurface drainage will be used in all reclaimed areas. Clay pipes are mainly used and plastic pipes will be used. In many reclaimed areas canal seepage is high and canals are bottomed by an interceptor, usually cement as in West Nubaria, Mariout and the sandy areas at Rashid. This reduces seepage and waterlogging in saline areas where there is soil salinity. In saline areas, gypsum is added to the soil to supplement the drainage system. Biological drainage using suitable trees would prove beneficial as in other countries.

Several practices to improve soil productivity of reclaimed soil can be developed by using fertilizers, and the proper choice of crop management.

In reclaimed areas both organic and chemical fertilizers are required. Organic fertilizers are more needed in the sandy soils to improve soil structure and reduce leaching of chemical fertilizers.

CROPS

Attention should be given to the choice and management of crops in the reclaimed areas especially in the first years after reclamation. Green manure crops such as alfalfa should be planted in the first years to improve soil and as a fodder for livestock. Legumes or crops which are ploughed under at the end of the growing season are preferred. In arid areas selection of water efficient plants such as barley, sunflower, melons and other species should be given consideration. Crop rotation is essential to preserve soil fertility. In some areas, as in the Western Desert, multiple cropping can be used, while annual crops can be planted under palm trees.

ROLE OF AFFORESTATION IN DESERT RECLAMATION

It has been known that considerable blocks of natural forests once existed in Egypt. Although forests can be produced in submarginal land and although submarginal land is always overexploited by man, it is necessary to devote at least a small proportion of agricultural land in special cases to the production of tree plantations. The most expensive crops (including forest crops) are those grown on the poorest land.

The threat of the desert is ever present in Egypt. Moving sands, hot winds and dust storms are reminders of the fact that deserts, like the sea, must be kept within bounds. Thus, some of the most valuable vegetation in the country is paradoxically that which appears the most worthless, i.e. the desert shrubs and shelter belts in lands adjoining agricultural areas. These prevent the movement of soil patches and the cumulative buildup of sand. Therefore, there should be insistance on an arbitrary percentage of afforested land in Egypt and on the establishment of shelterbelts and woodlots, as these provide the most economic answers to many of the problems confronting desert encroachment on newly reclaimed deserts.

The objectives of afforestation can be summed up as follows: (a) The conservation of soil, water and ground water separately or together in all aspects; (b) Protection from such calamities as floods, sandstorms, wind, dessication, moving sand dunes, to arrest deterioration of the environment; (c) Production of timber, poles, fuel, grass and minor forest products of all kinds; (d) Provision of amenities, shade, windbreaks, wildlife, sport, recreation, to enlarge and enrich the pattern of life in arid regions for Egyptians and for increased touristic attraction; (e) Provision of employment. Forests provide a way of life and an occupation for parts of the rural population. Proper utilization and harvesting of forest products will permit a renewable resource to be developed permanently and support a type of peasantry which is an asset to any country.

These are, in short, the aims of afforestation in Egypt. They apply to the activities carried out so far and even more so to the expanded programs for the future. There is an extensive program of afforestation and establishment of properly designed windbreaks being undertaken by the Department of Forestry at Alexandria University. The main objectives of this program are to protect the newly reclaimed desert areas in Northwestern Egypt and to provide wood and other tree products for the inhabitants.

Choice of Species: In recent times, several dozens of tree species from other regions of the world have been tried in Egypt, but the majority of those were merely ornamental trees. They are found in public parks and private gardens but have no importance as forest trees. Thus, the list of actual forest trees which have been used for establishing shelterbelts, windreaks and the very limited real stands comprise not more than 10 species: *Casuarina* sp, *Eucalyptus* sp, *Dalbergia sisoo*, *Poplar* sp, *Tamarix articulata*, *Albizia lebbek*, *Prosopis juliflora*, *Khaya senegalensis*, *Acacia* sp, and *Brachychiton pepulneum*.

The *Casuarina* species we find in Egypt are: *C. gluaca*, *C. equiset* and *C. cunninghamiana*. The most common *Eucalyptus* species are: *E. camaldulensis*, *E. citriodeora*, *E. gomphocephala*. The poplars unfortunately are mostly hybrids and clones of the euramerican type, and are obviously not suited for the hot Egyptian climate. After some apparently thrifty growth in nurseries, these unsuitable provenances start starving as soon as they are planted in the field. Much more research work needs to be done before we can say that poplar cultivation has a future in Egypt. The main *Acacia* used are *A. cyanoplylla* and *A. nilotica*. *A. cyanoplylla* has proven suitable in areas with less than 200 mm rainfall. Except *Cupressus*, which is now planted as a christmas tree, there is practically no conifer in Egypt though it can be assumed that certain pines would grow. Work on pines is, therefore, an urgent research need. There are two genetics programs at present to improve some of the fast growing species in Egypt. They aim at producing trees of desired characteristics, such as wind resistance and better wood quality, to be planted as windbreaks and woodlots. Growing drought resistant and

and salt tolerant trees are among the trials being given consideration.

Techniques of Tree Growing: Location for tree nurseries in Egypt is usually well chosen, availability of water being the main consideration. Most of them are permanent nurseries, but comparatively small in size as there are no ambitious afforestation programs yet. So far, all the work is done by hand. The irrigation system is usually based on gravitation in open ditches.

Plastic bags are used to some extent, though the more conservative method of using clay pots still prevails since it is an old technique and these pots are made cheaply everywhere. The planting itself always takes place in different locations, sometimes far away from each other, the reason being that there are no permanent workers for such work. It should be stressed, however, that the skill to plant is somehow innate to all Egyptians, and the results of the plantings are quite satisfactory as a rule. Eventual losses are caused mostly by damage due to lack of proper protection of plants.

Some species such as tamarisk and poplar are propagated by cuttings which are planted toward the end of the winter season.

Utilization of Afforestation Products: The dependence of the Egyptian economy on imported fibrous and wood products gives added values to afforestation activities. While such activities are essential for agricultural development in arid regions, the fiber shortage in Egypt also gives a higher value to the end products of afforestation projects. Assuming a short rotation in the shelterbelts of eight to ten years, great amounts of wood can be produced. This can be seen from the data collected from the reclamation areas in the northern coastal areas where casuarina for instance gave $0.3m^3$ volume after about ten years growth. Such volume is doubled in the Tahrir area due to the abundance of water. The consideration of the $0.3m^3$ per tree means considerable wood material when it is noted that one feddan supports 400-500 such trees planted as shelterbelts. Research activities at Alexandria University proved the suitability of windbreak eucalyptus trees grown in short rotation for pulp production and paper making. The paper quality of such pulp is at least double to triple that of rice straw or bagasse.

The possibility of using windbreak trees after short rotation in particle and fiber board production shows great potential. The thinning products of afforestation projects also provide additional raw material for both pulp or fiberboard industries in the newly reclaimed desert areas. There is thus high integration between environmental effects and industrial use of afforestation activities.

FUTURE STRATEGIES

The problems of combating desertification through the reclamation of desert areas are of great importance to Egypt. However, there must be a logical extrapolation of the ever-present Egyptian agricultural growth curve. Any strategy to reclaim desert regions should consider the time factor. For Egypt, a short run strategy of 25 years must be included as the first phase of a 100-year long run strategy.

Consideration of annual water resources shows that about 68.2 million m^3 are available. This volume comes from the Nile (55.5), drainage water (12.2) and underground water (0.50). With the current estimate of 51.41 million m^3 water requirement for the nation, it can be seen that a possible 2.5 million feddans can be added to the current cultivated area in Egypt. Within the next decades added water resources from the Nile projects in Africa would be around 9.0 million m^3 (same as Sudan's share). This would allow the addition of 1.5 million feddans to the cultivated land. Another 0.65 millions can be added through the use of underground resources.

Based on studies made by the Ministries of Irrigation, Reclamation, and Agriculture, 3.5 million feddans can be added to the cultivated land using available resources. Another 0.3 million can be added based on rain and underground water outside the Nile Valley.

Careful study of the surveyed desert areas for future projects along with analysis of water requirements for crops, based on their contribution to the agricultural economy, is needed. Such studies may alter current cropping systems in the newly reclaimed lands.

The introduction of the private sector to land reclamation activities should prove beneficial. This should be encouraged by the authorities along with local cooperatives, firms and companies. Current reclamation possibilities cover the following areas.

Sinai and East Suez: A pioneer working project started in 1960 with 15,000 feddans in that area. Land reclamation activities were reassumed in 1973 in the region and promise possibilities for a larger project. The establishment of green belt protection of these newly reclaimed lareas is of great importance to prevent advancing desert.

Western Desert: According to recent estimates of water and land resources, it can be estimated that agricultural development can be established on 505,000 feddans apart from the currently cultivated areas.

South New Valley: It is estimated that a large development project can cover 1.3 million feddans in that region. This project would include agriculture and industrial centers along with newly planned cities. Feasibility studies are needed for the cropping pattern and to evaluate the different productive parameters involved. This project includes areas for both agricultural and industrial development programs that would develop a huge area in the southern Egyptian desert. Carefully planned protection belts would go hand in hand with reclamation projects to guard against desert encroachment.

It is necessary to consider the details of establishment of desert communities along with the national agricultural and reclamation policy. The future strategy of the country should include a complete plan with a short run phase to the year 2000, part of a long term plan covering 100 years. A major part of these activities would go hand in hand with the establishment of new cities and urban communities in the newly reclaimed desert regions. This, along with a farsighted integrated reclamation plan should promise a breakthrough to face the huge population increase that will add 30 million to the current Egyptian population by the end of the century.

REFERENCES

Arab Republic of Egypt. United Nations Conference on Desertification Country Report. Nairobi, Kenya. 1977.

O.A. Badran (1978). Green belt project for North Africa: Feasibility Study. ALECSO, UNEP. Kenya.

M.M.D. Ezz (1972). An ecological and physiological study on some timber tree species grown in northwestern coast of Egypt. M.Sc. Thesis. University of Alexandria.

S.A. Kandeel, O.A. Badran, M. El Osta and H. Aly (1978). Complete tree pulping of young Eucalyptus windbreak trees. Chemistry and Pulp & Paper Session 32nd Forest Products Research Conference USA.

K.H. Oedekoven (1970). United Arab Republic, Afforestation in Arid Zones (R.N. Kaul).

United Nations Conference on Desertification Secretariat (1977). Desertification: its cause and consequences. Nairobi. Pergamon Press, Oxford.

THE DEVELOPMENT OF SINAI

RUSHDI SAID
Consulting Geologist, Cairo

ABSTRACT

The Peninsula of Sinai offers possibilities for development, having a temperate climate, a strategic location with a reasonable infrastructure, and a variety of natural resources. Water resources that could be used for agricultural development are limited to the northern coastal areas and around the mouth of Wadi El Arish, a trunk channel which drains almost two-thirds of the surface area of Sinai. The Gulf of Suez and its coastal strip is rich in oil and a variety of proven mineral resources, including manganese, glass sand, kaolin, gypsum, anhydrite, salt and a large assortment of building and industrial materials. These could form the base of thriving industries which could draw their energy from gas now wasted from the existing oilfields and desalinated sea water; the latter will have to furnish domestic needs. The Peninsula also offers possibilities for tourism and recreation, possessing rich scenery, beautiful beaches, spas and a variety of archaeological sites.

It is perhaps appropriate to begin our survey of the desert potential of Egypt with Sinai, which is not only in the center of the news today, but is unique in its potential for development and for the significance of its development. Sinai occupies a strategic position linking Asia and Africa, bordering the Mediterranean Sea and the Red Sea. Throughout history it was the gate through which waves of immigration, trade or invading armies entered Egypt. In recent years, Sinai has represented a threat to Egypt and its development was hampered by the fact that it was looked upon as a barrier that could block imminent invasions from the east. Today, the view has changed and the development of Sinai is looked upon not only as the best weapon for its defense, but if good will prevails, as a bridge rather than a barrier.

Because of its strategic position Sinai boasts of having the best infrastructure of any of the deserts of Egypt. It has a railway line which crosses the Suez Canal near Ismailia, follows the coast and joins Egypt with Palestine and other Arab countries. Today, work is in progress to construct several tunnels underneath the Suez Canal to facilitate movement to and from Sinai. Sinai also has a network of roads, about 3800 kms of macadamized roads, i.e., about 1 km for every 15 km^2. In other Egyptian deserts the density of macademized roads is 1 km for every 200 km^2 in the Eastern Desert, and 1 km for every 800 km^2 in the Western Desert. In addition, Sinai has a large number of airports and great potential for port development along its shores.

Another unique feature of Sinai is its climate, which is amiable, Mediterranean in aspect and less harsh than that of most other deserts of Egypt. Its scanty rainfall and topography allows easy movement and maneuverability. It was, after all, the land of the "wanderings."

Sinai also possesses ample natural resources. From the point of view of agriculture, its coastal western part across from the Suez Canal is an old flood plain of the Nile and represents a natural extension of the agricultural expansion of the Nile Delta. More than 25 years ago, plans were devised to irrigate about 250,000 feddans by Nile waters coming through a canal running in a tunnel underneath the Suez Canal. The project was studied by a group of UN experts. A survey has been made on every aspect of the project and it was found to be viable. The project was not implemented for political reasons; Sinai was then considered a barrier rather than a bridge.

In addition, potentialities for minor agricultural development exist along the coast where a 200 mm rainfall brings sporadic, but successful, cultivation.

Groundwater potential also exists in the delta of Wadi El Arish, a wadi which drains two-thirds of Sinai. Even though the rainfall is small, its volume gathers up until it reaches the delta of this wadi. Close to 200,000 m^3 of water can be drawn per year from this groundwater reservoir. This could reclaim roughly 40,000 feddans of land or, better still, it can be used for industry.

Sinai possesses ample energy sources. It has the only commercially extractable coal field in Egypt. In Gebel Moghara in northern Sinai half way along the Ismailia-El Arish road, coal has been discovered in Jurassic beds which crop out at the core of the anticlinal structure. When war broke out in 1967, this coal field was about to be developed. The plan envisaged the transport of the coal to Egypt to be used either as a thermal source or for partial coking. Today it may well be worthwhile considering that this coal be used as a source of energy for industries that could develop in Sinai itself. The proven reserves of the Moghara coal field, as worked out by the Geological Survey of Egypt, amount to 35 million tons.

The coastal strip of the Gulf of Suez and the offshore areas again offer great potential for the discovery of new oil and gas fields. Today, oil fields of Belayim, Belayim Marine and El Tor produce about 150,000 barrels/day. Offshore areas of Northern Sinai and the Mediterranean have great promise of oil. Recent reports of the discovery of gas fields in the sub-recent deltaic deposits in northwest Sinai are of interest. They may be related to similar discoveries in the deltaic plain of Egypt.

Among the mineral resources mention must be made of the sedimentary manganese deposits known in the Carboniferous rocks which crop out to the east of Abu Zeneima along the Gulf of Suez which have since the early years of this century been exploited. Most of the production was exported, although in recent years smaller quantities were used in the iron and steel furnaces of Helwan and plans were made to use it totally in Sinai and convert it to ferromanganese. Power was generated by the use of the gas flare of the Belayim oil fields transported along the coast to Abu Zeneima. The average yearly production until 1967 was 200,000 tons of low grade ores averaging 28.4% Mn and 17.4% Fe. Small quantities of high grade Mn ores were also won.

In addition, Sinai possesses a variety of resources ranging from gypsum, kaolin, glass sand and various building materials. Most of these are of very high quality and were extracted on large scales prior to 1967.

This note shows that Sinai possesses an industrial potential. It does possess energy resources and the basic raw materials that can be used to build a self contained industrial complex. The most suitable place is along the coast of the Gulf of Suez where most of these resources are concentrated. Here, sea water could be used for industrial purposes and desalinated later for domestic purposes.

Sinai has also unlimited possibilities for tourism and recreation. Its long beaches with palm trees along the Mediterranean, mountains, spas, historical sites and its climate all make Sinai one of the best areas for this industry.

In conclusion, it can be stated that Sinai has unique characteristics which permit its development on a scale that will give a reasonable standard of living for a number of people who must be trained in the use of the desert environment and who must be aware of the limits of desert development. It would be futile to begin any reclamation or development of the desert, the last frontier open to overcrowded Egypt, with people who have the current attitude of misusing the environment and proliferating without control.

DEVELOPMENT OF NATURAL AGRICULTURAL RESOURCES IN THE SINAI PENINSULA

ABDU A. SHATA
Director, Desert Institute, Mataria, Cairo

ABSTRACT

In the Sinai Peninsula, seven geomorphological districts are distinguished, each of which has its own potential of natural agricultural resources:

1. *Wadi El Arish - Wadi Ghaza Basins, including the Mediterranean littoral;*
2. *Wadi El Giraf basin, draining into the Red Sea;*
3. *West Sinai morphotectonic depression, including the Gulf of Suez littoral;*
4. *The littoral zone of the Bitter Lakes, including the eastern dune complex;*
5. *The northeast extension of the Nile Delta flood plain, including the littoral zone west of El Bardawil;*
6. *South El Bardawil dune complex;*
7. *Southeast Sinai morphotectonic depression, including the Gulf of Aqaba littoral pockets.*

The soil and water resources in each district are described and discussed briefly.

GENERAL OUTLINE

Certain facts must be mentioned when discussing the development of the natural agricultural resources of the Sinai Peninsula (62,000 km^2).

1. The location of Sinai within the arid belt of North Africa and southwest Asia and its extension, northward, into the temperate belt of the "Mediterranean Climatic Region." This explains the general occurrence of better land and water resources in the northern portion, particularly in the Mediterranean littoral zone, compared to the southern portion.

2. The distinction of thelandscape in Sinai into different "morpho-tectonic districts," comprising the elevated granitic horst in the south (+2600m), the calcareous homoclinal plateau in the center, the anticlinal ridges in the north and the faulted depression in the east and west. This accounts for

the formation of complex watershed areas and drainage basins. Some of these basins are steep, others are gentle and their uptake areas extend principally into the elevated horst and into the calcareous plateau of South and Central Sinai. Such basins become occasionally active during the rainy seasons (October to February). There is evidence that the initial phase in the formation of these basins took place during the Late Tertiary and Early Quaternary. Due to lack of information of the climatological records for most of South and Central Sinai, little can be mentioned about humidity and surface runoff, which are among the important factors affecting the landscape in this part of Sinai.

3. The occurrence in Sinai of wide tracts covered by non-consolidated deposits (provisionally termed "soils," at least 15,000 km^2). Such deposits represent the weathering products of the different rock units exposed and were formed under alluvial, lacustrine and aeolian conditions. Again, these deposits have a wide geographical distribution in the littoral zones, in the synclinal depressions and in the drainage basins. The thickness of such deposits varies from a feather edge to many meters, as evidenced by boring in the alluvial deposits of Wadi El Arish and by field observations in the areas covered by drift sand accumulations in North and West Sinai.

4. The occurrence in Sinai of major groundwater aquifers, which extend eastward into the rainy areas of the East Mediterranean Region. These comprise essentially the Nubian Sandstone, the Eocene and Cretaceous carbonate rocks and the Quaternary calcarenites. When explored adequately, such aquifers may contribute tremendously to the water supply of Sinai.

5. The location of West Sinai within reach of the surface water supply of the Nile. This water has been used, before the 1967 war, for the irrigation of limited areas of Northwest Sinai (opposite Ismailia), but it is expected that wider areas can benefit from this water in the foreseen reclamation projects.

For the convenience of treating this subject, the natural agricultural resources, and in view of the geomorphological setting, the Sinai Peninsula will be subdivided into seven hydrographic districts (Fig. 1).

WADI EL ARISH BASIN

This basin has an area of about 22,000 km^2 and extends beyond the limit of Sinai into El-Naqab area. From south to north, this basin has a length of about 400 km and the

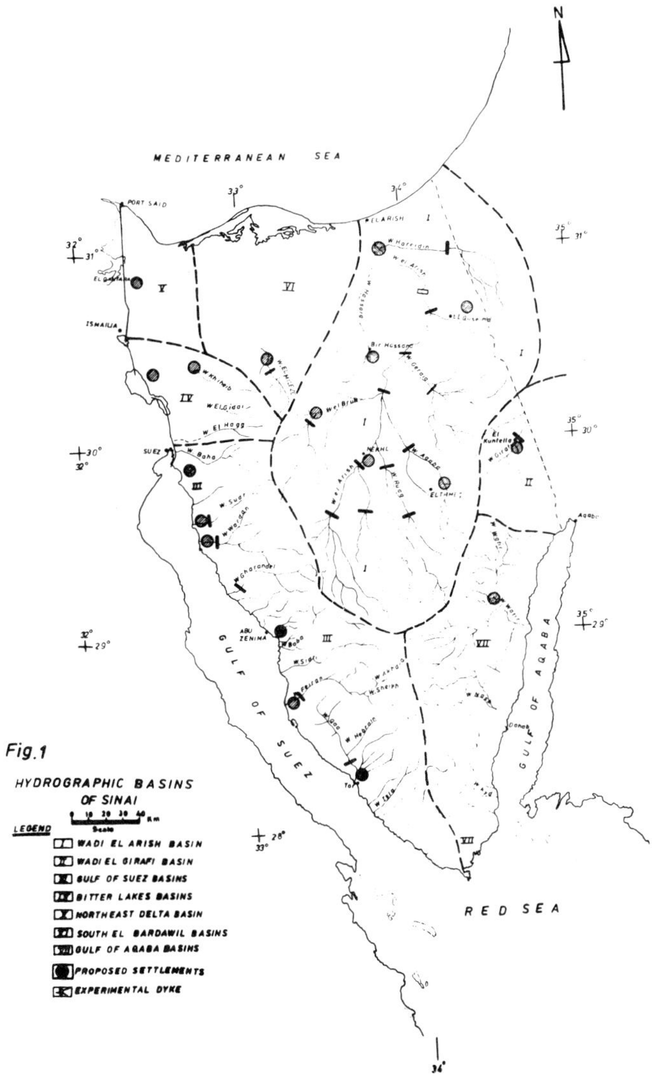

Fig.1

HYDROGRAPHIC BASINS OF SINAI

difference of relief is of the order of 1000 m. In Wadi El Arish basin, the nonconsolidated deposits cover an area of about 6000 km^2 and are differentiated into two main groups:

(a) The calcareous alluvium; this group dominates the present stream floods as well as the old flood plains. The alluvium is occasionally mixed with gravel and sands and is subjected to gully erosion in some places and to wind deflation in others. The calcareous alluvium covers wide low land areas around Nekel in Central Sinai, between El Mitmetni and El Daiqua in North Central Sinai and between El Rawafa and El Arish town in North Sinai.

(b) The aeolian deposits; this group dominates the northern portion of this basin, and is particularly developed along the littoral zone. Such deposits take occasionally the form of dunes, either oriented parallel to the present coast (3-W), or to the delta of Wadi El Arish (N-S). Again, such deposits are in many places mixed with the calcareous alluvium and account for the formation of a light textured soil, which is considered favorable for cultivation.

In Wadi El Arish basin, the strata of hydrogeological interest comprise the following:

(a) The Nubian Sandstone (Paleozoic and Mesozoic). This formation has a thickness of about 600m (increasing gradually to the north). This formation is either exposed or reported in the subsurface at variable depths (controlled both by the local as well as the regional geological structure). Boring in this basin both for oil and for water revealed the occurrence of artesian water in the Nubian Sandstone (in some wells the water rises to about 200m above sea level). The water is generally brackish and the salinity is of the order of 1500 ppm. The Nubian Sandstone aquifer has neither been explored nor exploited on any reasonable scale. There are indications from its wide geographical distribution, where it extends to the recharging areas in the East Mediterranean Region, that it can form an important source for groundwater in Wadi El Arish Basin and that it can contribute to the supply of other aquifers (principally the carbonate aquifers overlying it).

(b) The carbonate rocks (Secondary and Tertiary). These are least explored in Wadi El Arish Basin, but the association of a good number of natural springs (exemplified by El Gedeirat near El Qusaina) reflects the high ground potential of such rocks.

(c) The oolitic carbonate rocks or calcarenites (Quaternary). These are best developed along the littoral portion of Wadi El Arish Basin and in the delta of the wadi itself. The water in such rocks exists under subartesian conditions and depends for its recharge on local and regional precipiattion and on seepages from other aquifers. Although the water produced from this aquifer is brackish (1000-2000 ppm), it is used on a large scale for irrigation. Other aquifers in the Quaternary rocks are known in Wadi El Arish Basin but the potential is rather limited.

With regard to the surface water supply in Wadi El Arish Basin, this is composed of the complex system of the master stream and its derivatives, which include Wadi El Rowaq, Wadi El Bruk, Wadi El Aqaba, Wadi Geraia, Wadi El Hassana and Wadi Haredein. This complex system becomes occasionally active during the rainy seasons and the amount of surface runoff has been estimated at 60 million m^3. From this point of view, Wadi El Arish Basin forms an ideal catchment area for the conservation of surface water, and detailed studies are needed.

From the agricultural point of view, the water and soil occurrences in Wadi El Arish Basin may be differentiated in two, unequal, divisions:

(a) The northern division: this comprises the littoral zone as well as the delta of the main stream. Although this division is adequately developed, it needs better management of both land and water resources.

(b) The southern division: this is almost undeveloped. Conservation of surface runoff water and establishment of pilot areas around deep test water wells are foreseen as recommendations for that portion of Wadi El Arish. A number of experimental earth dykes for water spreading and a number of settlements (e.g., at Nekhel, El Hassana, Sudr El Hetan, El Qudaeima, Lihfin) can provide useful information that can be implemented in the further development of this important portion of Sinai.

WADI EL GIRAF BASIN

This basin has an area of about 2000 km^2 and its surface slopes from west to east. It extends into El Naqab area and ends in the Dead Sea depression. The uptake areas of this basin are located in the eastern uplands of El Tih Plateau.

The surface area of this basin displays landforms similar to Wadi El Arish Basin, but the nonconsolidated deposits have a higher content of sands and gravels. In this basin, the surface runoff has been estimated at 4 million m^3 and the

Nubian Sandstone is the main aquifer (many details are lacking, however).

No organized agricultural development is practiced in this basin, and it is expected that the establishment of a small number of earth dykes and at least one settlement at a test well tapping the Nubian Sandstone aquifer, can answer several questions about its potential.

GULF OF SUEZ BASINS

These comprise the series of hydrographic basins, at least 20 in number, which drain West and Southwest Sinai into the Gulf of Suez depression. The drainage basins occupy an area of about 14,500 km^2 and their uptake areas extend into the water-divide in South and Central Sinai. Such basins, best exemplified by Wadi El Qaa, Wadi Feiran, Wadi Sidri, Wadi El Taiba, Wadi Gharandel, Wadi Sudr, etc., took on most of their present shape in the Late Tertiary and Early Quaternary, and they act as active drainage arteries during the rainy seasons in some wet years.

In the courses of the drainage basins of West Sinai, alluvial deposits and occasional lacustrine deposits are distinguished and take the form of well developed terraces, rising above the present stream floor. The downstream portions of such drainage basins are marked by the occurrence of deltaic deposits dominated by alluvium intermixed with sands and gravels.

In West Sinai fresh water is not available and brackish water is reported in various formations including the Jurassic, the Cretaceous, the Neogene and the Quaternary. In some of these formations the water exists under artesian conditions and the head is attained from the recharging areas located at higher altitudes, in and outside Sinai, presumably to the east. The groundwater occurrences of this portion of Sinai has not been explored on any reasonable level. The Desert Institute has already started a program of hydrogeological investigations which is expected to cover much of that area.

The present agricultural development in West Sinai is restricted to patchy areas located at El Tar, Abu Zeneima, Sudr and Ayoun Musa. It is expected, however, that wider areas can be cultivated provided adequate management of available natural resources is supplied. While exploring such resources, a number of experimental earth dykes across a selected number of the drainage basins and a number of small settlements is proposed for this portion of Sinai.

BITTER LAKES BASIN

This district occupies an area of about 3000 km^2 and is bounded on the east by the elevated calcareous plateau of Umm Khesheib. This plateau is broken by the channels of Wadi El Giddi and Wadi Umm Khesheib. These two wadis are directed from east to west and are lost in the area covered by sand dunes which stretch between the plateau and the Bitter Lakes. The eastern fringes of the Bitter Lakes are dominated by mud flats, which are subjected to salinization processes.

No information is available at present about the ground water possibilities and also about the surface runoff. The amount of water received annually by the area during the rainy season has been estimated at 150,000,000 m^3. It is, therefore, hard to estimate the agricultural potential of this portion of Sinai. Portions of the mud flats opposite Ismailia, however, placed under reclamation in 1966 by using Nile water syphoned underneath the the Suez Canal, are presently being evaluated by the Desert Institute.

Due to the geological setting, there should be aquifers of the Nubian Sandstone, and a deep test well and small settlement can be recommended for the eastern portion of this district, along Wadi Umm Kheshib or Wadi El Giddi.

NILE DELTA NORTHEAST EXTENSION

This district has an area of about 2000 km^2, is dominated by salinized alluvial plains of the Nile Delta and is densely covered by drift sand accumulations. This portion of Sinai displays salient features similar to the southern fringes of El Manzala Lake and other peripheral lakes in the Nile Delta region. The evaluation of the agricultural possibilities of this portion of Sinai depends on how much water can be syphoned to it from the Nile.

SOUTH EL BARDAWIL BASINS

This district has an area of about 6000 km^2 and extends from the anticlinal ridges of El Maghara to the coast at El Bardawil Lake. The surface is dominated by complex dune sand accumulations, and this accounts for its inaccessibility. At the approach to El Bardawil Lake, the surface is developed into salinized mud flats.

With the exception of the dense date palm groves scattered in this district, almost no signs of agricultural activities are seen.

In view of the scanty information about the deep groundwater aquifers in this district and in view of the nonoccurrence of an adequate soil cover, little can be stated about future agricultural development. However, along the course of Wadi El Mileize, located in the southern portion, an experimental dyke and a deep test well for water is initially recommended within the framework of an intensive program aiming at the evaluation of existing natural agricultural resources.

GULF OF AQABA BASINS

The natural agricultural resources are least understood in this portion of Sinai (12,500 km^2). Field investigations are necessary, and with special emphasis on hydrogeology, morphopedology and plant ecology. Because the area is rugged, use of remote sensing techniques in mapping can be recommended.

THE WESTERN DESERT OF EGYPT, ITS PROBLEMS AND POTENTIALS

FAROUK EL-BAZ
Research Director, Center for Earth and Planetary Studies, Smithsonian Institution, Washington, D.C. U.S.A.

ABSTRACT

The Western Desert occupies two thirds of the area of Egypt. It is basically a platform of sedimentary rocks crossed by numerous belts of sand dunes that inundate whatever stands in their way. The flat terrain is broken by numerous depressions enclosing oases that provide the only respite from the prevailing harsh conditions. Early exploration in the 1920's by Hassanein and Kamal El-Din and in the 1930's by Sanford, Peel, and particularly Bagnold provided the basic data on this desert. However, to this date much of the desert is relatively unexplored. Space photographs and images provide a useful tool in the selection of areas for detailed field study.

The most potentially valuable aspect of the Western desert is the occurrence of clayey, fertile soils in many locations. Some of these lie above a reservoir of fossil underground water, which if carefully used could supply enough water for irrigation. Additional possibilities for development include deposits of iron, phosphates and other valuable minerals as well as a vast reservoir of building materials, particularly those necessary for cement production.

INTRODUCTION

The Western (Libyan) Desert of Egypt lies on the northeastern border of the North African Sahara. It occupies 681,000 km^2 or more than two thirds the land area of Egypt. Early travelers, geographers and explorers are credited with contributing to our knowledge of this desert and its ways.

The ancient Egyptians are known to have traveled extensively in the Western Desert and settled in many parts of it. The interest in this desert continued through the time of the

Romans as indicated by their numerous remains. However, the first truly scientific account of part of this desert was that of Mohamed Ibn-Abdulla Ibn-Batuta. Starting in 1325, he spent twenty years traveling in North Africa and the Arabian Peninsula meticulously recording his observations.

After Ibn-Batuta little progress was made until the end of the eighteenth century when the "Association for Promoting the Discovery of the Interior Parts of Africa" was established by Sir Joseph Banks in 1788. For nearly a century, the exploration of Africa captured the imagination of westerners much as spaceflight did in the 1960's. Although the main expeditions were dedicated to finding the source of the Nile and charting its course, much information was gathered on the Sahara Desert.

Scientific exploration of the Western Desert began with Gerhard Rolfs. He was a German soldier who, in order to prepare himself for desert travel, joined the Foreign Legion in Algiers. He made an expedition in 1874 traveling over 650 kilometers from Sallum on the Mediterranean coast to the heartland of the desert crossing the Great Sand Sea. Zittel[1] and Beadnell[2,3,4] described parts of the desert. Also, two major expeditions were conducted in the 1920's by two Egyptian explorers, Mohamed Hassanein Bey and Prince Kamal El-Din, who reached the remotest corner of Egypt at the border intersection with Libya and Sudan. Another notable geomorphological contribution was made by Ball[5].

Ralph Bagnold of the Royal Corps of Signals pioneered a new era of desert exploration in the 1930's. Using Ford automobiles, his select team of explorers covered vast tracts of the Western Desert reaching most of its remote corners. His explorations and their accounts (Bagnold)[6,7,8,9] provided much of the basic information on the Western Desert in general and on the movement of dune sand in particular. Other classical works were written during the same period such as those by Sanford[10] and Peel[11].

More recent and detailed geological studies of the Western Desert were reviewed and summarized by Said[12]. The regional geology of this desert has also been summarized in the map that was published in 1971 by the Geological Survey of Egypt. During the last decade, orbital photographs obtained by astronauts of the Gemini, Apollo, Skylab, and Apollo-Soyuz missions (e.g., El-Baz)[13] and images transmitted by the unmanned Landsat missions (e.g., El-Shazly et al)[14] have been used in studying the Western Desert and its features.

TERRAIN CHARACTERISTICS

The Western Desert is a physiographic region of Egypt that stretches westward from the Nile Valley to the Libyan border and northward from the border of Sudan to the Mediterranean Sea (figure 1). It is basically a rocky plain which slopes gently to the north. The generally flat plateau desert of sedimentary rocks is broken by depressions, which enclose oases, and by granitic mountains in the Oweinat region at the border intersection between Egypt, Libya and Sudan. These mountains and the Gilf Kebir plateau just to the northeast constitute the only prominences that are higher than 1000 m above sea level. The rest of the central and southern parts of the desert average 200 to 500 m in elevation, and rocks of the northern belt are less than 200 m above sea level.

The seven major depressions in the Western Desert are bounded by escarpments which appear to be erosional features. However, the boundaries of some indicate control by intersecting faults. This suggests that the original cause of the depressions was the fracturing of the sedimentary platform. Fracturing allowed erosional mechanisms to excavate the rock along the zones of weakness.

The largest of the Western Desert depressions is the Qattara, which occupies, below sea level, approximately 19,500 km^2, or 1/15 the land area of Egypt (figure 1). This depression also contains the lowest point on the continent of Africa, being 134 m below sea level. Approximately 5,800 km^2 of the floor of the depression is covered by *Sabkha,* which is made of water saturated salt that is thinly covered by sand. The rest of the floor is covered by sand, pebbles, clay, and exposures of limestone. Much like other depressions in the Western Desert, its northern borders are clearly marked by a distinct scarp, whereas the depression grades into the terrain to the south.

The Qattara Depression figures prominently in plans for development of the Western Desert. A project now under study aims at the generation of electric power using two natural phenomena: the flow of water from higher to lower levels, and its evaporation by the sun's heat. It will entail bringing water from the Mediterranean Sea to the edge of the depression via an open canal or an underground pipeline. The gravity flow of water will be used to run turbines to generate electricity. The high rate of evaporation will assure removal of water until a layer of salt is formed in the depression. It

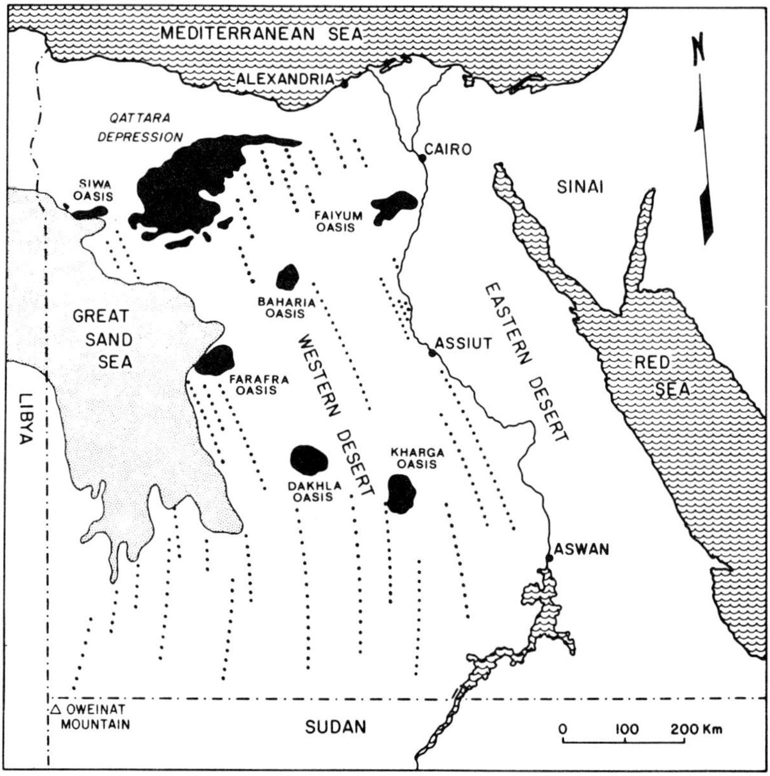

Figure 1: Map of Egypt showing the location of the seven major depressions in the Western Desert. Dunes of the Great Sand Sea occupy much of the west central part of the desert. In addition, numerous longitudinal dunes occur throughout. Some of these are made of one enormous dune like Ghard Abu Muharik, which stretches from Baharia to Kharga. Others are made of dune bundles, e.g., near Qattara.

is calculated that the electric power to be generated from this project would be up to a maximum of 10,000 megawatts, which is nearly five times that generated at the High Dam south of Aswan. Apparently the required electricity output in Egypt by the year 2000 will equal approximately 15,000 megawatts. It must be stated that the effects of the "Qattara Salt Lake" on the underground water reservoir in the northern Western Desert are not yet known. Also the project plan requires detailed topographic maps of the region which are not available.

The depression in which Siwa Oasis occurs is irregularly-shaped and generally elongate in an east-west direction. It is the farthest depression from the Nile Valley and is approximately 50 km long and covers 300 km^2. The irrigated area within the depression is 90 km^2 and parts of it are 60 m below sea level. About 15 km east of the Siwa Depression is another depression about two thirds its size; both are bounded by faults that intersect in an "X" pattern. In both places the lowermost stratum is a non-porous shale layer. This layer prevents easy drainage of irrigation water, which is allowed to stagnate and evaporate leaving a crust of salt on the land. Potentially valuable rocks in the Siwa Depression include marble, highly pure crystalline limestone, and good quality Egyptian alabaster.

The southern borders of the Siwa Oasis blend with the northern boundary of one of the largest dune fields of the Sahara, the Great Sand Sea. Near Siwa, the dunes of this sand mass are reminiscent of the Algodones dunes southeast of the Salton Sea in California. However, the Great Sand Sea displays some of the largest dunes in the world and covers an area that is 200 times that of the California dunes. The Great Sand Sea's formidable zone is 500 km long and 200 km wide. The main sand mass is characterized by enormous whaleback dunes with superposed sharp-crested longitudinal dunes. The gently sloping whaleback dunes are relatively static, whereas the longitudinal dunes shift as the wind blows the sand southward.

The Faiyum Depression is closest to the Nile Valley and occupies a circular area of approximately 1,700 km^2, much of which lies below sea level. The Bahr Youssef canal carries Nile water to the depression where it drains into Lake Qaroon (200 km^2) in addition to El-Gharaq, a small enclosed basin on the south side of the depression. Herodotus, who visited the area in 450 B.C. described an artificial lake 3,600 furlongs in circumference and 50 fathoms deep. According to him, water from the Nile flowed into the Lake during six months of the year and drained back to the Nile the other six months. This Lake Moeris is now believed to have existed in the southern part of the depression where large embankments exist. The cliffs of the Faiyum Depression expose limestone of good quality for use as building material and/or in a cement industry.

The Baharia Depression is spindle-shaped and is elongate in a northeast direction. It is approximately 95 km long and 45 km wide and covers an area of 1,800 km^2. The escarpment that bounds the depression enclosing it on all sides, is approximately 100 m high. In addition to this feature the depression is unique among others in the Western Desert in that hills form a discontinuous line in its center. Many of these hills such as Gebel Hefhuf and Gebel Mendisha are partly made of the black-colored dolerite rock or capped by it. Also aligned with these hills are basaltic intrusions, particularly in the northeast part of the Depression.

The Baharia area has two essential elements for economic development. First, iron ore occurs in the area; it is mined at El-Gedida and is being evaluated in several other localities. El-Gedida ore is shipped by rail to the steel mill at Helwan south of Cairo. Second, the floor of the depression is composed of sandstone and variegated shale; many patches of it constitute arable soil. Also ground water exists in sandy subsurface horizons. Wells occur along the central line of the depression and near its eastern boundary.

The Farafra Depression is roughly triangular in shape and is bounded by high scarps on the east and west and a low scarp on the north. It is open towards the south where it meets the Dakhla escarpment some 200 km away. Most of the exposed rock in the depression is chalk of Maastrichtian age (Said).[12] The eastern half of the depression is covered by dune sand. Field observations indicate that this sand mass is similar to that of the Great Sand Sea. The large dunes are low-sloping whaleback dunes, several kilometers across. Superposed on these dunes are sharp-crested, partly sinuous longitudinal dunes.

Because of the prevalence of thick sand deposits in the Farafra Depression, the only areas where soil is suitable for agriculture is in the vicinity of Qasr El-Farafra, and particularly to the south near Abu Minqar. This is significant because in the latter location the ground-water reservoir is larger, less saline and under higher pressure (up to 20 times the atmospheric pressure) than farther southeast of Dakhla and Kharga.

The Dakhla Depression lies south of an escarpment that trends west-northwest from Kharga for approximately 250 km. South of the escarpment the exposed rock belongs to the Nubia Sandstone formation, which covers most of the southern half of the Western Desert. However, the erosion of the chalk and shale layers that are exposed on the scarp provides a layer of arable soil. For this reasons, agricultural communities are lined parallel with the scarp.

In addition, east of Dakhla at Abu Tartur Plateau, a phosphate layer occurs that is up to 6 m thick and 15 km long. Experimental mining of this deposit has begun and it promises to be one of the major phosphate deposits in North Africa. This is particularly true since low-grade phosphates cropout in a nearly continuous layer, 60 to 100 cm thick, all along a scarp that stretches from Abu Tartur in a west-southwest direction for some 300 kilometers.

The Kharga Depression is the only one in the Western Desert that is aligned in a north-south direction. The succession of rocks from the floor upward begins with Nubia Sandstone, which is overlain by purple and variegated shales, phosphatic beds, Dakhla shale, chalk, Esna shales, Thebes limestone, and travertine and loess on top. The depression is open to the south and southwest. The southern opening made the Kharga Oasis easy to reach by camel caravans from the interior of Africa along the celebrated Darb El-Arbein road. The width of the depression varies from 20 km in the north to 80 km in the south. The communities in the Kharga Depression occur along a line that trends in a north-south direction. This line is a major fault, which allowed the settlers to reach underground water that used the fault as a channel way for upward movement. The underground water reservoir at Kharga is smaller, more saline and is under less pressure than those farther to the west. It is believed to be made of fossil water (old and not rejuvenated) that is, some 30,000 years old (E. Degens, pers. comm.). Less is known about the underground reservoir of the region due south of the Kharga Depression, which is known as "Ganoub El-Wadi" or the South New Valley.

Bundles of active sand dunes hamper the development of Kharga and to a lesser extent Dakhla. The sand atop the high plateau to the north accumulates in broad dunes of the longitudinal type. The prevailing sand-carrying winds are towards the south as indicated by the numerous north-south-trending lines (figure 1). As these broad dunes reach the scarp, the sand creates patterns that are reminiscent of waterfalls. After the sand is carried south of the scarp, it forms small and relatively fast moving barchan dunes. The speed of the Kharga barchan dunes has been measured to vary between 20 and 100 m per year (Embabi).[15]

These barchans inundate roads, telephone lines, agricultural fields, and houses along their way. In some instances, entire villages in the Kharga Depression have been completely buried by the shifting sands, for example Ginah village whose inhabitants had to abandon it in 1970 and settle elsewhere. Barchan dunes are also encroaching on the Kharga-Assiut road which is the life line of communications between Kharga and Dakhla oases and the Nile Valley.

Beyond Kharga and southwest of the other oases, the terrain of the Western Desert is flat to undulating; it is interrupted by two major highs in the southwestern corner of Egypt. The Gilf Kebir plateau is made up of dark, resistant layers of Nubia Sandstone, which Issawi[16] prefers to designate the "Gilf Sandstone." The sandstone is locally interrupted by basaltic intrusions up to a kilometer across (Peel).[17]

Farther to the south, the region consists of flat-lying, Cretaceous sandstone overlying Paleozoic sandstone that rests unconformably on basement rocks. This sequence is interrupted by 800 to 1400 m high mountains. The most prominent of these is the Gebel Oweinat that is composed of a Precambrian complex of granites and metamorphic rocks, which have not yet been studied in detail. It is conceivable that the Oweinat area has economic concentrations of rare or radioactive elements such as uranium. These usually occur in granitic plugs or as weathering products of granites.

Detailed study of the Gilf Kebir-Oweinat area is hampered by the remoteness as well as by the lack of detailed maps. The existing maps of the southwestern part of the Western Desert are at 1:1,000,000 scale with 100 m contours and poor feature portrayal.

DEVELOPMENT CONSIDERATIONS

The population of Egypt has been increasing at the alarming rate of nearly 1,000,000 persons per year. This increase has been accompanied by a rise in food consumption beyond the present output of the Nile Valley. Dependence on foreign loans to augment the food that is consumed yearly in Egypt has disrupted the country's economy and promises to cripple it if the situation is not remedied.

This situation has increased the interest in the development potentials of the large expanses of unused land in the Western Desert. Emphasis is placed on agricultural land, particularly in view of the fact that the output of the Nile Valley lands has been decreasing due to three main reasons: (1) migration of people from farms to urban areas; (2) decrease of silt that is held behind the High Dam; and (3) encroachment of sand on arable land at the western borders of the Nile Valley (figure 2).

Reclamation of land from the desert is not new to Egypt. It is known that the Ancient Egyptians regulated the flow of water along the Nile for use in irrigation. In fact the agricultural land of the Faiyum Depression (figure 3, top) is due to a successful land reclamation project that was started by Amenemhet III who died in 1801 B.C. The present irrigation system in Faiyum made use of canals dug during the reign of this pharoah. The land of Faiyum is among the most fertile in Egypt, and some areas have been reclaimed in recent years at the oasis' northern boundary at Kom Oshim.

Also a strip of land just west of the Nile Delta was reclaimed from the Western Desert (Shata and El-Fayoumy).[18] This reclamation project encountered several problems, particularly because it was started in a sandy zone in the southern Tahrir Province. However, its continuation in the north was more successful since it began in a clay-rich zone (El-Baz and El-Etr).[19] Comparison of two orbital photographs taken ten years apart by the Gemini and Apollo-Soyuz astronauts indicated that nearly 1,100 km^2 of land have been reclaimed from the desert (figure 3, bottom).

These reclamation projects, however, were started close to the Nile and benefited from its water and human resources. Clayey soils elsewhere in the desert are very far from the Nile except for the area west of Kom Ombo. These areas must be reclaimed either by underground water or Nile water transported via pipelines or both. In addition, because of the sparse population in the desert, these areas lend themselves to large agricultural schemes that must be economically attractive for agribusiness investors, local and/or foreign.

From the above it follows that any agricultural development scheme in the Western Desert must be based on detailed study of water resources. The selection of areas for such

Figure 2: Aerial view of dunes that are encroaching on agricultural land west of the Nile Valley town of El-Minya, 200 km south of Cairo.

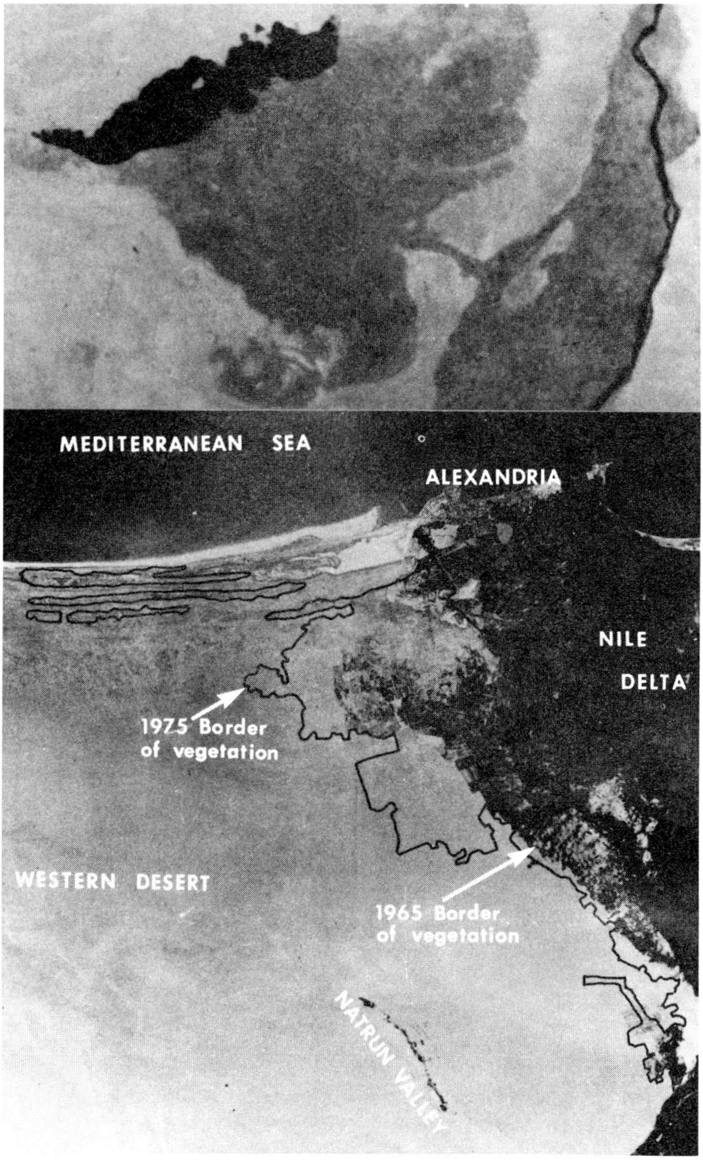

Figure 3: Examples of successful reclamation projects in the Western Desert at Faiyum (top) and west of the Nile Delta (bottom).

development must be based on the effective utilization of these resources. This includes use of modern irrigation technology such as spray or drip irrigation. Water in the Western Desert is far too valuable a resource to waste by the presently used methods of irrigation.

It is also important to consider agricultural potentials in areas of mining and industrial development. The development of the Western Desert resources must be thought of in terms of self-sufficient settlements, where food and water is plentiful and services are adequate for the community of developers. Over-dependence on products shipped from the Nile Valley increases the cost of living in the desert and hampers its development.

One factor that must be considered in the development of the Western Desert resources is the motion of sand dunes. Space photographs and images have recently been used to map the general distribution patterns of dunes (Gifford et al.)[20] 1978). However, more detailed aerial photographs are necessary to map and monitor the migrating sand dunes.

In dealing with shifting sand dunes, one could avoid them, try to halt their advance, or learn how to live with them. Avoidance of migrating dunes can be done by planning in advance. For example, if a dune belt is moving at the rate of 50 m per year, then a settlement should be built at least 25 km farther to the south. This would allow the settlement to exist without the danger of inundation for about 500 years.

The most popular method of halting the advance of dunes is planting trees in their path. In the Western Desert oases, *tamarisk* trees are usually used for this purpose, although some farmers plant palm trees instead. The trees in this case serve as a physical barrier downwind. They are effective until the dune becomes tall enough to engulf the trees themselves. In Tunisia and Algeria, higher eucalyptus trees are being planted. The eucalyptus trees work as wind breakers rather than physical barriers. Their height (up to 100 m) disturbs the wind regime, forcing the wind to unload the sand before it approaches the wall of trees.

Other methods of halting the advance of dunes include spraying them with petroleum, as was successfully but expensively done in Saudi Arabia (Kerr and Nigra),[21] or with chemicals such as "Sand Seal," a non-toxic water emulsified blend of liquid polymers and silicate derivatives. Stabilizing dunes by seeding them with grass would not work in the Western Desert because of the scarcity of water and lack of humidity.

However, it must be stated that inserting dry palm fronds on dune crests has been successfully applied in reducing the migration rate of dunes at Kharga from 30 to 4 m per year.

From nature we know that rock is eroded by sand until the aerodynamically stable prominences called "yardangs" are formed. These yardangs are spindle-shaped with rounded ends toward the wind and tapering tails downwind. It is conceivable that yardang-shaped frames made from appropriate materials can be placed in the path of advancing dunes in rows allowing the sand to pass between them. Gradually changing the orientation of such man-made yardangs in these rows might result in directing the shifting sand away from the area requiring protection. This method, however, has not yet been tried in the laboratory or in the field.

THE DEVELOPMENT BELT

The idea of a restricted zone for economic development in the Western Desert was studied by Gen. Mohamed Mohsen.[22] It proposes a development belt stretching from the western shores of the High Dam's Lake Nasser in the south to the Mediterranean coast in the north, and includes all the oases and other depressions.

This belt is bounded by rocky plateaus to the east and sand accumulations of the Great Sand Sea to the west. As shown in figure 4 the parts of this belt that are adjacent to water bodies are most suitable for developing tourism and fish industries. The rest of the belt encompasses all the land that includes patches of arable soil; clayey soil also exists near the southwest border of Egypt, but it is covered by a 15 to 25 cm layer of sand (A. El-Barqouqi, pers. comm.).

It is interesting to note that the most promising region in this belt is the one closest to Lake Nasser. The area west of the Nile between Aswan and Kom Ombo includes "nilotic" sediments as far as 50 km west of the present-day river course (Issawi and El Hinnawi).[23] These silts are mixed with sandy clay and gravel that form a plain that extends for nearly 200 km long and 10 to 40 km wide. The general slope of the plain is toward the north with a drop of about 50 m from south to north.

This area should receive more attention than other parts of the development belt because it is close to the fresh water lake and also to the densely populated Nile Valley. A development scheme may be recommended for this area in which silt-

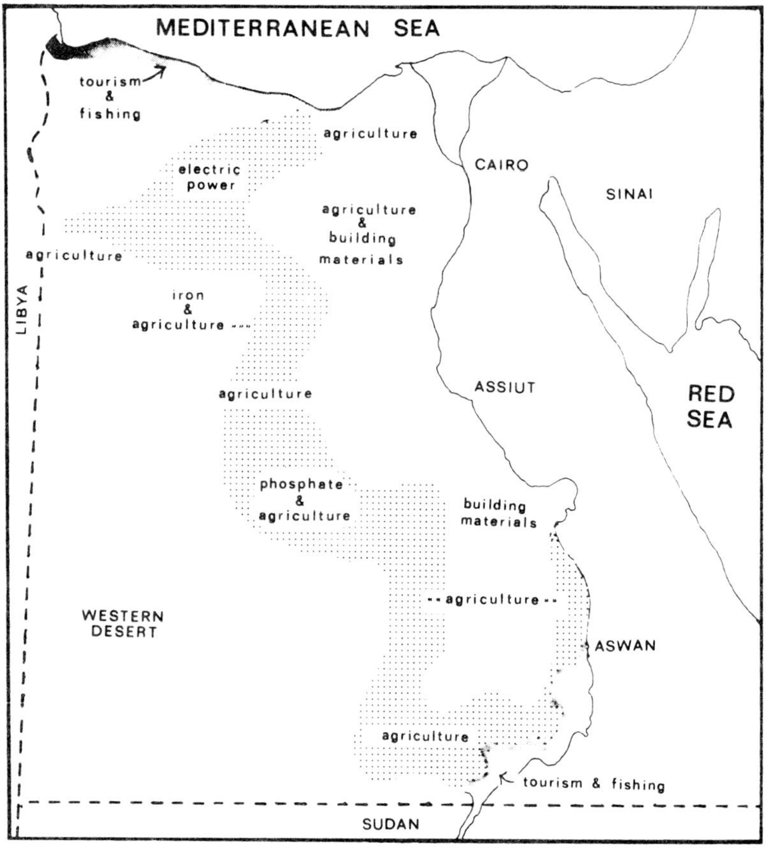

Figure 4: Map of Egypt showing the maximum extent of "the development belt" in the central part of the Western Desert (patterned area). This belt includes the seven major depressions and the South New Valley region. It also includes the coastal zones of the Mediterranean Sea and the High Dam Lake.

laden water is pumped from the lake and transported via a pipeline on the edge of the western scarp. Water may flow eastward by gravity via branch pipelines to reclaim the arable patches in the region. However, the effects of such a project on the narrow zone of agricultural land adjacent to the Nile River need to be thoroughly studied.

The southernmost segment of the development belt, the South New Valley, is estimated to contain 1,300,000 feddans surrounding the Toshka Depression. If confirmed as suitable for agriculture and later developed, it would increase the total cultivated land of Egypt by 22%. However, the problems of undulating topography, underground water availability, necessity for elaborate drainage systems, and high costs of operation are likely to reduce this figure by a factor of ten, to about 130,000 feddans.

The central segment of the belt includes the depressions of Kharga, Dakhla, Farafra, and Baharia. These oases have their own resources of good soil in patches that vary from 100 feddans (suitable for individual or family ownerships), 1,000 feddans (suitable for agricultural cooperatives) or 100,000 feddans (suitable for large-scale agricultural development). All three forms of development are necessary in solving the problems of redistribution of Egypt's population and the production of enough food for local consumption and conceivably for export.

In addition to agricultural development of the belt's central segment, its mining potentials are confirmed. In addition to phosphates and iron, the existence of limestone and shale strata in the same areas allows their exploitation for cement industries. Such industries would serve the development projects in the oases as well as the needs of the Nile Valley to the east.

The northern segment of the development belt includes the depressions of Qattara, Siwa and several smaller ones. These depressions include land that can be reclaimed for agriculture if the drainage problem is resolved. Also in Siwa, layers of marble, crystalline limestone and alabaster can be mined, particularly after the road from Siwa to Mersa Matruh is completed for shipping of products to national or international markets.

The full development of this northern segment will depend on whether or not the Qattara project will come into being. If electric power is to be generated via the flow of sea water into the depression, this will open the way for numerous industries. The "Qattara Lake" would also be used in fish industries, salt and other chemical industries, and tourism. The activities in the area would also attract part of the population from the overpopulated Nile Delta.

CONCLUSIONS

The Western Desert will figure prominently in the future economic development of Egypt because of its many potentials. These include areas of arable land that can be used to produce badly needed food for a growing population. They also include mineral deposits such as iron, phosphates, carbonates, clays, etc. However, the desert must be more thoroughly studied to allow thoughtful planning.

Self-sufficient settlements of various sizes appear to be the most suitable way of developing this desert along the median "development belt." Agricultural development may require the use of modern irrigation technology such as mechanized spray or drip irrigation. Also, the underground water reservoir may require augmentation by fresh water from the Nile. The latter may best be obtained south of the High Dam, because of the need for silt, and carried to reclamation areas via pipelines to minimize water loss by evaporation and seepage.

Based on the above it is clear that although the Western Desert has several potentials in agriculture, mining and industry, many of its features must be studied in more detail. The necessary studies include the making of topographic base maps, at 1:50,000 scale or larger. These maps would allow thorough study of areas being considered for development projects, and exploration of the yet unexplored tracts of the Western Desert.

REFERENCES

1. Zittel, A.K. (1883). Beitraege zur Geologie und Palaeontologie der Libyschan Wueste und der angrenzenden Gebiete von Aegypten. Palaeontolg. 30(1):1-112.
2. Beadnell, H.J.K. (1901). Dakhla Oasis, it topography and geology. Egypt. Survey Dept., Cairo.
3. Beadnell, H.J.K. (1909). An Egyptian Oasis: An Account of the Oasis of Kharga in the Libyan Desert. Murray, London.
4. Beadnell, H.J.K. (1910). The sand dunes of the Libyan Desert. Geog. Jour. 35:370-395.
5. Ball, J. (1927). Problems of the Libyan Desert. Geog. Jour. 70:21-38; 105-128; 206-224.
6. Bagnold, R.A. (1933). A further journey through the Libyan Desert. Geog. Jour. 83:103-129.
7. Bagnold, R.A. (1935). Libyan Sands. Hodder and Stoughton, London.
8. Bagnold, R.A. (1939). An expedition to the Gilf Kebir and Oweinat, 1938, Part 1: Narrative of the journey. Geog. Jour. 93:281-287.
9. Bagnold, R.A. (11941). The Physics of Blown Sand and Desert Dunes. Methuen, London.
10. Sanford, K.S. (1935). Geological observations on the northwest frontiers of the Anglo-Egyptian Sudan and the adjoining part of the southern Libyan Desert. London Geol. Soc. Quart. Jour. 91:323-381.
11. Peel, R.F. (1941). Dedudational landforms of the central Libyan Desert. Jour. Geomorph. 5:3-23.
12. Said, R. (1962). The Geology of Egypt. Elsevier, New York.
13. El-Baz, F. (1978). The meaning of desert color in earth orbital photographs. Photogramm. Engin. and Remote Sensing, 44(1):69-75.
14. El Shazly, E.M., *et al.* (1975). Geologic interpretation of Landsat satellite images for west Nile Delta area, Egypt. Remote Sensing Project, Cairo.

15. Embabi, S.N. (1978). The movement of barchan dunes and its effect on development in the Kharga Oasis depression (Arabic). In press.

16. Issawi, B. (1973). Contribution to the geology of Uwein-ate-Gilf Kebir area, Western Desert, Egypt. Geol. Survey Egypt. Inter. Rep.

17. Peel, R.F. (1939). An expedition to the Gilf Kebir and Oweinat, 4: The Gilf Kebir. Geog. Jour. 93:295-307.

18. Shata, A. and El-Fayoumy, I. (1967). Geomorphological and morphopedological aspects of the region west of the Nile Delta with special references to Wadi Natrun area. Bull. Inst. Deser. Egy. 17(1):1-28.

19. El-Baz, F. and El-Etr, H.A. (1978). Color zoning in the Western Desert of Egypt. In: Apollo-Soyuz Test Project Summary Science Report, Vol. 2: Earth observations and photography. NASA SP-412 (In press).

20. Gifford, A.W., Warner, D.A., and El-Baz, F. (1978). Orbital observations of sand distribution in the Western Desert of Egypt. In: Apollo-Soyuz Test Project Summary Science Report, Vol. 2: Earth observations and photography. NASA SP-412 (In press).

21. Kerr, R.C. and Nigra, J.). (1952). Eolian sand control. Bull. Amer. Assoc. Pet. Geol. 36:1541-1573.

22. Mohsen, M.A. (1977). A brief study of major proposed projects for the development of Egypt's deserts. Mil. Sur. (Arabic).

23. Issawi, B. and El Hinnawi, M. (1978). Contribution to the geology of the plain west of the Nile between Aswan and Kom Ombo (In press).

SALINITY HAZARD AND WATER LOGGING IN THE AREA WEST OF NOUBARIA CANAL

M. NAGUIB HASSAN, S. MAREI, M. MOTAGALLI, AND H. EL ABD
UNDP/FAO West Noubaria Project

Z. ARNAOUT
The General Co. for Land Reclamation

ABSTRACT

This area is a part of the Egyptian Western Desert on the fringe of the Nile Delta. Reclamation of the area started in 1956 through five main canals taking water from the Noubaria main canal.

The yield of crops at the beginning was startling but within a period of 5 years deterioration was observed. A study has been carried out to assess the cause or causes of this deterioration. It was concluded that waterlogging, secondary salinization and inseepage of saline ground water into the irrigation canals were the main causes.

Surface samples of irrigation water at the intake of the main canals at Noubaria have been analysed monthly since January 1975 till the present date, a total of 42 months.

Interpretation of the data with projections of the future situation in the years 1980, 1985, 1990, 1995, 2000 were calculated by means of simple linear regression.

THE NOUBARIA CANAL

All the water supplied to the area is pumped from the Noubaria canal. The sampling program for this canal consisted of two different parts: (a) The sampling program for the main irrigation canals in the area included sampling at their intakes from Noubaria canal and then at every kilometer of the branch canals including the suction and discharge of the pumping stations. This sampling is carried out monthly. (b) Samples were taken along the Noubaria canal itself about once every three months. The important sites along the Noubaria canal are shown in Table 1. The salinity of Noubaria canal comes from two main sources: saline inseepage from the ground water, and drainage water inflows into the canal.

Table 1: The important sites along the Noubaria canal

Site	Km
D3	53.0
D1	55.5
Nasr canal intake	57.5
M-F canal intake	65.0
New Nour drain	74.5
Thawra canal intake	79.4
Mariut canal intake	95.7
Navigation lock	100.0

SALINE INSEEPAGE FROM THE GROUND WATER

As mentioned earlier, the Noubaria canal was constructed to irrigate most of the western part of the Delta. The elevation of the canal is between 6 - 3.3 m MSL in the section under study. As irrigation extends to the upper lands westwards up to 30 m MSL, the saline ground water flows into the Noubaria canal.

The canal had been excavated for powered navigation, and a removal of the mud cake, which had built up on the bottom and the bank slopes, was effected. This mud cake was acting as a barrier against groundwater inflow into the canal. As a result of the movement of boats in the canal and the resulting waves most of the banks collapse into the canal and this helps the saline ground water to seep in.

DRAINAGE WATER INFLOWS INTO THE CANAL

All the areas which are irrigated by Nasr canal drain their water into the Noubaria canal, through D1 and D3 drains at km 53.0 and 55.5 respectively. The outfall of those drains is in the upstream side of Nasr canal intake. The water salinity of those drains is about 4500 ppm and this high saline water causes the contamination of Noubaria canal water.

Also, a new drain (New Nour drain) was constructed in 1975 to drain the water of about 28,000 feddans, 10,000 of the Mechanized Farm and 18,000 from North Tahrir company. This new drain discharges into the Noubaria canal at Km 74.50 on the upstream side of Thawra canal intake.

Small drains from private farms along the Noubaria canal also drain their water in the canal causing more pollution to the canal water.

SALINITY IN 1975

The average water salinity of Noubaria canal at the beginning of our program was 210 and 250 ppm in May and June, respectively, at the upstream side of D3 outfall into the Noubaria canal (Km 53.00, Table 2). This drainage water caused increases, in water salinity upstream of D1 to 260 ppm in May and to 280 ppm in June.

Due to the outfall of D1 into the Noubaria canal the salinity went up to 390 and 425 ppm in May and June, respectively, at Nasr canal intake. It means that due to the outfall of the drains (D3 and D1) into Noubaria canal, the salinity has increased about 70-80%.

It is certain that most of the drainage water goes to Nasr canal intake, since for the same months (May and June) the water salinity of the Noubaria canal at the Mechanized Farm canal intake decreased (310, 290 ppm, respectively, see Table 2). Another reason for this decrease in water salinity at Mechanized Farm canal intake compared to Nasr canal intake, is that the chances for dilution of the drainage water with fresh water in the reach between the outfall of the drains and the Mechanized Farm canal intake are higher.

The salinity of the Noubaria canal at Thawra canal intake is always higher than the salinity at Mechanized Farm canal intake and even higher than Nasr canal intake. The salinity increase was due to the construction of the New Nour drain, which at that time was draining a part of Thawra command area into the Noubaria canal. The drain was completed in June 1976. The salinity of the Noubaria canal water at Mariut canal intake is the highest of all canal intakes. The water salinity in Noubaria canal at this point was affected by the cancelled Tahrir canal which is acting as a drain now, draining its water into the Noubaria canal. Also, some other small drains for private farms discharge into Noubaria canal.

Table 2: Monthly salt content of Noubaria canal water (ppm)

Sites	Jan	Feb	Mar	Apr	May	Jun	Jul	Aug	Sep	Oct	Nov	Dec
					Year 1975							
U/S D3					210	250						
U/S D1					260	280						
At Nasr	700	560	430	360	390	425	470	490	520	510	500	350
At M-F*		760	250	280	310	290	320	450	370	400	350	460
At Thawra	980	820	370	400	400	460	510	580	580	560	540	530
At Mariut	590	990	500	420	580	430	530	555		530		670
At N.Lock**								770				
					Year 1976							
At Nasr	460	450	450	530	560	590	510	510	630		580	780
At M-F*	490	460	420	390	465	530	365	360	530		480	520
At Thawra	580	500	450		670	675	735	580	670		630	700
At Mariut	910	540	570	560	850	810	690	770	830	760	770	800
At N.Lock**		510			1025		790					
					Year 1977							
U/S D1				640						352		
At Nasr			580	690	580	672			944	896	640	450
At M-F*		270	450	420	510	420			480	640	380	510
At Thawra		630	680	540	630	690			640	960	736	
At Mariut		690		710	770	750			1070	1280	755	
					Year 1978							
At Nasr		512	544		640	736						
At M-F*	704	320	448	544	365	450	570					
At Thawra					770	820						
At Mariut		544			992	832						
At N.Lock**												

* M-F : Mechanized Farm

** N.Lock: Navigation Lock

SALINITY IN 1976

As seen in table 2, the salinity of the Noubaria canal water increased yearly due to more drainage water inflow into the canal and the inseepage of saline ground water into the canal. The water salinity at Nasr canal intake increased following the increase of the amount of drainage water from D1 and D3 drains.

The New Nour drain was completed in June 1976, so the Noubaria canal water became more saline, especially at Thawra canal and Mariut canal intakes. The salinity at Mariut canal intake in the last months of this year reached more than 750 ppm.

SALINITY IN 1977 AND 1978

The increase in salinity at different sites followed the same trend as in 1976 (Table 2). It was observed that close to the end of Noubaria canal, at Mariut canal intake, the salinity increased to a critical value, about 900 ppm.

DATA ANALYSES

It has been assumed that the present data can be used through linear regression to forecast the situation of the salinity of the Noubaria canal at these intakes in the years 1980, 1985, 1990, 1995, 2000 (Table 3), assuming that the present conditions continue and no drastic measures are taken to prevent the pollution by salts of the irrigation water of the canals.

For example from Table 4 and Figures 1, 2, 3 and 4 it can be concluded that: (1) At Nasr canal intake the salinity will reach 1500 ppm in October 1989 and 2000 ppm 7 years later in November 1996; (2) At the Mechanized Farm canal intake the salinity will reach 1500 ppm in 1993 and 2000 ppm in the year 2002; (3) In the Thawra canal intake the salinity will be 1500 ppm in May 1985; and (4) In the Mariut canal salinity of the irrigation water will reach 1500 ppm in August 1984 and 2000 ppm in June 1989.

Naturally, the project UNDP/FAO/EGY 73/048 has done voluminous work in studying the causes and the measures that should be undertaken to remedy the hazard of salinity on the area through four main lines of approach namely, ground water, drainage structures, irrigation structures and irrigation practices. The economics of the present losses due to the prevailing conditions and the projected future improvement in production together with the costs have also been evaluated. A comprehensive groundwater model was established to permit programming of new cropping patterns and irrigation schedules that will make optimum use of water available and maximum output. It has been found that:

a) Tiled drainage at a depth of about 2 meters should be installed in the area as soon as possible, and related main drains must comply with the requirement of this network.

b) Greater use of furrow irrigation, and introduction of sprinkler and drip irrigation have greatly improved water management efficiency under the specific soils and crop conditions. For new lands, these findings favour wider application of sprinkler and drip irrigation, while surface irrigation should be in controlled furrow systems rather than in basin-flood systems.

c) Making use of the aforementioned water model will help reach the optimum use of the water available and in reaching the maximum agricultural outputs.

Table 3: The expected salinity on the Noubaria canal (ppm)

Locations/years	1980	1985	1990	1995	2000
Nasr canal	785	1149	1513	1876	2240
M-F canal	638	947	1257	1566	1875
Thawra canal	944	1454	1964	2474	2984
Mariut canal	1021	1535	2048	2561	3074

Table 4: Expected occurrence of 1500 and 2000 ppm in irrigation water at different sites

Locations/salinity	1500 ppm	2000 ppm
Nasr canal	October 1989	August 1996
M-F canal	November 1993	January 2002
Thawra canal	May 1985	April 1990
Mariut canal	August 1984	June 1989

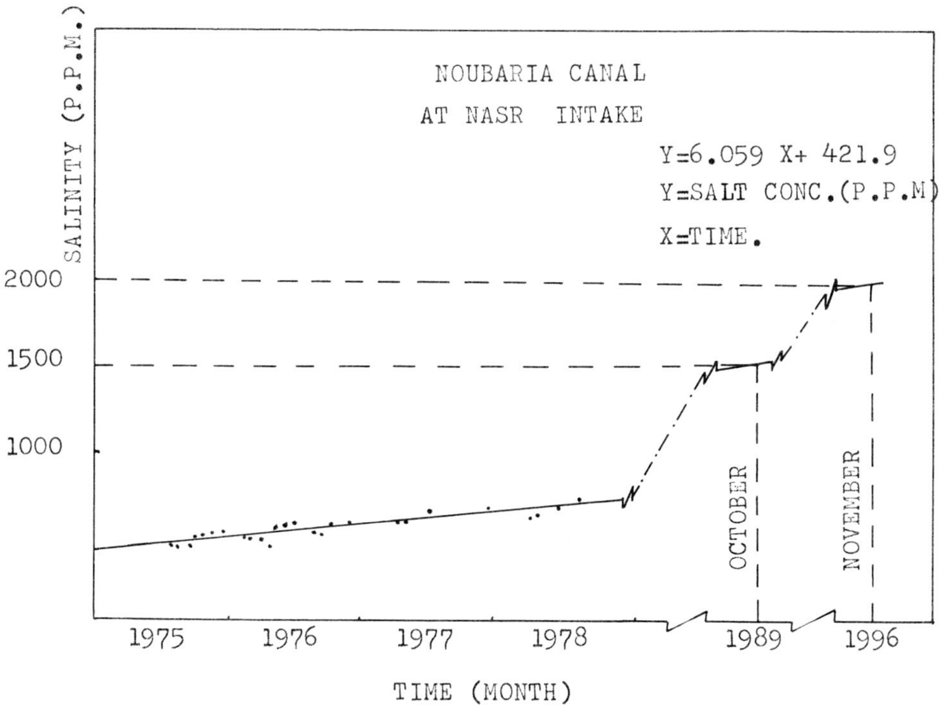
NOUBARIA CANAL
AT NASR INTAKE
Y=6.059 X+ 421.9
Y=SALT CONC.(P.P.M)
X=TIME.
SALINITY (P.P.M.)
2000
1500
1000
OCTOBER
NOVEMBER
1975
1976
1977
1978
1989
1996
TIME (MONTH)

Figure 1

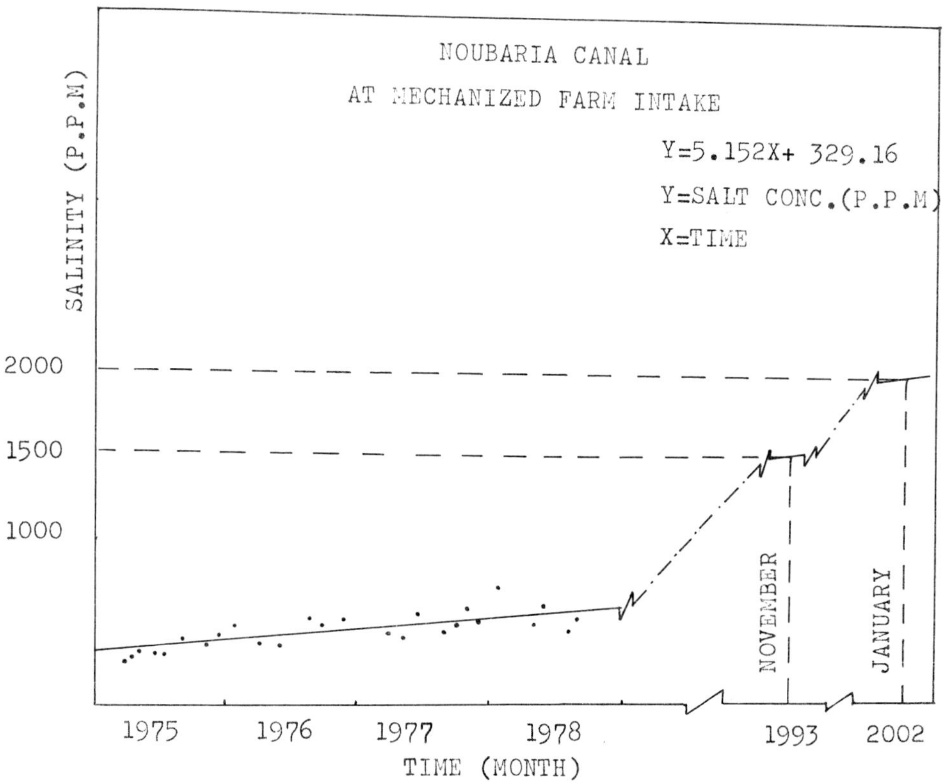
NOUBARIA CANAL
AT MECHANIZED FARM INTAKE
Y=5.152X+ 329.16
Y=SALT CONC.(P.P.M)
X=TIME
SALINITY (P.P.M)
2000
1500
1000
NOVEMBER
JANUARY
1975
1976
1977
1978
1993
2002
TIME (MONTH)

Figure 2

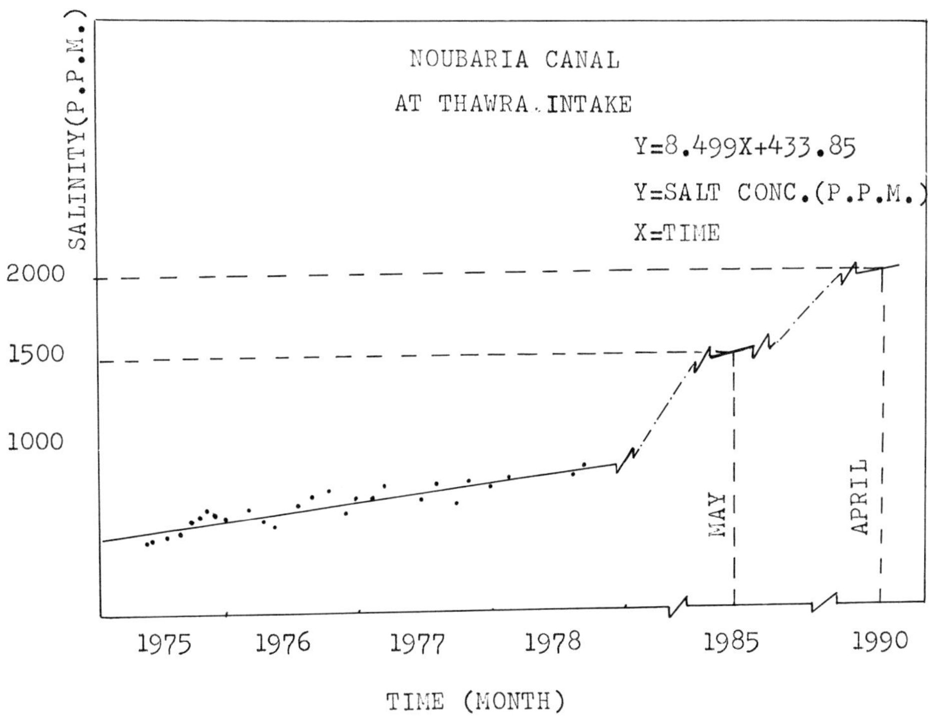
SALINITY(P.P.M.)
NOUBARIA CANAL
AT THAWRA INTAKE
Y=8.499X+433.85
Y=SALT CONC.(P.P.M.)
X=TIME
2000
1500
1000
MAY
APRIL
1975
1976
1977
1978
1985
1990
TIME (MONTH)

Figure 3

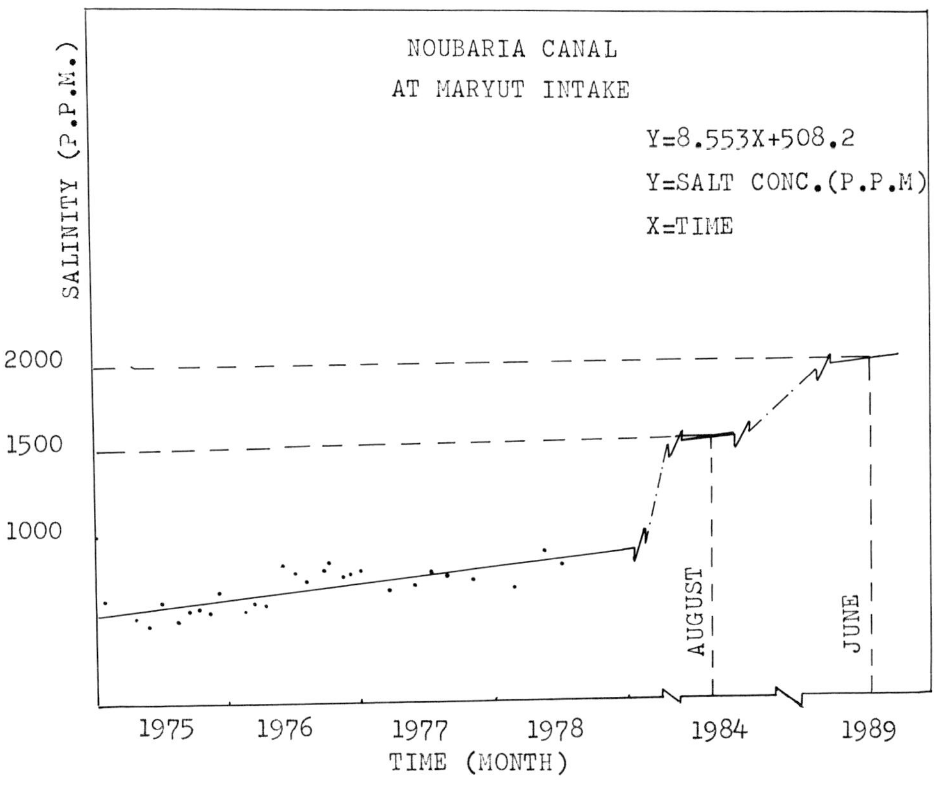

NOUBARIA CANAL
AT MARYUT INTAKE
Y=8.553X+508.2
Y=SALT CONC.(P.P.M)
X=TIME
SALINITY (P.P.M.)
2000
1500
1000
AUGUST
JUNE
1975
1976
1977
1978
1984
1989
TIME (MONTH)

Figure 4

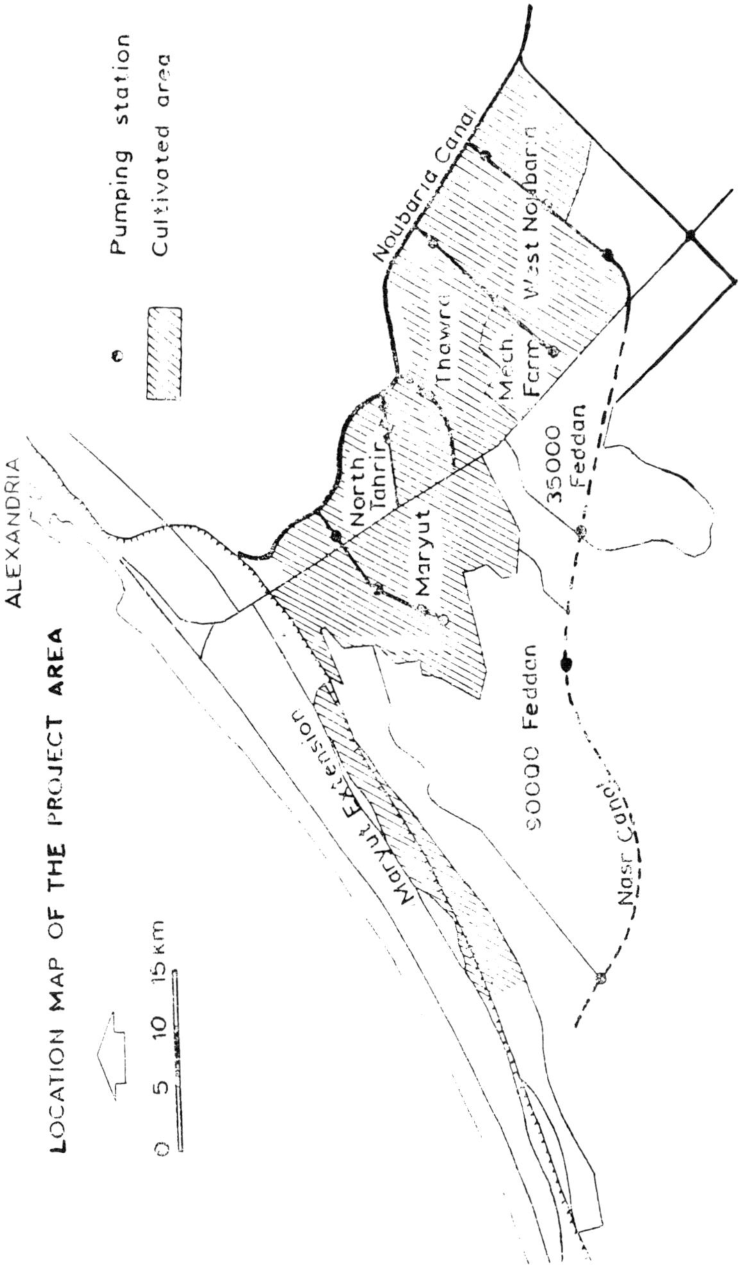
LOCATION MAP OF THE PROJECT AREA
0 5 10 15 km
ALEXANDRIA
Pumping station
Cultivated area
Noubaria Canal
West Noubaria
Thawra
Mech. Farm
North Tahrir
Maryut
35000 Feddan
50000 Feddan
Nasr Canal
Maryut Extension

AN OVERVIEW OF THE SAMDENE PROJECT

MOHAMED A. AYYAD
Principal Investigator, SAMDENE Project, University of Alexandria

ABSTRACT

SAMDENE (Systems Analysis of Mediterranean Desert Ecosystems of Northern Egypt) project was initiated in 1974 to collect and synthesize information that could be utilised in future management plans in the Mediterranean coastal desert of Egypt. This project is sponsored by University of Alexandria and is supported by grants from the U.S. Environmental Protection Agency and the Ford Foundation.

INTRODUCTION

An environmental management policy can only be successful if it is formulated with a basic understanding of the full complexity of the ecosystem: its structure and function, a knowledge of its properties - in particular the properties of stability and resilience, and a consideration of the history of manipulations and perturbations that have induced changes in its components. If, for whatever pragmatic reasons any of these criteria is ignored, there should be full appreciation of the consequences on environmental qualities. This is particularly true with arid and semi-arid ecosystems which are fragile and of low stability and resilience. These ecosystems have limited carrying capacities: limited potentials of material cycling and energy flow, and limited abilities to absorb changes, particularly those effectuated by inputs of materials and energy. If land-use practices exceed the carrying capacity of an ecosystem, the dynamic balance that gives it apparent stability will be upset, the ecosystem functioning will be impaired, and damage in the form of soil erosion, salinization, outbreaks in phytophagous insects, rodents, and weeds ... etc., may be irreparable.

SAMDENE (Systems Analysis of Mediterranean Desert Ecosystems of Northern Egypt) project was initiated in 1974 to fulfill these criteria, and to lay a scientific basis that could be utilised in future management plans in the Mediterranean coastal desert of Egypt. This project is sponsored by the University of Alexandria and is supported by grants from the U.S. Environmental Protection Agency and the Ford Foundation. Its specific objectives are:

1) to foster research for broadening the base of ecological information, which would directly or indirectly contribute to the practical development needs of the area;

2) to construct simulation models of the ecosystems in the area that would help in crystallizing the collected information, pointing out the gaps in knowledge about these ecosystems, assessing relative priorities for further research, and ultimately providing recommendations for formulating rational land-use schemes;

3) to provide training for junior ecologists in integrated and interdisciplinary research, and in a variety of advanced ecological techniques including laboratory and field methods and simulation modeling.

PROGRESS

Four major types of habitat in the study area were identified; of these, two were selected for study: (a) littoral sand dunes and (b) non-saline inland depressions, since these have better land-use potentialities. One site was selected at Gharbaniat (about 50 Km west of Alexandria) representing the first type of habitat, and another at Omayed (about 80 Km west of Alexandria), representing the second type of habitat. A plot of 8 hectares was fenced at each of the two sites in August 1974. This constituted the area of the validation study, while studies which created disturbances, such as vegetation clipping and animal collecting, were conducted outside the fenced area.

As a first step, a survey of literature and of collections was conducted to provide existing information about the biotic and abiotic components of the two ecosystems. Validation measurements included an initial survey of the spatial distribution of biota on each site, broken down by species, life form, and age, all expressed in terms of biomass.

During the last four years, a major step has been made towards filling the gaps of knowledge about the structure and functions of the major ecosystems in this region, and the integration of information on these ecosystems. During the

1974-1976 period, the general emphasis was directed more to the state-of-the-system measurements, since information about most components of the ecosystems was scanty at the time when the project started; but studies on some processes were also initiated. Periodic measurements of most of the state and exogenous (mostly climatic) variables thus provided a more or less complete picture of the monthly (or seasonal) fluctuations in the state of the system. These measurements included: biomass and phenology records and estimations of the chemical constituents for plant species representing different life forms, population and biomass determinations of vertebrates, invertebrates and micro-organisms, and records of soil moisture content and climatic factors. Starting in 1976, process studies were carried out in various fields besides the state-of-the-system measurements. Of these studies, mention may be made of growth, germination and transpiration of common plant species, feeding of common vertebrates and invertebrates, and decomposition of litter, nitrification and ammonification by micro-organisms.

By the end of 1976, SAMDENE reached a turning point in its activities. While the state-of-the-system measurements and the process studies mentioned above were continued, others were initiated, such as soil respiration, humification and mineralization of soil organic matter, feeding, respiration, defaecation and water relations of soil animals, and nutrient cylcing between the soil and common plant species. Efforts have been also directed towards two main themes: evaluation of the impact of various types of land use on the structure and functions of ecosystems, and the socio-economic structure and flow of energy in bedouin societies. The state-of-the-system measurements and process studies were continued to serve three major objectives: (a) enriching our knowledge about the structure and functions of desert ecosystems, thus broadening the scientific base for future management-oriented research; (b) providing information on temperal variations in state and exogenous variables as reference for the validation of ecosystem models; and (c) furnishing quantitative means for the assessment of the values of different model parameters that have been estimated previously on the basis of scanty information, intuition or mere guesses.

Studies on the impact of common types of land use on the structure and functions of ecosystems were started in May 1977, according to a scheme which ensured the orientation of data collection for serving the major objectives of SAMDENE. The first item in this scheme was to define the most common types of land use in the Mediterranean desert of Egypt, and then to select sites representing such types. It was agreed that the following types were the most common, and were most likely to have the greatest impact on the dynamic balance of the ecosystem: (a) grazing and cutting; (b) dry-farming represented by fig and olive plantations; and (c) irrigated farming which started a few years ago after the introduction of Nile water to the region.

Uncontrolled cutting and grazing have dominated North Africa since the eleventh century, and have been generally blamed as one of the major causes of degradation of desert ecosystems in this region, exhibited by the reduction of perennial cover, impoverishment of flora, soil erosion, dust storms, formation of mobiles and dunes, and establishment of desert pavement. To that end, the impact of overgrazing and cutting has been one of the main concerns of SAMDENE project since the outset of its activities. It will be recalled that areas were fenced at the two project sites representing two of the ecosystems subjected to uncontrolled cutting and grazing, in order to protect samples of ecosystems in these areas from such activities. This served three purposes: (a) furnishing locations for validation measurements where the effects of all but the state and exogenous variables of the ecosystem were excluded; (b) demonstrating the variation that would take place in the ecosystem by protection for different lengths of time after a long history of cutting and overgrazing that has culminated in the present state of disclimax stability; and (c) providing a means of describing the state of the ecosystem at one end of the gradient of grazing intensity in any experiment dealing with the effect of this factor.

It is interesting to note that at Omayed, although the vegetation inside the exclosure flourished after two years of protection and reached a peak in cover, biomass and number and abundance of species, its vitality declined during the third and fourth years in comparison with the much less abundant vegetation outside the exclosure. Meanwhile, there was a building up of large insect populations as a result of protection, which may be the reason for deterioration of vegetation inside the exclosure, and which may constitute a source of infection to neighbouring cultivations. Whether the loss of vitality of plants was due to an increase in phytophagous insects, or to an increase in competition for moisture and/or nutrients between plants after attaining larger cover and biomass, or to any other reason, remains to be investigated. But this provides a clue to an issue of practical implication: in arid lands with long history of overgrazing, complete protection of ecosystems for extended periods might lead to further deterioration of their productivity, and perhaps controlled grazing and/or cutting would be a better management policy.

Conceivably, the grazing intensity, in terms of time of grazing and number of grazing animals, that would ensure the maximum possible productivity and does not upset the balance of the ecosystem, varies from one ecosystem to the other, depending on its structure and geographic location. In the SAMDENE program, additional areas were fenced in September 1977 at Omayed in which grazing intensity was fixed at certain limits, with the aim of providing a means of describing the state of the ecosystem at different points between the two ends of the gradient of this factor: one end representing no grazing (furnished by the area inside the old exclosure) and the other end representing free grazing (furnished by the area outside the exclosure). In addition to the records on different state variables in these areas with different grazing intensities, investigations of the feeding habits of domestic grazing animals and palatability of plant species were started.

In the meantime, several plots within the old fenced area were delimited for experimenting on the effects of the increase in soil moisture content by adding varying amounts of water, the increase in the content of certain nutrients, artificial cutting, and scratching of the soil surface. It is premature at this stage to draw any conclusions, but preliminary results of some experiments indicate varying degrees of increased vitality and phenological activities due to supplemental irrigation.

For evaluating the impact of some types of farming manipulations on the structure and functions of ecosystems in the project region, five sites representing these types were selected: (a) fig plantation under dry farming on the coastal sand dunes at Omayed; (b) fig plantation under dry farming in the inland depression at Omayed; (c) olive plantation under dry farming in the inland depression at Burg El-Arab; (d) olive plantation which was turned to a crop field under irrigation with Nile water, in the inland depression at Burg El-Arab; and (e) vine plantation under irrigation with Nile water, in the inland depression south of Gharbaniat. Investigations were extended to these agroecosystems, in an attempt to determine the changes that might have occurred in soil physical and chemical properties, populations and activities of insects, due to different manipulations taking place in these ecosystems in comparison to unmanipulated ecosystems.

A reservoir of populations of parasitoids of our common insect pests was found in the irrigated farms. These parasitoids have almost disappeared from the Nile Delta farms due to intensive use of insecticides. It is suggested that these parasitoids could be reintroduced in the Delta. Agricultural authorities are advised not to use chemicals extensively in the newly irrigated fields in order not to annihilate the last refuge of these parasitoids. Soil fauna in rain-fed and irrigated agroecosystems was found not significantly different; nevertheless, earthworms appeared in irrigated agroecosystems in greater numbers. It was also observed that prolonged farming operations and irrigation lead to lower infiltration rates, less available nutrients and higher salinities. Increased soil compaction prevented earthworms and other useful soil animals from maintaining their initial invading populations.

Modelling activities were started at an early stage of the project. At first, a preliminary version of a whole-ecosystem model was prepared, incorporating the best existing knowledge of the systems. This served as a valuable guide in collecting field data and in assessing relative priorities for further research. It was so written that the system could be defined at execution time by specifying the various organisms present, the soil characteristics, etc., together with initial values of state variables and parameters in the system. Where data for some parts of the system were not available, they were replaced temporarily by informed guesses, or data from similar systems elsewhere.

Modelling activities were then continued towards the improvement of the conceptual framework of the whole-ecosystem (DESERT 2) model, and the estimation of more realistic values for parameters, with the aim of getting simulations which would be a closer representation of real-world variations in state and exogenous variables. In the meantime, the accumulation of data on the impact of different types of land use (grazing, dry farming, and irrigated farming) on the ecosystem would allow the orientation of the DESERT 2 model to answering questions pertaining to management problems. It may, therefore, be expected that the last version of this model will be largely management-oriented. But within the constraints of time available for SAMDENE till February 1979 when it is scheduled to end, one can only hope that the last version of DESERT 2 will provide a framework which may be useful as a guide for more extensive research on management problems in the Mediterranean desert of Egypt. The modelling activities also included the construction of carbon- and nitrogen-cycling models of *Thymelaea hirsuta*, one of the prominent shrubs common in the two validation sites, as well as a more elaborate grazing module for the whole-ecosystem model.

ACHIEVEMENTS

The achievements of SAMDENE may be summarzied in the following main points:

1) Collection and synthesis of a substantial amount of ecological information about desert ecosystems of the western Mediterranean coastal region of Egypt, which could be used to develop further research-based resource management inputs to environmental and land-use problems in this region, and which would help build increased knowledge of world coastal desert ecosystems.

2) Preliminary evaluation of the impact of human manipulations and environmental perturbations on the structure and functions of desert ecosystems of Egypt's western Mediterranean region, as a step toward more encompassing studies for laying the foundations of sound land-use schemes in this region.

3) Construction of simulation models which could be used to help solve some land-use decision-making problems, and to guide further work of wider scope dealing with environmental resource management of the western Mediterranean desert of Egypt.

4) Fostering an interdisciplinary research program with a closely linked interinstitutional working team of ecologists for the first time in Egypt. Both investigators and research assistants of this team also benefited from contacts with their colleagues in American, Australian and European ecological centers.

5) Enabling Egyptian ecologists to work together on a single ecological problem of national concern, thereby giving them a chance for self-fulfillment and a sense of achievement.

6) Training of junior Egyptian graduates in modern techniques of interdisciplinary ecological research and systems analysis. A number of these students were able to submit their M.Sc. and Ph.D. theses as a result of their work in SAMDENE.

ARID LANDS DEVELOPMENT STRATEGIES: RESEARCH NEEDS

J. ELEONORA SABADELL
National Science Foundation

ABSTRACT

When mapping development strategies for arid regions in relation to present and future water resources, two options may be followed: (1) increasing water supply, and (2) controlling demand. In describing and assessing these alternatives the technological, economical, social and institutional problems must be considered not only separately but weighed relative to one another, and "trade-off" analyses are necessary for balancing the water supply and demand in the projected development. Some of the research topics in need of investigation for the better management of natural resources in arid regions are identified. Due consideration should be given to: energy and the environment; the particular socio-economic needs of specific regions; the coordination needed between the institutions conducting this research; the scientific and technological cooperation needed between nations; the communication between researches and the policy makers.

INTRODUCTION

The rational and effective planning and management of a nation's water resources is fundamental for its present and future development. Population growth, food production, mineral and fuels extraction, energy conversion and industrialization are all activities supported by the natural resources of a country but especially by its water resources. This is true throughout the world, in developed and developing countries alike, but more so where arid regions constitute a large part of a nation. It is evident that being in a constant water shortage situation, not only in quantity but also in quality and subjected to low, erratic and often unfavorable rainfall distribution, the planning for the usage, allocation and management of the available water becomes critical.

One point to be stressed is the increasing economic and resources interdependency and integration between nations.

Taking food production as an example, it has been calculated that during the global drought of 1972 the food produced was approximately 888 million tons, 2% less than in 1971 and 9% less than in 1973. During 1973-74 the Sahelian drought and crop failures in India and the USSR occurred. If drought had taken place in the USA at the same time the consequences of such a combination of circumstances would have been staggering. It is clear that as the world moves closer to the limits of its capability for producing food, the apparently marginal effects of climate behavior and local or regional water resources mismanagement become correspondingly significant. It might be added that the probability of such events happening is increasing.

The intent of this paper is to present some considerations that should be examined when mapping development strategies for arid regions in relation to present and future water resources, and the research needed to aid the formulation and implementation of such policies. In describing and assessing alternatives the technological, economical, social and institutional problems must be considered not only separately but weighed relative to one another, and "trade-off" analyses are necessary for balancing the water supply and demand in the projected development. In general there are two options to be followed: (1) increasing water supply, and (2) controlling demand.

INCREASING WATER SUPPLIES - PROBLEMS

In the past, increasing the system capacity meant the buildup of new hydraulic networks for the diversion of surface fresh waters and the mining of ground waters, so as to convert an undesirable distribution of the water stock, in space and time, to one more in harmony with the needs. The rising costs of exploration, construction, storage, transmission and treatment, especially of surface waters, as well as the energy consumed in every operation are limiting factors that have to be taken into account at the planning stage of large waterworks. Environmental controversies have arisen in relation to major surface water engineering projects, especially in their health hazards aspects since waterborne diseases are one of the main causes of death and decreased productivity in developing countries. Other environmental concerns of large irrigation, impoundment and distribution projects have to do with: erosion, sedimentation and related pollution problems; the physical, ecological, economic and social impacts on the places where the water is taken from, on the irrigated areas and on the contiguous lands. Evaporation, siltation, eutrophication and seepage are also some of the serious problems present in reservoirs that need to be solved.

Many and abundant aquifers exist throughout the world, even beneath deserts, and mining ground water is one of the most commonly used methods for augmenting the water supply in arid and semiarid regions. It must be realized, though, that the capacity of the system can only be manipulated up to a point. Overdrafts will result in the depletion of the water stock, land subsidence, saline intrusion, pollution of aquifers, larger energy requirements and growing capital outlays for more and deeper wells, and possible economic dislocations. Hydrologic information is usually limited and is often obtained only during drilling and pumping. In general there is not a complete understanding of the relationship between surface and ground water, as well as of the chemical, physical and biological processes affecting ground waters, especially when related to the agricultural activities and wastewater disposal.

Other methods to increase freshwater supplies are: desalting of sea and brackish waters, weather modification, iceberg towing, and wastewater reuse and recycling. Even though desalinization methods have been studied for a long time and are based on diverse phenomena (distillation, ion exchange, reverse osmosis, electrodialysis, freezing, foam separation) they are still expensive, energy-intensive, and present problems in disposing of large amounts of the brine produced.

Weather modification has also been examined for many years but producing rain by cloud seeding poses many serious questions that are still unresolved: imperfect knowledge of the precipitation-causing physical processes, environmental effects of cloud seeding, disruption of weather patterns, engineering problems in the production of seeds, legal rights to cloud moisture, etc.

The towing of icebergs is a technology that is still in the speculative stage. Tapping some of the 85% of the world's unused freshwater entrapped in its ice caps is a very attractive prospect. Preliminary estimates show that the idea merits serious consideration but the problems to be examined are many: melting losses; towing methods; collision dangers due to manageability limitations; legal rights and political issues; techniques for iceberg melting; pumping of the water produced from sea level to the storage sites or distribution systems; environmental effects on the polar regions; changes in the microclimate of the region receiving the water imports, etc.

Recycling and reuse of wastewater can be done at the domestic, municipal, agricultural and industrial levels and it is in the latter where high recycling rates have been achieved. These practices as well as the utilization of cooling systems effluents can have a large impact on the overall water budget. Reuse of municipal effluent and domestic wastewaters have been used in many places for a long time and is becoming a common practice in some parts of the U.S.A., especially when used in land application that provides supplemental irrigation water. The added attraction is that it partially solves the economic problem of wastewater treatment and it returns nutrients to the food production cycle. The main problems in these practices are: the presence of pathogenic agents, particularly viruses, in the recycled water; the presence of toxic elements, especially when industrial effluents are used, that could limit the productivity and decrease the suitability of the produced food and animal feed; the possible contamination of soils, aquifers and adjacent bodies of water; the costs and energy requirements for the collection and distribution of wastewaters.

CONTROLLING DEMAND - PROBLEMS

In general, controlling the demands for water is equated to restrictions or rationing of water usage, rate schedules and pricing policies. But it also means managing water resources in such a way as to satisfy the fundamental needs of a region, allowing for development to an extent that will not overpower its carrying capacity and will minimally impact its environment. This last point is of particular importance for fragile arid, semiarid biomes. Conservation in irrigation and other activities, improvements in land use, in aquifers and watershed management, and improving the efficiency of water systems are methods that aid in achieving the best possible usage of scarce water resources. In effect these practices help in allocating water more efficiently and in satisfying highly varied demands.

Agricultural irrigation is the best example of a major freshwater consumptive use. Conservation in irrigation can be achieved by: reducing evaporation from soil and water surfaces; reducing seepage and percolation losses; reducing transpiration losses; developing plants that use less water and crops that are salt-tolerant; developing novel irrigation methods, e.g., subsurface irrigation; adopting less consumptive irrigation methods, e.g., trickle or drip systems. These types of irrigation techniques have the advantage of minimizing the waterlogging and salinization of soil and water problems that the standard methods of irrigation may produce.

There are limitations to all these conservation practices that in general relate to their costs, the available technologies not yet well developed nor validated, and the scarcity of data on the short- and long-term effects of these conservation methods on the total system.

Improvements in land use and in the management of water and other natural resources can be achieved by better understanding the interactions occurring between these resources and the stresses created by human activities in the region. Each proposed land use plan should be supported by the available resources and the carrying capacity of the regional system. Conversely, the effect that the spatial and functional distribution of varied land uses have on the amount and quality of the water and all other resources in the region and even in adjacent lands should be examined carefully. Another important point to consider is that arid and semiarid regions vary greatly from one another not only in physical but in socio-economic-political conditions, each presenting a different set of problems, constraints, and possible management schemes for their development. The synergistic effects produced by development between the physical, social, economic, ecological conditions of a given arid region and the interactions between the resources being utilized, extracted, imported and exported, are still not well understood nor quantified. It is worth mentioning that the amount and variety of resources found and used in arid regions, as well as their economic value, are of substantial magnitude and affect 13% of the world's population. Any perturbation in the balance of these fragile ecosystems can be devastating, as many past experiences have indicated.

One example of the interactions mentioned above and the problems generated by development are those encountered in the exploitation of the mineral and fuel resources of an arid region. On the one hand we have the essential role water has in the mining and processing activities and in the production of energy; on the other hand the effects these activities may have on the water in the system, on other resources, on the socio-economic structures; etc. Some of the problems related to these activities are: adverse changes in the regional ecology; unfavorable alterations of well-established land use practices; social and economic dislocations; degradation of soil, air and water quality; groundwater depletion; unfavorable trade-offs between existing and new water uses; land subsidence; saline intrusion in aquifers; increasing energy budget and costs for land rehabilitation and water treatment. Conservation programs are essential in balancing the mineral processing and energy-related water demands with those demands by agriculture, livestock and other uses.

The efficiency of water systems can be improved by eliminating or minimizing water losses in storage and distribution systems. Evaporation and seepage losses have already been mentioned in regard to large reservoirs and canals. The use of diverse types of linings and covers is still unsatisfactory and costly. Leakages in water distribution systems are difficult to locate and to seal. The losses are seldom measured and water quality deterioration in the water lines may occur due to infiltration and inflow.

RESEARCH NEEDS

The issues presented throughout this document are only a sample of the large number of complex, multiple-nature problems to be addressed in selecting suitable policies for the management of water and other natural resources in the arid regions. Basic and applied research is needed to: (a) achieve a better understanding of the impact of human activities on the dynamics of arid and semiarid biomes; (b) solve the scientific, technological and managerial problems raised by the development and usage of the natural resources in the arid regions; (c) help in the formulation of natural resources policies that take into account the special needs of the arid areas; and (d) facilitate the implementation of these policies by helping to attain specific objectives and keeping open alternative options. It must be realized that science and technology have very few ready-made short-term solutions to offer. In general, results of research take some time to be applied and their effects are usually felt in the medium and long term. Therefore it is necessary to readily develop research programs where needed and to maintain and upgrade programs already in place to support developmental efforts. New research approaches are also necessary that will look upon the natural resources of an arid region as all limited in supply, closely interrelated, and as parts of a complex system in which the relationships between energy, materials, food and the environment are fundamental.

Basic research in arid lands needed to support all these activities range from physical to life and social sciences. Some examples are: improving our understanding of climatic changes and climatic trends; studies on forecasting climatic changes; studies on the impacts of pollution on microclimates; development of better methods to measure aquifer boundaries, thickness, transmissiveness, quality; studies on sources and transport processes; pollutants in ground and surface waters; investigating saline-induced changes in the physiology of plants; conducting taxonomic studies of arid, semiarid flora and fauna; studies on ecological indicators for drought; better methods to detect, identify and measure diverse disease vectors and other pathogenic materials; development of better ecological, hydrological, socio-economic models; studies on demographic trends; new approaches for calculating cost-benefit ratios and for benefit assessment; development of new biological conversion processes; anthropological and archeological studies of diverse land uses and settlements; etc.

Applied research is usually problem-oriented and multi- and interdisciplinary in nature. Some of the topics in need of investigation for the better management of natural resources in arid regions have been mentioned throughout the text of this paper and many more will be identified during this conference. In general, applied research is geared towards the identification, assessment and control of problems in the resources system and to the development of new and/or better processes for extracting, converting, utilizing and recycling resources. It is important to point out that in undertaking studies on arid lands development and its impact due consideration should be given to: energy and the environment; the particular socio-economic needs of specific regions; the coordination needed between the institutions conducting this research; the scientific and technological cooperation needed between nations, to profit from each other's experience and to avoid delays and unnecessary overlapping of efforts; the communication between the researchers and the policy makers required for improving the transfer of scientific and technological knowledge to the management and decision-making process. Science and technology cannot by themselves solve all the problems and set the course of the necessary changes in developing and managing arid lands natural resources but they are the foundations on which realistic objectives can be set and sound policies can be built.

DESERTIFICATION AND PRODUCTIVITY

DONALD H. BAEPLER
Chancellor, University of Nevada System, U.S.A.

ABSTRACT

As a function of world-wide population growth, man has increasingly invaded semi-arid lands in an effort to increase total world food supply. As a consequence, and frequently correlated with natural phenomenon, the process of desertification has accelerated. Man is becoming increasingly dependent upon arid and semi-arid lands, yet understands very little about his role as a causitive agent in accelerating the deterioration of these lands.

Long-term agricultural productivity must take into account the delicate and fragile ecological balance that exists in these lands, and the consequences of disturbing this balance must be understood. Agricultural productivity must be considered as only one possible use of such lands, and must be weighed against the economic benefits of alternative uses such as energy production and recreational use.

This paper examines the process of desertification in terms of man's relationship to the process and analyzes the limited choices available for arid and semi-arid land development. Examples illustrating the process of desertification as well as examples indicating the potential for agricultural productivity are presented from a variety of semi-arid and arid regions of the world.

It is very appropriate that we are holding this international workshop on the "Applications of Science and Technology for Desert Development" in Egypt, a country with tremendous potential, but a country also with tremendous problems. As a result of population increase, Egypt must explore ways of becoming self sufficient in terms of agricultural productivity and must turn to its arid lands and to careful land management practices and policies in an effort to obtain this goal. Egypt, however, is not unique in having problems of population growth or in having the need to increase agriculutal productivity and, therefore, it is highly advantageous to bring people together from a variety of countries to discuss common problems.

Desertification, the process whereby semi-arid and subhumid land is degraded to impoverished desert as a function of drought and the activities of man is accelerating today on a global basis. As the world's population increases, man looks to semi-arid and arid lands as a potential resource for agricultural production and frequently, in exploiting such lands, increases the rate of desertification. Additionally, man tends to build cities in the most fertile and ecologically favorable portions of semi-arid lands. As populations increase in these cities, thousands of acres of productive land are removed from agricultural utilization and, from the perspective of productivity, become impoverished, a process that might be referred to as "urban desertification." Examples of expanding cities removing productive land from agricultural usage are common the world over. Perhaps nowhere is this process more evident than in Southern California where urban sprawl has greatly reduced the amount of acreage available to agriculture and has forced a dependency on vast irrigation projects in less favorable regions such as the Imperial Valley. Thus, population increase not only makes us more dependent on semi-arid and arid lands for grazing and for agricultural productivity, but frequently it removes valuable agricultural land from production.

My interests in deserts stem from personal observation in the deserts of the American Southwest, including Baja California (Mexico); the North American Arctic; arid regions of South America, including the Galapagos Islands; and the Egyptian Sahara. As a biologist, one factor is obvious to me as I observe different deserts, and this factor must be uppermost in our minds as we consider desert productivity: Deserts represent very fragile and delicate ecosystems. Once disturbed, it is very difficult for deserts to replenish themselves and to recover from the disturbance. Mistakes made in attempting to improve desert ecosystems are not easily corrected and take inordinately long periods of time for recovery, if indeed recovery is ever achieved.

In attempting to utilize arid lands, we must, through fundamental research, attempt to understand the parameters of arid land development. We must analyze the potential of such lands for agriculture (particularly for plants adapted to arid or saline conditions), for energy production (solar, wind, or production of plants for conversion to ethanol, methane, or petroleum derivatives), and for recreational and tourism usage.

Historically, we have often abused semi-arid and arid lands and only recently have we approached the problem of arid land utilization on an international basis. In the western part of the United States, much of the available lands with noted exceptions such as the coastal regions of the Pacific Northwest and the higher elevations of the major mountain ranges, can be classified as semi-arid or arid. Much of this land is characterized by low rainfall (which often occurs sporadically in the form of thunderstorms), rocky soils, pockets of salinity, and by a topography that is more suitable to grazing than to agriculture. Additionally, these lands are subject to sporadic droughts which, together with historical grazing practices, have greatly damaged the natural vegetation and, as a consequence, have promoted basic research in land management.

Although the ecological characteristics of this part of the United States differ tremendously from that of the Egyptian Sahara, there are some problems relating to man's usage of these lands that the two regions share historically. Perhaps the most obvious problem is that of overgrazing, and until recently in the United States, at least, no effective policies and regulations existed to control the number of animals utilizing such land or the times of year that these lands might be grazed.

Until the middle 1930's all public lands in the American West were available to anyone without any government restrictions. Cattlemen and sheepherders exploited the land to the extent that massive livestock losses occurred in times of drought, and the native vegetation of these federal lands was greatly depleted. In the Great Plains region of the United States, hundreds of thousands of acres of native prarie were put to the plow for agricultural usage ultimately, in a time of drought, resulting in the great dust-bowl era.

As a result of severe economic losses and the gradual realization that native lands were becoming increasingly nonproductive, the federal government finally passed legislation in the 1930's to regulate grazing on these lands. In the intervening years, land management practices in these semi-arid Western lands have been a source of controversy. Many federal programs intended to improve these lands have been met with limited success because of public attitudes or because of an insufficient knowledge of the basic ecosystem that was to be managed.

Although the lands will never be restored to their original composition, particularly with respect to the distribution of native plants for forage, there has been a marked improvement in the grazing conditions as a function of effective range management. The basic improvements in the Western United States semi-arid range lands came about as a function of restricting the number of animals permitted on any given grazing district, and by restricting the times of year that grazing was permitted. Range improvements included not only an increase in vegetational coverage, but a decrease in erosion with a concomitant decrease in run off during times of heavy rains which reduced flooding and added to the ground water reservoirs. Although many problems still exist in this region, the government is continuously attempting to analyze and change its regulations in an effort to upgrade the quality of its Western lands. It is obvious that management practices enacted since the 1930's have slowed or reversed the process of desertification in these Western lands and have permitted cattlemen and sheepmen to operate in a more consistent and predictable manner without experiencing the vast fluctuations in herd or flock size as a function of periodic wet cycles and periodic droughts. Although extensive regions in the semi-arid Western regions of the United States have been brought into agricultural productivity through massive irrigation projects, and although many millions of dollars have been expended in improving watersheds and in the removal of certain plants such as sage-brush in an effort to upgrade grazing conditions, the basic improvement in the range lands was brought about not through a massive infusion of dollars, but by the establishment of management policies which, in time, gave nature itself the opportunity to at least partially rectify the abuses to which the land had been subjected.

A similar pattern of land-use management in Australia is documented by Ian Douglas.* Much of the Australian continent can be classified as semi-arid or arid land, and much of this land has been subjected to grazing by both cattle and sheep. Periods of drought alternate with wet cycles and, like arid and semi-arid lands the world over, this part of Australia is

---

* Desertification: Process, Problems, Perspectives, The University of Arizona, Tuscon, 1976, p.p. 87-107.

subject to infrequent heavy rain and flooding. The exploitative use of grazing lands has increased the process of desertification in many parts of Australia.

Early policy decisions towards the utilization of these lands demanded, in an effort to increase animal productivity, that herd and flock sizes be maintained at a level which proved to be disastrous to the ecology of these lands. Recently, policy changes have dictated a reduction in the number of animals per grazing district, but competition with such introduced forms as the rabbit has greatly limited the ability of the range lands to recover. Although policies governing the utilization of such range lands in Australia are constantly being reevaluated, and although much basic research is currently under way in the areas of range-land management and irrigation, the ecology of the continent with its extensive periods of drought will make recovery a long process, if, in fact, recovery is possible at all.

Similarly, in Mexico, particularly in arid regions such as Baja California, one can commonly observe misuse of land through overgrazing resulting in the classic patterns of erosion, destruction of native vegetation, and livestock that is obviously undernourished. Land management policies, so desperately needed in these desert regions of Mexico, are either missing or are poorly enforced and there is little hope that these desert lands will ever recover to the point where animal husbandry can be a productive industry.

One also sees in Baja California that in an effort to improve agricultural productivity and to encourage settlement, underground aquifers, developed over thousands of years, are rapidly being depleted in an attempt to make the desert bloom. Thus, we face the prospect of increasing numbers of people becoming dependent upon a finite resource which ultimately will be depleted, since withdrawal of water from the aquifers is obviously taking place at a rate which is much more rapid than the ability of the aquifer to recharge itself.

It can be observed that in a variety of arid and semi-arid regions of the world, policies concerning land management and research into the fundamental nature of the delicate ecosystem of arid lands are needed in an effort to obtain a sustained yield in terms of animal husbandry. In the Mediterranean we see the effects of uncontrolled grazing, particularly by cattle; in Australia the combined over-exploitation of grazing land by sheep and by cattle; and in the United States, the over-exploitation of public lands by cattle and by sheep.

In places where sound land-management practices have been implemented, time alone tends to improve range conditions leading to a sustained yield.

In addition to the utilization of arid and semi-arid lands for animal husbandry, we must also turn our attention to the utilization of these lands for agricultural productivity. In certain parts of the world where soil conditions are suitable and where water is available, massive irrigation projects have been undertaken and semi-arid lands have been farmed in a manner similar to that used in more humid regions of the world.

In many semi-arid or arid regions, however, soil conditions are only marginal for agriculture due to the lack of organic matter or the problems associated with salinity, and water is not abundantly available. In these areas, traditional methods of farming have typically failed. Only recently have we come to the realization that the proper utilization of these arid lands must involve basic research to locate plants that are suitable for life in the desert. Such plants may fall into two categories. First of all, one must examine the flora in the deserts of the world in an effort to locate native plants heretofore unused in agriculture, but which might have potential for cultivation in the production of food, fiber, oil, or some product useful to man. Secondly, we must continue our basic research into the genetics of traditional agricultural plants in an effort to produce plants that are drought resistant or salt tolerant.

Water, of course, is the limiting factor in the utilization of arid lands. Too often we have approached this problem only from the perspective of bringing more water to our lands from distant sources or by exploiting the subterranean aquifers in an effort to treat arid lands as though they were, in fact, humid lands. Not only have underground aquifers been depleted in many parts of the world, but the infusion of water on the surface of arid lands from these aquifers and from distant water sources has often created problems, particularly in the area of increased salinity, that has proven ultimately to be detrimental to agricultural pursuits.

Through the utilization of drought-resistant and salt-tolerant plants, the problem of desert agriculture centers on the conservation of water and should emphasize a decrease of water loss through reduction of evaporation and increased and more efficient use of run off in an effort to conserve the limited ground water available in arid regions. Growing plants under cover, utilization of drip irrigation methods, and irrigating with saline water are methods that are currently being used with great success in various parts of the world.

We must also consider the uses of land for purposes other than animal husbandry and agriculture. It is obvious that arid lands can become productive and can greatly augment the world's food supply, but they nonetheless will always play a secondary role to the more humid regions of the world in terms of total world food production. As a consequence, all alternative uses of these lands must be explored. The vast potential for energy production in arid lands must be considered as one of the most valuable resources of deserts, particularly as the world's supply of fossil fuels becomes exhausted. On a global basis, arid lands must be viewed in such a way that they complement the humid regions of the world rather than attempting to imitate these regions. The search for mineral deposits should be accelerated, but once found, care must be taken in the methods used in exploiting these resources in an effort to preserve the quality of the desert environment.

Finally, the question of tourism and the resort industry must not be overlooked as we consider uses for arid lands. In the United States, Las Vegas, Nevada serves as a classical example of the interrelationship between a desert area which is not agriculturally productive, but which has been extensively developed for tourism. Although there are other attractions besides the desert in Las Vegas, unquestionably the dry climate with the unusually high percentage of sunny days and a year-round climate conducive to outdoor activities has added greatly to the attractiveness of the region for tourism. The Southern tip of Baja California and the cities on the desert coastal plains of Western Mexico also provide excellent examples of the utilization of arid lands for recreational usages and for the development of tourism.

In Egypt, the prospects for efficient and effective utilization of desert land would appear, on the surface, to be bright and one should be optimistic. Egypt has miles of extraordinarily attractive Mediterranean coastline which could be developed into resort areas, attracting tourists on a global basis. The Nile River constitutes an enormous potential for traditional methods of irrigated farming, and the desert might even provide the energy necessary to move this water from the River onto the more arid regions of the Sahara. Energy from the desert also might be utilized to power desalinization plants along the Mediterranean to generate water for high-intensity farming utilizing desert adapted plants and strict water conservation methods. This type of farming often requires a very large labor force, but Egypt also has the manpower available to implement high-intensity farming practices.

In conclusion then, utilization of arid lands in an effort to increase productivity requires a variety of approaches. Sound land management policies and practices, rigidly controlled and enforced, are necessary regardless of the ultimate utilization of such lands. Further, research and experimentation in the utilization of desert-adapted plants should be accelerated based on recent successes in the development and discovery of such plants. Water conservation, rather than water exploitation, is absolutely essential for long-term development and productivity in all of our arid regions, and finally, every effort should be made to utilize arid lands for non-agricultural pursuits in an effort to complement the more humid regions of the world.

ECOSYSTEM DETERIORATION IN THE INDIAN DESERT: RETROSPECT AND PROSPECTS

H.S. MANN
Director, Central Arid Zone Research Institute
(CAZRI), Jodhpur, India

ABSTRACT

Once the seat of Indo-Ghaggar civilization, the Indian Desert, constituting the eastern-most fringe of the Sahara-Rajasthan desert basin, has until recently been subjected to over exploitation and neglect with the result that the desert ecosystem has deteriorated. Combined with climate, deterioration is due to escalation of human and livestock population and their activities. Already a densely populated desert of the world, the human population in the Indian Desert has increased from 3.4 million in 1901 to 8.84 million in 1971. Marginal lands have been brought under the plow, reducing grassland and exposing the land to erosion. Sparse vegetation, trees, shrubs and roots are indiscriminately cut for fuel.

Paradoxically, though grazing area is decreasing, the livestock population has increased from 9.4 million in 1951 to 15.5 million in 1972. Within this period, the 1961-1971 drought decreased cattle population by about 11 percent, while the proportion of goats increased substantially to 34 percent, a further stress on sparse vegetation. Studies reveal that over 9000 square kilometers of Indian Desert have already been affected by desertification processes and another 1,620,900 square kilometers is highly vulnerable to desertification.

Fortunately, the Indian Desert problem is now receiving attention of the Government. The Central Arid Zone Research Institute at Jodhpur in Rajasthan, established in 1959 with the collaboration of UNESCO and the Government of Australia, has studied the desert ecosystem. With the aid of scientific and technological information now available, a major irrigation project to irrigate 2.4 million hectares (Rajasthan Canal project) is now half completed. A Drought Prone Area Program is under implementation with World Bank assistance. The Rajasthan Government now has proposals involving Rs. 110 crores for Desert Development Programs to be implemented in the next seven years.

INTRODUCTION

The Indian or Thar desert is spread over four Indian states, the Punjab, Haryana, Rajasthan and Gujarat, situated in the west of India. Of the total arid area of 3,170,000 km^2 62 percent lies in the Rajasthan State. The average annual rainfall varies from 150 to 500 mm with a coefficient of variation as high as 60-70 percent. The distribution of rainfall is erratic, occurring mostly in the period July to September. The mean daily maximum temperature in summer is generally 40°C and in the winter, 22°-28°C. The mean minimum temperature varies from 24° to 26°C during summer and 4° to 10°C during winter. The mean diurnal temperature variation ranges from 12° to 17°C. The mean relative humidity during summer varies from 36 to 50% and from 66 to 78% during monsoon in the morning hours. In the afternoon, however, there is an expected drop and the mean values range from 20 to 35% during summer and 48 to 60% during monsoon. The mean evaporation during summer months exceeds 10 mm per day. The potential evapotranspiration values during summer (April to June) vary from 7 to 9 mm per day. The mean daily wind speed is recorded to be highest during summer and monsoon seasons, 8-20 km/h.

MAJOR FACTORS OF DETERIORATION

Besides the climatic and geomorphic features, the major factors of ecosystem deterioration within the desert are the escalation of human and livestock population and their activities. The human population in the Rajasthan desert has increased from 3.42 million in 1901 to 8.84 million in 1971, a growth rate of 158 percent (Table 1). The percent increase in population exhibits an inverse relationship with the amount of annual precipitation. In the regions receiving an annual rainfall of 30 cm, 30-40 cm and more than 40 cm, the growth rate during the last 40 years has been of the order of 143.8, 132.1 and 107.2, respectively.

In spite of the low productivity of the arid lands, the Rajasthan sustains a fairly high population of livestock. Paradoxically, along with the reduction in the grazing area during the last two decades, the livestock population has displayed an alarming increase, from 9.4 million in 1951 to 15.5 million, or almost double, in 1972 (Table 2). The livestock density per 100 hectares rose from 72 in 1951 to 175 in 1971 in the desert districts; but in the adjoining districts this

Table 1: Decinnial per cent variation in population in Rajasthan Arid Zone, Rajasthan & India (1901-1971)

Year	Arid Zone of Rajasthan (11 districts)		Rajasthan		India	
	Population (Millions)	Growth rate (percentage)	Population (Millions)	Growth rate (percentage)	Population (Millions)	Growth rate (percentage)
1901	3.42		10.29		236.28	
1911	3.67	+ 7.23	10.98	+ 6.70	252.12	+ 5.42
1921	3.41	- 7.27	10.29	- 6.29	251.35	- 3.03
1931	3.94	+15.67	11.75	+14.14	279.02	+11.00
1941	4.74	+20.40	13.86	+18.01	318.70	+14.22
1951	5.53	+16.58	15.97	+15.20	361.13	+13.31
1961	6.95	+25.87	20.16	+26.20	439.24	+21.64
1971	8.84	+27.22	25.77	+27.83	547.37	+24.66
Overall increase from 1901 to 1971	+5.42	+158.22	+15.48	+150.47	+311.09	+131.65

Table 2: Livestock Population (millions) and Growth Rate in Rajasthan and Arid Districts during 1966 and 1971

	1956		1971			
	Rajasthan	Arid districts	Rajasthan		Arid districts	
	Nos.	Nos.	Nos.	Growth rate(%)	Nos.	Growth rate(%)
Cattle	12.1	3.8	12.5	3.3	5.8	52.6
Buffaloes	3.4	0.7	4.6	35.3	0.9	28.6
Sheep	7.4	4.6	8.6	16.2	5.1	10.9
Goats	8.7	3.6	12.2	40.2	5.7	58.3
Camels	0.4	0.3	0.7	75.0	0.5	66.7
Total	32.0	13.0	38.6	20.6	18.0	38.5
Density per 100 ha of grazing land	160	104	238	49	175	68

enhancement was not so spectacular (25% as compared to 293% in desert districts). The goat and sheep population ranged from 57.1% to 69.3% during this period. Owing to continued droughts during 1967-71 with the consequent migration and mortality, the number of cattle fell heavily (10.8% decrease), whereas the number of hardy animals (e.g., goat) increased substantially (34% increase). The data thus not only reveal a preponderance of the goat and sheep populations in the arid Rajasthan but also point to their increase during the years of drought.

Field rodents (gerbils) are well adapted for arid conditions. In various habitats the rodent population varies from 5 to 600 animals per hectare. Estimates show that the merion gerbil alone can dig out 61,500 kg stabilized soil/day/km^2 during its burrowing activities, leaving it loose as mounds at the surface exposed to wind action. On a rangeland, field rodents can consume 90% of the edible fodder, thus severely competing with grazing animals. A population of 477 rodents/hectare required an annual feed of 104 kg/ha as against the total production of 110 kg/ha.

CROPPING MARGINAL LANDS

In the sandy tracts, quite large areas are ploughed and millet and pulses, which have a relatively low water requirement, are sown soon after the first effective rainfall. After the crop is harvested the land is left exposed to high velocity winds. This annual disturbance of the soil not only robs it of sand-binding vegetation but increases soil erosion, especially the semi-stabilized sand dunes. In fact, in certain areas of the Indian arid zone, rainfed crops should be substituted by perennial ranges, but as a consequence of the high growth rate of population, the sub-marginal and marginal lands have been brought under cultivation. In 1951, the net sown area in the arid zone of Rajasthan was 28.6 percent. It increased to 44.6 percent in 1951-61 and by an additional 9.5 percent during 1961-71 (Table 3). This increase in the net sown area has caused a reduction of about 25 percent in the pasture and rangelands. This substantial shrinkage of rangelands is adversely affecting the fodder and livestock production, besides increasing soil erosion hazards and inducing desertification. Moreover, it has been observed that the area under different crops increased in 1954-70 while production per unit area considerably decreased (Table 4), indicating that soil fertility is decreasing.

Table 3: Land use pattern in arid zone of Rajasthan (Excluding Ganganagar)

Land use	Percengage of total geographical area of western Rajasthan			Percentage change		
	1951-52	1961-62	1970-71	1951-61	1961-71	1970-71
Forests	0.85	0.95	1.09	+ 12.13	+ 14.35	+ 28.22
Land put to non-agricultural use	1.99	2.79	3.02	+ 41.13	+ 8.36	+ 52.92
Barren waste permanent pasture + culturable + current fallow	71.03	57.26	53.67	- 18.89	- 6.18	- 23.90
Net area sown	26.12	39.02	42.23	50.31	+ 8.34	+ 62.85
Area sown more	0.32	0.77	1.31	141.45	+ 70.03	+312.50

Table 4: The annual linear growth rate of the area, production and productivity with respect to various crops (1954-70)

Crop	Area	Production	Productivity
Bajra	+ 2.41	+ 1.34	- 0.66
Jowar	- 3.41	- 5.42	- 4.88
Wheat	- 0.82	- 0.24	+ 0.60
Barley	- 1.16	- 0.05	+ 0.68
Gram	- 6.31	- 1.67	- 1.59
Sesamum	+ 3.24	- 3.57	- 4.23
Other pulses	+ 1.91	- 1.08	- 2.70

A case study conducted by CAZRI for UNESCO in Luni Block revealed that over a period of 18 years (1958-76), due to human interference, sand movement activity leading to further accentuation of undulation has taken place over an area of 166 km^2 or 8.4 percent of the total area of the block. Further, it was observed that recent sand activities have led to an increase in the thickness of sand on previously created fence line hummocks by 15 to 30 cm and to have enhanced their width by 1-2 meters. Even greater increase in height (1-2m) has occurred in parts of the Lunni Block. This phenomenon has concurred with an increase in thickness of sand sheeting. The area so affected is 163.3 km^2 or 8.2 percent of the total area of the block. The stabilized sand dunes in the extreme northwest show an increase of sand piling by 1 to 2 meters on the flank and 3 to 5 meters on the crest.

Saline ground water up to a salinity level of over 10,000 mmhos (6-7 g/liter soluble salts) is commonly used for growing barley and "kharchia" wheat, a salt tolerant variety. An area of 0.4 million hectares is under such an irrigation system. Lands irrigated with saline waters are left fallow for one, two or more years for leaching down the salts by natural precipitation. Continuous use of saline water has rendered the soil saline or sodic, resulting in lower crop yields.

The increasing human population is a serious stress, even on the natural vegetal resources of the desert. The trees and shrubs and even their roots are indiscriminately cut by the rural population for fuel, top feed, thorn fencing and the construction of thatched hutments. It has been estimated that the requirement of the people in the desert with respect to woody biomass for fuel purposes has increased from 1.85 million tons in 1951 to 3.33 million tons in 1971. The desert people have developed peculiar food habits. All the available air-dried seeds and pods of the trees are used as delicacies. The seeds of *Acacia senegal* (kumat), the fruits of *Capparis decidus* (karir) and the pods of *Zizyphus nummularia* (ber jhadberi) growing in accessible parts of the desert are harvested for human consumption. The seeds of grasses, e.g. *Panicum turgidum, P. anti?otable, Cenchrus biflorus* and *Echinochloa coloum,* are mixed with millet for making "capatis" (unleavened cales) especially during drought years. The grass seeds are supposed to add to the nutritive value of the food. The intensity with which seed collection is made for direct human consumption throughout the desert region serisouly affects the natural process of regeneration of the desirable plant species in the inhospitable terrain. Along with efforts to evolve technology for increasing vegetable production locally, the problem has also to be tackled by formulating a suitable educational program for the masses.

An important consequence of overgrazing is the interference in natural regeneration and succession of vegetation, especially the grass species. The lush monsoon vegetation in the desert may seen an inexhaustible source of foraging material for the very large number of domestic livestock. But most of the species, e.g. *Tephrosia purpurea, Indigofera cordifolia, Tribulus terrestris, Crotalaris burhis, Cyperus* sp., are not only inedible or unpalatable to livestock, but compete severely with the perennial species of nutritive grasses and consequently the landscape is infested by low-ranking, unpalatable annuals, thus decreasing vegetation production and affording poor coverage for soil conservation.

The net result of these various activities in the desert ecosystem is that about 9,920 km^2 of 4.35 per cent of the western Rajasthan has been affected by the processes of desertification. Moreover, 162,900 km^2 or 76.15 per cent of the area has been categorized as highly and moderately vulnerable and 41,692 km^2 or 19.5 percent of the area is moderately to slightly vulnerable to the various processes of desertification.

PROSPECTS

The rulers of the erstwhile Princely States were well aware of combating the deteriorating conditions of the desert and did take steps which, if reintroduced, are still viable. They exploited the religious sentiments of the people by establishing temples on hill tops and devoting the hill to the deity (usually *Jogmaya*); tree felling from the hill was religiously banned. The Princes declared *orans* (protected forests, *birs* (protected, sometimes forced, grasslands) and *shikargahs* (wildlife conservation paddocks, where in only the Maharajah had the right to shoot). These measures resulted in protected areas where vegetation and wildlife developed. Another important activity was bringing surface water into the desert. In the pre-independence era, two major endeavors of the Maharajahs to bring surface water from the non-desert areas into the desert are exemplary: the Ganga Canal and the Jawai dam scheme. These attempts did bring considerable areas under irrigated crops and made drinking water available to desert people.

Soon after independence, the Central and State Governments became aware of improving the xeric conditions and the importance of the arid region to the national economy. A number of steps were initiated. Probably the most important was the establishment of the Central Arid Zone Research Institute, at Jodhpur, which has completed 25 years of productive research. Besides CAZRI, there are the Central Sheep and Wool Research Institute (Avikanagar); Central Soil Salinity Research Institute (Karnal); Indian Grassland and Fodder Research Institute (Jhansi), and Exploratory Tube Well Organization at Jodhpur. The Botanical, Geological and Zoological Surveys of India have their regional stations to assess and properly exploit the natural resources of the desert. Agriculture, animal husbandry, forestry and soil conservation groups, the Ground Water Board and a number of other State Departments, are also actively engaged in desert management. Several agricultural universities sprang up in the arid region to advance the well proven agricultural practices to to transfer technology to the grass root level. For the speedy implementation of the programs, a Desert Development Commission has been established by the Rajasthan Government.

CAZRI and other organizations have by now developed sufficient technology to reclaim desertic areas and improve production and desert management in general.

To assess the natural resources and to check their overexploitation, CAZRI has standardized techniques for integrated natural resource surveys, based on the interpretation of aerial photographs. Surveys have been completed for an area of 45,267 km^2 in Rajasthan, Haryana and Gujarat. These surveys provide comprehensive information on (a) land use capability classes based on land use data, erosion hazards, both wind and water, types and intensity of sand dunes, soil texture and fertility; (b) vegetation types, pastures; (c) water resources both surface and ground water, water quality (salinity), water potential zones; (d) socioeconomic conditions of the desert dwellers, and human-animal-vegetation relationships.

On the basis of geomorphological investigations, the sand dunes have been classified as parabolic, longitudinal, transverse, barchan and shrub-coppice. Unstabilized sand dunes shift their locations not only to encroach over the roads, railway lines, villages and clog irrigation canals but the drifting sand also covers the saplings of food crops, occasionally in vast stretches of cultivated desert lands. Techniques for sand dune stabilization have evolved. The rehabilitation of marching sand dunes through revegetating involves three distinct processes: (a) protection against biotic influence by fencing the area; (b) treatment of the shifting sands by fixing brushwood micro-windbreaks from the crest of the dune to the heel of the dune, and (c) revegetating by reforestation and establishment of grasses, shrub and tree species. This technology has been successfully demonstrated in Bikaner, Jhunjhunu and Barmer districts of Rajasthan. The cost of such technology is on an average about Rs. 860/hectare with an anticipated return of Rs. 1000/ha. The cash flow analysis revealed that cumulative streams of receipts from fuel, fodder, seeds and top feed will start "outwitting" costs from the 13th year and onwards, yielding a net benefit cost ratio of 31.65 percent for the 13-year period.

Unstabilized sand dunes contain some moisture that permits growth of adapted vegetation. Properly exploited, this moisture could be used for establishing vegetation on the dunes for their stabilization and for providing fodder for livestock.

In the field of afforestation, a large number of species of *Eucalyptus*, *Acacia* and other genera from similar iso-climatic zones of the world were introduced. Of the different exotic tree species, *Acacia tortilis*, *A. aneura*, *A. radiana*, *Eucalyptus cmaldulensis*, *E. melanopholia*, *E. terminalis*, *Cassia phyllodenea*, were found to be promising for planting in the Indian arid zone. Nursery techniques for raising saplings for plantation in large numbers have also been evolved.

The rangelands of the Indian arid zone support palatable, adapted and productive grasses and shrubs for top feed. Unfortunately, however, mainly due to over-exploitation, overgrazing and poor management, the grasslands are deteriorated to a great extent. Intensive research work on the various aspects of range management and livestock production has been in progress for the last 20 years. The perennial grasses, *Cenchrus ciliaris*, *C. setigerus*, *Lasiurus sindicus* in the northwestern desert, and *Dichanthum annulatum*, *Sehima nervosum* and *Chrysopogon fulvas* in the southern region are directly sown in the fields after a complete soil working and grubbing out of the weeds and annuals in lines of 50-75 cm apart at a depth of 1-2 cm. Planting of 2-3 seedlings per hill followed by one weeding gave 60 percent establishment of a pasture.

Fertilization with nitrogen further increases the forage yield. The average amount of nitrogen recommended is 2 kg per hectare for *D. annulatum* and *Panicum antidotale*, 2 to 3 kg/ha for *C. ciliaris* and *C. setigerus* and 5 kg/ha for *L. sindicus*.

Intercropping of the grasses with legumes (*Phaseolus acentifolius, P. radiatus*) gave 20 to 30 percent increase in forage production as compared to pure culture of grasses.

The livestock production in relation to improved grass-ands indicate a high level of production as the body weight gains of yearling heifers range from 92 to 1964 kg per year in different pasture types. Likewise, gains in thc body weight of ram lambs is also superior to those grazing on depleted natural ranges.

Important crops in the area are millets (*Pennisetum typhoidium*) and legumes (*Vigna radiata*). Adapted varieties of these crops have been developed and agronomic practices including water-harvesting, mulching, fertilizer use, alternative crop choice (mid-season corrections) have been developed. New crops such as sunflower and castor bean have been found better under late-sown conditions.

Techniques such as sprinkler and drip irrigation have been tried and found suitable for optimum water use efficiency. With drip irrigation use of saline water is also possible.

The fuel requirement of the relatively dense population of the Indian arid zone is usually met with locally available brushwood and trees and cow dung. This is a severe strain on the natural vegetation resources. Most of the fuel wood comes from *Prosopis cineraria, Acacia jaguemontii* and *Prosopis juliflora* and, deeper into the desert, from the best sand-binding shrub, *Calligonum polygonoides*. Even the roadside plantations of *Albizzia lebbek* and *Azadirachta indica* and shelter belts are not spared by the local populations. Research in solar energy utilization has shown that low cost solar water heaters and solar ovens can be used and if they were adopted widely, the shrubs and trees could be spared to some extent and the process of deterioration could be partly checked.

On the basis of the studies conducted on bait preferences, seed consumption and lethal dosages of various rodenticides, effective methods for the control of field rodents have been recommended. The rodent control operation should be taken up prior to crop sowing, during summer, when the reproductive activity of the rodents is lowest during the year, their population density is minimum and acceptability of baits maximum.

It is, however, felt that despite useful technologies, India has not been able to achieve a significant breakthrough on the development front. The available scientific information on the national use of ecosystem resources are yet to be converted into action programs. With this in view, the National Commission on Agriculture has made important recommendations for improving conditions in the desert and a State Desert Development Board is implementing the recommendations. A Working Group on desert development has also announced its recommendations regarding formulation of action programs. Operational Research Projects have also been initiated in the arid areas of Rajasthan to expedite effective transfer of technology and for involving the desert people themselves.

Mention must be made of the two important programs in India for development of the desert ecosystems:

(a) Rajasthan Canal Project: This project, which is about half completed, is bringing water from Harke barrage at the junction of the Sutlej and Beas Rivers via a canal traversing western Rajasthan in a north-south direction about 20 miles inside the western border. It will feed a series of distribution canals providing irrigation to the west of the main canal. The project covers a gross area of 5.9 million acres. It has been estimated that 3.36 million acres could be the command area and 2.6 million acres actually irrigated. This project will have considerable influence on a large area of western Rajasthan.

(b) Drought Prone Area Program: The Drought Prone Area Program (PAP) was initiated five years ago with the assistance of the World Bank. As the name suggests, the program covers the most chronically drought prone areas in India. It is in operation in 74 districts in 13 states, and started initially as a rural works program. It has now been restructured as a comprehensive area development program in the arid and semi-arid areas of the country. Its thrust is for optimum utilization of land, water and livestock resources consistant with the ecological balance. Development programs are undertaken on a watershed basis to improve agricultural productivity and to provide employment opportunities.

REFERENCES

CAZRI, Jodhpur (1964). Recent Development in Rajasthan. Souvenir Volume.

CAZRI, Jodhpur (1974). Summer Institute on Desert Ecosystem and its Improvement (In press).

CAZRI, Jodhpur (1978). Arid Zone Research in India. Silver Jubilee Volume.

Garg, H.P. (1975). Solar Energy Utilization Research. CAZRI (Jodhpur) Tech. Monograph No. 3.

Gupta, R.K. and Prakash, I. (1975). Environmental Analysis of the Thar Desert. English Book Depot, Dehradun.

Kaul, R.N. (1970). Afforestation in the Arid Zone. Dr. W. Junk, B.V., Publishers, The Hague.

Mann, H.S., Malhotra, S.P. and Jagdeesh C. Kalla (1974). Desert spread, a quantitative analysis in the arid zone of Rajasthan. Ann. Arid Zone 13(2):103-113.

Ministry of Education, India (1964). Proc. Symp. on Problems of Indian Arid Zone, Jodhpur, Nov. 23-Dec. 2.

National Commission on Agriculture, India (1974). Interim Report on Desert Development. New Delhi.

National Institute of Science (1952). Proc. Symp. on Rajputana Desert. New Delhi.

National Institute of Science, India (1969). Symp. on Planning for Drought Areas. New Delhi.

Roy, B.B. and Pandey, S. (1970). Expansion or contractions of the Great Indian Desert. Proc. Indian Nat. Sci. Acad. 36(6):331-343.

Sen, A.K. (1972). Agricultural Atlas of Rajasthan. ICAR, New Delhi.

COMBATING DESERTIFICATION IN JORDAN

KAMAL I. TADROS
Head, Range Division, Ministry of Agriculture,
The Hashemite Kingdom of Jordan

ABSTRACT

In arid and semi-arid regions throughout the world, human over-use, and misuse, of fragile ecosystems has led to increasing desertification. Desertification in the Kingdom of Jordan has caused serious changes in desert ecosystems, as grazing animals are displaced by agriculture onto areas unsuitable for grazing and soil erosion and groundwater depletion follow in the wake of agricultural practices unsuitable to the environment. Examples of recent processes of desertification in Jordan are given and the Government's programs aimed at combating desertification are briefly noted.

INTRODUCTION

The total area of the East Bank of Jordan is approximately 9.26 million hectares. About 91.4% of this area is desert and semi-desert, as classified by the Five-Year Plan, with annual rainfall of less than 200 mm and yellow or grey desert soils. These areas are not suitable for economic agriculture and their optimal use is grazing.

The main causes of the formation and extension of deserts is a combination of harsh environment and human misuse of natural resources, and within this combination, man is the most important cause of desertification. Throughout the world's arid and semi-arid areas, the following factors have increasingly been responsible for desertification of once productive and/or inhabitable lands. Cultivation is rated as the major cause of accelerating desertification, i.e., ploughing of erodable soils and unstable soils without taking appropriate soil and water conservation measures, and the ploughing of range lands with annual rainfalls of less than 200 mm. Natural vegetation cover is destroyed, and the soil is loosened, with greater areas exposed to wind and water erosion, leading to desertification. Overgrazing in fragile desert ecosystems occurs when cultivation displaces livestock from range lands onto arid and desert regions. Vegetation is destroyed, especially on steep slopes, leading to loss of rainfall runoff and erosion. Overgrazing around new, unplanned wells and bore holes aggravates the problem. The use, and overuse, of fuel wood to provide people's energy demands is a current cause of desertification throughout the centuries. When this fuel harvest reaches of proportions of tree felling, uprooting of shrubs and grasses and the gathering of any vegetative material available, desertification becomes imminent. Fires in hot and dry seasons develop enough heat to destroy seeds, and hence the regenerating capacity, of many trees and shrubs. Overuse of ground water in arid lands leads to lowering of the water table and increased salinity. Use of this saline water for irrigation leads to greater salinities and decreased plant production.

Desertification in Jordan has caused serious changes in ecosystems, as the following examples illustrate: (1) Secondary plant associations that have low economic value and are weak in soil conservation took the place of climax vegetation. For example, oak forests were replaced by *Poterium* species, and these were replaced by *Asphodelus*. (2) Under the effects of cultivation, overgrazing and uprooting of shrubs, range lands deteriorated so that plant associations having low nutritive value and less palatability took the place of original associations; for example, *Artemisia* and *salsola* were replaced by *Anabasis*. (3) Soil destruction can be seen in the low organic matter of the soils, appearance of mother rocks at the surface, and decreased rainfall absorption capacity, increasing runoff and soil erosion. (4) Extensive use of underground water for irrigation in the drylands has increased the salinity of this water, as in the Wadi Dhuleil. (5) Some wild animals in Jordan have vanished, such as the cheetah and oryx, and others have greatly decreased (gazelle, hare, houbara bustard).

COMBATING DESERTIFICATION

The Government of Jordan has recongized the dangers of desertification and has established development plans for the rational use of natural resources. The difficulties that face this rational development are more social than technical.

Efforts to halt and reverse desertification include the following:

a) Every year, 3000 hectares are afforested with forest trees such as *Pinus* sp., *Eucalyptus* sp., *Acacia* sp., *Tamarix* and *Causaurina*. Five range stations (2200 hectares) in various range areas were protected to give natural vegetation a chance to grow, and palatable species time to establish themselves. These areas were improved by applying soil conservation techniques such as check dams and were planted with fodder shrubs such as *Atriplex* sp. and *Colutea* sp. Some range stations, such as Khanaseri and Fujij, were utilized by sheep, depending upon the estimated carrying capacity. Recently, 10,000 hectares in Shaumari and 1054 hectares in Sabha were registered as range lands to be protected and improved.

b) During the last ten years more than 2 million fruit trees, primarily olive trees, were planted on earth terraces in rainfed areas. Trees have been planted along major roadways to improve both soil conservation and the general environment. The Royal Society for Protection of Nature established a national park in Shaumari (2200 hectares) to protect wildlife and the environment. Windbreaks are planted around orchards and row crops in the Jordan Valley and other areas to reduce wind velocity and conserve soil and water.

c) Contour ploughing on gentle (0-9°) slopes is practiced in dryland farming areas, especially for wheat. Watershed protection has been enhanced by catchment dams (e.g., King Talal, Wadi Shuab, Kafrein, Ziglab dams). There is a project for sand dune fixation in Ram and Desi in an area of 5000 hectares.

d) A Bedouin settlement project was begun in 1962 to use available underground water for agriculture, to train the bedouin in agriculture and to improve their social, economic, educational and health conditions. Nine project sites covering 695 hectares were reclaimed and planted, and 139 bedouin have been trained and given the land.

WATER CONSERVATION

Under an overall plan for water conservation, the following works have been undertaken in Jordan: Digging of wells (859 up to 1977); construction of 32 runoff catchment dams; use of overhead and drip irrigation in the Jordan Valley; protected vegetable cropping.

The Government of Jordan, aware especially during the past ten years of the processes of desertification, and its deleterious effects, plans the extension of some current projects and implementation of new projects in a direct effort to combat desertification. These include:

(1) Extension of the rangelands improvement schemes, whether by protection or plantation of fodder shrubs that are drought resistant and palatable (*Atriplex nummularia*).

(2) Proposals for a 60,000 hectare project for wild life protection and range management being prepared by the Ministry of Agriculture, the University of Jordan and the Royal Society for Protection of Nature will be submitted to the UN Environmental Protection Agency.

(3) The Royal Scientific Society, in cooperation with the Ministry of Agriculture, is preparing a project for collecting and analyzing range plants to determine their feed value, economic use, and the variations of feed value with different seasons.

(4) In arid areas of Jordan where runoff management is an available agricultural technique, the Ministry of Agriculture is in close cooperation with the Mennonite Central Committee in rainfall management studies.

The Ministry of Agriculture in Jordan takes as first priority in arid areas the improvement of pasture and grazing control.

REFERENCES

1. Desert creep and range management in the Near East. Mission Report. RNEA, Cairo, 1976.

2. Desertification in Jordan. Report to the United Nations Conference on Desertification. Nairobi, Kenya, 1977.

SELECTING THE OPTIMAL TECHNOLOGY LEVEL FOR DEVELOPING DESERT RESOURCES

MARTIN M. FOGEL
Professor, School of Renewable Resources and Department of Hydrology and Water Resources, University of Arizona, Tucson

ABSTRACT

In the context of development, much has been written on the concept of "intermediate" technology as differing from "high-level technology" employed in the Western world. While development is the application of technology to improve the quality of human life, more technology does not necessarily mean improvement. This paper will focus on the use of a systems approach for determining the optimum technology level within the framework of developing land and water resources of arid lands. It will be shown that system analysis, which consists of the application of economic concepts to the problem of resource allocation, forces decision makers to look at an entire system, at a number of alternatives to achieve a goal, and at the consequences of implementing the alternatives. Inasmuch as most large scale developmental projects are undertaken in large part for social, environmental, military or other non-economic objectives, a means is needed for handling multi-objective projects, for trading economic returns against effectiveness of meeting non-economic goals.

INTRODUCTION

Evidence is accumulating that questions such widely held beliefs as, "more technology is better," or, "high level technology is better than low level technology," with "better" referring to the "quality of life" in developing countries. What I am alluding to is that in the development of desert resources, more or higher technology is not *always* the best option to choose for a developing country. Circumstances may indicate that a relatively unsophisticated level of technology will, on occasion, meet the goals of development better than more or higher level technology when considering the overall system, the trade-offs and the long term effects. The objective of this paper is to present what may be called a systems approach for choosing the best set of options for effectively utilizing the desert resources of a desert country.

Reviewing the titles of papers to be presented at this conference, we see that the majority refer to technology associated with the development of deserts for agricultural purposes. Since my experiences, both in the United States and the Middle East, are along the same lines, my comments will be limited to the development of desert resources for the production of food and fiber. In addition, inasmuch as the major components or attributes of a desert region are basically lots of sand and sun, water has to be introduced to the scene in some manner in order that this production take place. Thus, all the problems of developing and using limited quantities of water are applicable in this instance.

DEVELOPMENT PROCESS

In simple terms, desert development is contingent on technological possibilities, social requirements and resources available. The socio-technological development process can be subdivided into (1) the information process, (2) the decision process and (3) the realization process.

The initial step in desert development is gathering the necessary information on the possibilities, the requirements and the available resources (capital, manpower, natural resources, etc.). In discussing development of deserts for agriculture, thoughts are immediately turned to the subject of water. Assuming that surface waters are either non-existent or are fully utilized, the only source of water for an inland, large-scale project would be groundwater. Pertinent information needed includes the economics of developing the supply (a variety of technologic levels are certainly available), the interrelationship between supply and demand and the scheme for managing the land (e.g., will it be subdivided into small farms or will large corporate farms manage the land).

Making the appropriate decisions is the second step in the development process. An example of a decision-making process that required some adjustment was a project in Saudi Arabia that I helped initiate. Originally called the King Faisal Settlement Project in Haradh, the project was designed to provide for the settling of 1000 Bedouin families on 4000 hectares of irrigated land.

While the technical problems were great, due in large part to the harsh climate with high temperatures of 50°C in summer and a grand total of 50 mm rainfall, and to the remoteness of the area, the sociological problems were even greater. No amount of governmental subsidies could entice sufficient numbers of Bedouin to settle anywhere, much less on the Haradh Project. Without the Bedouin settlement that was counted on, the project was operated by the Ministry of Agriculture and Water, then by a large American firm, and is now back with the Ministry, I believe. I am not sure what has happened in the last two to three years.

The original intentions were to deliver water to a number of small units, about 5 hectares each. Under corporate farming however, the fields should be sufficiently large to allow the efficient use of machinery. The point is that, somewhere in the early planning stages, a number of possible alternatives should have been determined, either another scheme to settle the Bedouins (if that was the objective) or a scheme, other than by Bedouins, for managing the land.

The conclusion to be drawn from this project is simply that any proposed level of technology requires acceptance by local people if the third step, the realization process, is to be reached. The Saudi Arabian project was only a relatively small-scale irrigation, however.

In contrast to the above, Iran has problems of a similar type with the 1,000 km^2 Dez Irrigation Project. More than half the project, including the infrastructure, was assigned to four international companies. Preparing the land for mechanized farming consisted of reshaping the landscape to the extent that numerous villages were destroyed and villagers were resettled. According to Worthington,[1] resettlement has caused "social disintegration and social and economic decline of the population."

LEVELS OF TECHNOLOGY

In recent years, whenever technical assistance to developing countries is discussed, terms such as "intermediate technology" and "appropriate technology" have been common descriptors. *Intermediate technology* usually refers to a technology level somewhere between the traditional or that now employed in a developing country and that which is characterized by large-scale, capital-intensive, modern technology as that found in sophisticated, developed countries. *Appropriate technology,* as defined by the Agency for International

Development of the U.S. State Department,[2] is one "which is as responsive as possible to the capital, manpower, natural resources, institutional and social realities of a particular country environment."

Important characteristics of intermediate technology are that it can be produced by the nonmodern, rural sector with labor intensive production methods and local resources. It is relatively inexpensive in terms of production cost and selling price, and is simple to fabricate, operate, maintain and repair (OECD).[3]

Appropriate technology is a broad concept referring to technology at any level of sophistication appropriate to local circumstances (Becker and Ulinski).[4] This can be interpreted to mean that in the development of desert resources for agriculture, or any other purpose, there are a number of technology levels from which the most optimal can be selected.

In the production of food and fiber from deserts (or semi-deserts), examples of intermediate technology would include the traditional grazing practices, the use of primitive irrigation systems where water is readily diverted from nearby streams, the lifting of water from surface sources by animal or human power, such as by the "shadouf."

An example of such technology -- which is currently undergoing change, and not necessarily for the good -- is the large date-palm oases found in Saudi Arabia.

The development of the groundwater supply for the Qatif and Al Hasa oases in the Eastern Province appears to have taken place over 2000 years ago when many hand-dug wells (or 'ains) supplied flowing artesian water to a total of about 15,000 ha of date palms. The wells, which required chiseling through some 3 meters of rock, flowed for many centuries to provide a major share of the support for a relatively large population.

Irrigation practices were primitive as the fields were very small, labor was intensive and with no control devices, water flowed continuously, the surplus going to the Gulf. With the discovery of oil, water well-drilling technology was transferred to the local people who then proceeded to drill many wells which also flowed continuously. As might be expected, artesian pressures dropped and flows ceased. Responding to the resulting complaints from the people, the Saudi Arabian Government had valves installed on the drilled wells. Unfortunately, not all the necessary technology was transferred, and the land at the lower elevations became waterlogged and saline. What was not apparent from the ground surface was that the drilled wells were cased for the upper 5 meters where sand was encountered and were uncased the remaining depth, which was in a limestone formation. Once the three meters of rock were passed through, artesian water forced its way through crevices in the limestone, up through the sand and then to the surface. Without adequate drainage, some of the land became unproductive.

In contrast to the above intermediate technologies, higher levels of technology would consist of large-scale irrigation projects with water being supplied either from that sotred behind huge multipurpose dams (such as the High Dam at Aswan) or from a large number of pumped wells. Such projects would prepare the land, provide a network of canals and/or pipelines to distribute the water and would apply the water to the land in a number of ways utilizing a relatively small amount of labor. It is of interest to note that most of the projects in which water is applied by surface methods have irrigation efficiencies (ratio of water required to water applied) of less than 50 percent. The overall efficiency, when storage and transportation losses are considered, is even less, perhaps in the 25 to 33 per cent range. Many of the so-called more primitive systems, ironically, would have higher irrigation efficiencies than the more sophisticated irrigation projects.

Where annual rainfall approaches 300 mm, some form of rainfed agriculture may exist such that adapted technology could improve the quality of life of local people. The introduction of new plants, some of which are to be discussed at this workshop, offers a possibility. Another technique that bears consideration is water harvesting, which had its roots in the Middle East perhaps 4000 years ago. By clearing and smoothing the land at the higher elevations, runoff from rainfall was induced with the resulting water then channeled in contour ditches to agricultural fields at lower elevations.

Water harvesting is returning to the scene, this time with improved methods and materials for increasing, collecting and storing runoff water (Cluff).[5] Under the right conditions, common table salt applied to the land will increase runoff substantially. Storing the water in plastic-lined reservoirs reduces water losses from seepage and evaporation and allows the water to be used at the most beneficial times.

More complex systems for the production of food and fiber that are currently available are those that require the water, either sea or brackish, to be treated such as by large-scale desalinization plants. The use of controlled environmental systems, a subject discussed at this conference, is an example of one of the more sophisticated methods for agricultural production.

With this range of possibilities, decisions will have to be made on which set of components make up the most appropriate system. The question, then, becomes: Which set of components, which system or which technology should be chosen for a particular set of local circumstances? A very much related question would be, What are the criteria for making this selection?

OBJECTIVES AND CRITERIA

In conjunction with generating a set of alternatives for desert development, the objectives must be clearly defined and criteria for making the decisions must be firmly established. Without these two highly important decision-making elements, alternatives cannot be compared nor can an optimal decision be made.

Whatever it may be called, goals, aims or objectives, no logical analysis can proceed far without precise, explicit statements of purpose (de Neufville and Stafford).[6] For the development of deserts the following are examples of what might be considered broad, general objectives:

1) Pure economic efficiency, the maximization of returns on investments regardless of to whom they may accrue or at whose cost.

2) Redistribution of income, the promotion of the welfare of particular interests or the expense of others.

3) The fulfillment of desires which cannot be justified through their economic benefits, but which are considered worthwhile or essential, such as education, creation of parks, or defense.

Some people confuse the above with criteria for selecting a particular alternative. The above are objectives to be achieved and how well an alternative achieves these goals is a criterion or measure of effectiveness. Looking at the first of the above general objectives, economic efficiency, the measure of effectiveness may be quite simple, such as either the benefit/cost ratio or the difference between benefits and costs. In either case, the higher the number, the greater the measure of effectiveness; that is, the greater the return for the investment.

Most projects, however, are much more complex than the above simple example. More often than not, they will have several objectives. Expanding on the above three general objectives, the following is a more detailed list of objectives that may be used in desert development: (1) Maximization of output (food and fiber; (2) Maximization of the availability of consumption goods; (3) Maximization of the rate of economic growth; (4) Reduction in unemployment; (5) Regional development; (6) Reduction in balance of payments deficits; (7) Promotion of national self-reliance; (8) Improvements in the quality of life.

With the development of desert projects most likely having more than one objective, each objective should have its own measure of effectiveness. To go even further, each performance measure should be quantitative, if at all possible. Since quantification is not always possible, the problem of selecting an optimal alternative may be formidable.

TECHNOLOGY ASSESSMENT: EVALUATION AND SELECTION

Defined as a formal, systematic, interdisciplinary examination of technology that is existing, emerging or prospective, technology assessment seeks to estimate costs and consequences over an expanse of time (Madden).[7] As a new line of social accounting, it should evaluate expected impacts and should include consequences not anticipated.

According to Madden,[7] technological development tends to leave out the economic cost calculations of such humanistic values as abolition of poverty, preservation of the ecosystem and enhancement of human dignity. Not all development is positive as seemingly implied by Madden. As one voice of dissent for example, Michael G. McGarry[8] of the Health Sciences Division of the Canadian International Development Research Center has stated that sanitary sewers is not an appropriate technology for providing urban wastewater management services for developing nations. His study showed generally that: (1) Construction periods are necessarily long because of budget constraints; (2) Benefits accrue from sewerage in the distant future, because sewer mains and laterals must be complete before many house connections can be made; (3) Large investments are required during the first stages of sewer construction, such that they are not matched by early revenue returns and by benefits from improved health; (4) To invest in sewerage, a very high value must be placed on intangible environmental and amenity benefits.

McGarry's conclusions certainly are not universally acceptable. Much will depend on the objectives to be achieved and the measures of effectiveness that are used in determining how well the objectives are met. Different conclusions may be drawn if the objectives include developing an industrial base in a city. Outside capital and management may not be attracted if the city is sewerless. For desert development, the value of the water, both raw and treated, may alter the evaluation. In any case, specific wastewater management alternatives have to be compared in a like manner before the most cost-effective system is chosen.

The evaluation of alternatives, as stated by de Neufville and Stafford,[6] consists in associating each alternative system with its effects, i.e., costs, benefits, impacts on human resources and on the environment. Wherever possible, these are estimated by using various types of mathematical models so that the evaluation is based on the full range of conditions that may occur. Optimal water supply development is contingent on having a good estimate of the relative frequency of the occurrence of drought periods, of anticipated water requirements of supply and demand functions for the entire spectrum. This need is supplied by stochastic or probabilistic models which can be used to generate long-term synthetic records which are then used to obtain optimal designs and operations of water control structures.

Mathematical models also have the advantage in that it is often much easier to adjust a parameter in a computer model than it is in obtaining factual information about the parameter's effect on the system. For example, in those instances where water supplies are limited, it is necessary to know which crop should be limited, when and by how much. Unfortunately, such information is not readily available anywhere in the irrigated world. What is generally known is the amount of water required to produce a crop so that water is not the limiting factor to maximum production. However, to maximize the *economic returns* to the farmer and the country, an amount less than this would generally be the optimal application. Validated models could provide this economically-optimum water requirement information and could also determine which crops to limit, how much and when.

COST-EFFECTIVENESS: A SUMMARY

Desert development for agriculture would most likely be a multi-objective project in which all the measures of effectiveness could not be quantified or estimated in economic terms. How is "the quality of life" determined? Most complex projects represent a trade-off between economic returns and protection of both humanistic values and the physical environment. While recognizing technology's essential role in development, decision makers must be given all the information pertaining to this trade-off in order to make the best possible decision, in order to select the most appropriate technology. The following material presents a logical, step by step, standardized approach to making a cost-effectiveness evaluation. The methodology is adapted from Kazanowski,[9] and in effect, summarizes the foregoing material.

1. Define the objectives, goals or purposes that the project is to accomplish. Principal classes of objectives include economic efficiency, redistribution of income and attainment of other desires which cannot be justified solely through their economic benefits. Examples of this latter group of benefits are those associated with the enhancement of the quality of life and the preservation of the environment.

2. Identify measures of effectiveness which establish the criteria for evaluating a particular project. By providing some means for specifying to what degree objectives are met, alternative designs or projects can be compared. Examples of effectiveness measures relating to economic efficiency are simply the benefit-cost ratio or the difference between benefits and costs, while income redistribution can be measured by regional increase in income or reduction in unemployment. Criteria for quality of life indicators might be employment opportunities, use of local resources, necessity for resettlement and reliability of project. One of the best indicators of environmental protection is the quality of water that leaves a project on an area that has been disturbed by man's activities.

3. Develop alternative designs, management schemes or operational policies with different technology levels that can meet objectives. At least two distinctly different systems must be conceived to use cost-effectiveness evaluation.

4. Determine capabilities of alternative technologies or systems. Once the appropriate criteria have been identified, the next step is to express the abilities of these alternatives

in terms of measures of effectiveness, quantitatively if possible, qualitatively if not.

5. Perform sensitivity analysis: In many cases the outcome of a cost-effectiveness evaluation is very sensitive to the assumptions made. Conclusions reached under these circumstances may be unknowingly but significantly biased by assumptions essential to the analyses. If such is the case, either sound justification for the assumptions must be developed or at the very least, sensitivity of the conclusions should be indicated.

6. From a display of costs and effectiveness measures for each alternative, the dicision makers can select the most appropriate technology level for desert development.

CONCLUSIONS

The following represents some of the writer's principal conclusions on the technology for developing desert resources for agriculture:

1. In considering the development of deserts for agricultural purposes which involve the utilization of large quantities of water, the objectives a nation desires to achieve must be clearly stated, the measures of effectiveness must be defined and other alternatives for meeting the objectives should be considered.

2. In selecting the most appropriate technology level for desert development, social accounting and environmental protection should be considered along with economic evaluations.

3. To meet a national objective of increasing food production within an economically-efficient framework, improving irrigation practices on existing projects will generally bring greater returns than developing virgin land using any technology levels.

REFERENCES

1. E.B. Worthington (1978). Irrigation's benefits: miracle or mirage? Civil Eng. 48(8):59-63.

2. Agency for International Development (1972). Appropriate technologies for international development. Office of Science and Technology, Agency for International Development. Washington, D.C. p. 30.

3. Organization for Economic Cooperation and Development Center (1975). Low cost technology - an inquiry into outstanding policy issues. OEDC, Paris. p. 10.

4. A.L. Becker and C.A. Ulinski (1976). IRRI small agricultural machinery project: U.S. technology transfer to resource-poor developing countries. In, Technology Transfer and U.S. Foreign Policy, Henry Naw, Ed. Praeger Publishers. New York. pp. 208-264.

5. C.B. Cluff (1975). Engineering aspects of water harvesting research at the University of Arizona. Proceedings, Water Harvesting Symposium, Phoenix, Arizona, March 26-28, 1974. ARS-W-22, Agric. Res. Serv., USDA. pp. 27-39.

6. R. de Neufville and J.H. Stafford (1971). Systems Analysis for Engineers and Managers. McGraw-Hill Book Co., New York.

7. J.P. Madden (1978). Agricultural mechanization, farm size and community development. Agric. Eng. 59(8):12-15.

8. M.G. McGarry (1978). Sanitary sewers for undeveloped countries - necessity or luxury? Civil Eng. 48(8):70-75.

9. A.D. Kazanowski (1968). A standardized approach to cost-effectiveness evaluations. In, Cost Effectiveness: The Economic Evaluation of Engineered Systems, J. Morely English, Ed. Wiley and Sons, New York. pp. 113-150.

LAND USE IN ARID AUSTRALIA

RAY PERRY AND DENI GRAETZ
Division of Land Resources Management,
CSIRO, West Australia

ABSTRACT

Three-quarters of Australia is arid, of which about half is unpopulated desert. The other half is mainly used for grazing cattle for beef and sheep for wool. The grazing industry is export oriented, capital intensive, and uses very little labor. Small areas are used for mining, irrigation and tourism, all of which are also capital intensive.

INTRODUCTION

In this contribution we have set ourselves the following goals: (a) to present a synoptic overview of the land resources of arid Australia; (b) to outline the present pattern of land use and to list the factors which have determined this pattern; and (c) to examine the pressures for and to predict the direction of future changes in arid land use.

The arid lands of Australia and her management of them is unique when compared to climatically similar areas in the rest of the world. However, we have tried to summarize the national experience and to set out both the ecological and societal constraints that must be considered whenever and wherever arid lands are to be newly developed.

THE LAND RESOURCES OF ARID AUSTRALIA

(a) Size

The semi-arid and arid lands of Australia are together represented by a single, vast central area of the continent extending to the western and southern coasts. The various climatic classifications of arid and semi-arid can have little meaning over such a large land mass for they do not give sufficient attention to the particular seasonal atmospheric circulation patterns and they ignore the role that soils may play in moderating or exacerbating atmospheric drought. Hence we use as our definition of arid lands (collectively the arid and semi-arid) those areas in which *rainfall effectiveness precludes cropping or pasture improvement manipulations*. Thus the boundary is primarily determined by the climatic factors influencing rainfall effectiveness, i.e. seasonal distribution and reliability and not just annual totals alone, and include as well the hydrological characteristics of soils where they limit agriculture in an otherwise climatically favourable area.

The dividing line between the cultivated lands and the arid lands separates off only a small corner of the south west and eastern coastal arc of the continent as not arid. The remaining arid lands occupy some 5.3 x 10^6 km^2 or approximately 70 per cent of the total land survace area of Australia (Fig. 1).

The boundaries drawn on Fig.1, whilst being largely determined climatically, are also determined by historical experiences. In southern Australia, with a century of agricultural occupation, cropping is largely confined by the 250 mm median annual rainfall isohyet whereas in northern tropical regions the 600-700 mm isohyets best define the boundary between the cultivation and pastoral enterprises. This latter boundary remains the most fluid since agricultural experience in this area is young (30 years) and rather intensive pasture improvement is proceeding (Perry).[1]

The arid zone is more than 3,200 km from east to west and 2,000 km from north to south extending as it does from latitude 17°S to 35°S and from longitude 113°E to 148°E. Whilst the total area of arid land is large, it is second in size only to the North African-Middle East deserts. But as we shall see, some 26 per cent of these lands, the central and western spinifex sand plain deserts, are not used by European man on any permanent basis. Even so the area used for pastoral purposes, the range-lands, is roughly equivalent to the total land area of India. Within these rangelands all pastoral enterprise is based on the native vegetation.

On Fig. 1 we have included political boundaries, i.e. the state borders, to emphasize two points. The first is that the central core of arid lands of mainland Australia

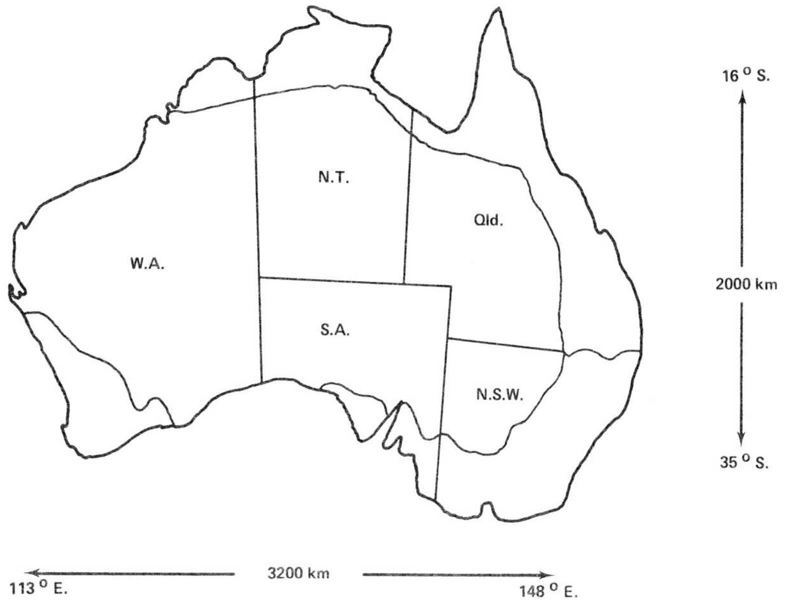

Fig. 1. A map of Australia showing the boundaries of the arid lands and the political boundaries separating the five mainland states. The abbreviations used on this and other maps are:

W.A.	Western Australia	N.S.W.	New South Wales
N.T.	Northern Territory	S.A.	South Australia
QlD	Queensland		

are effectively divided by these political boundaries and secondly these boundaries also demarcate differing policies and controls on land use, essentially through variations in land tenure legislation. Unlike the U.S.A. there are no large areas of federally owned lands; each state largely controls all the arid lands within its boundaries. There are some minor exceptions to this and these will be pointed out later.

Each state controls the arid lands within its boundaries because the state retains ownership of it; the majority of the land is only leased to pastoralists or other users. In the case of pastoralism these leases grant rights to graze the land but they do not grant rights to mine, or to cultivate or commercially harvest the timber standing on the lease. Access to these resources are granted and administered separately to ensure access for development.

In one state, Queensland, a small section of the arid lands has been alienated - i.e. converted to a conditional freehold; no other state allows this. Finally, some 9 per cent of the total arid land area is reserved to the government. These areas are the special purpose areas, the aboriginal homelands, the national parks and some defence areas. A complete summary of the distribution of arid lands between the five mainland states and the allocation of this to the various land tenures is set out in Table 1.

TABLE 1

The distribution of arid lands and the allocation of this land to the various tenures within the five mainland states of Australia

State*	Area of Arid land ($x10^3$ km^2)	Proportion of state (%)	Proportion of state arid lands: Alienated (%)	Leased (%)	Reserved: Aboriginal home lands	Reserved: National parks	Unoccupied arid lands (%)
W.A.	2,185	85	1	48	8	0	43
Qld.	967	55	9	88	0	0	3
N.T.	1,042	76	0	60	12	4	24
S.A.	836	84	4	58	9	4	25
N.S.W.	345	43	1	97	0	2	0
Australia	5,376	70	3	62	7	2	26

* These are respectively Western Australia, Queensland etc.

These political boundaries follow no natural feature of the landscape however, by and large the areas enclosed within each state roughly correspond to separate ecological regions of the arid zone, e.g. W.A. has most of the unused sand deserts; or to different herbivores, e.g. in N.S.W. only sheep are grazed; or to different periods of occupation and development. They certainly signify different controls over the intensity of use and of ownership, e.g. whether an individual or a company can hold land. The five states represent five current experiments in arid land management -- we have not yet evaluated the results.

(b) Climate

No continent, with the exception of Antarctica, is drier than Australia, for no continent is more dominated by the sub-tropical high pressure belt (Gibbs).[2] It is therefore most likely that no continent has less reliable rainfall for there is an almost universal association of high variability with low annual totals. Rainfall characteristics are paramount in the prescription of management for arid lands.

In broad scale we have marked wet summers and dry winters with median annual totals of $\leq$ 650 mm at the northern edges to a comparatively small area that has either uniform or winter maximum incidence with median annual totals of $\leq$ 250 mm at the southern edges. In comparison with the rest of the continent, the arid lands have moderate to high variability of annual rainfall with a large central area, roughly coincident with the Lake Eyre drainage basin, that experiences high to extreme variability (Fig. 2). A computed index of drought probability follows, though not closely, the continental distribution of rainfall variability. Of interest is that an area of moderate to high risk extends well down into the winter rainfall south eastern parts of the rangelands (Fig. 3).

The functioning of arid ecosystems is "water-controlled" or "rain-driven". Thus we can expect that all of the above-mentioned rainfall characteristics are reflected in the functioning of and management strategies for the vegetation and the animals in the arid lands. For although the rainfall is both low and variable, winter temperatures are not low enough to inhibit plant grownth. Therefore each rainfall tends to produce a finite period or pulse of plant growth;

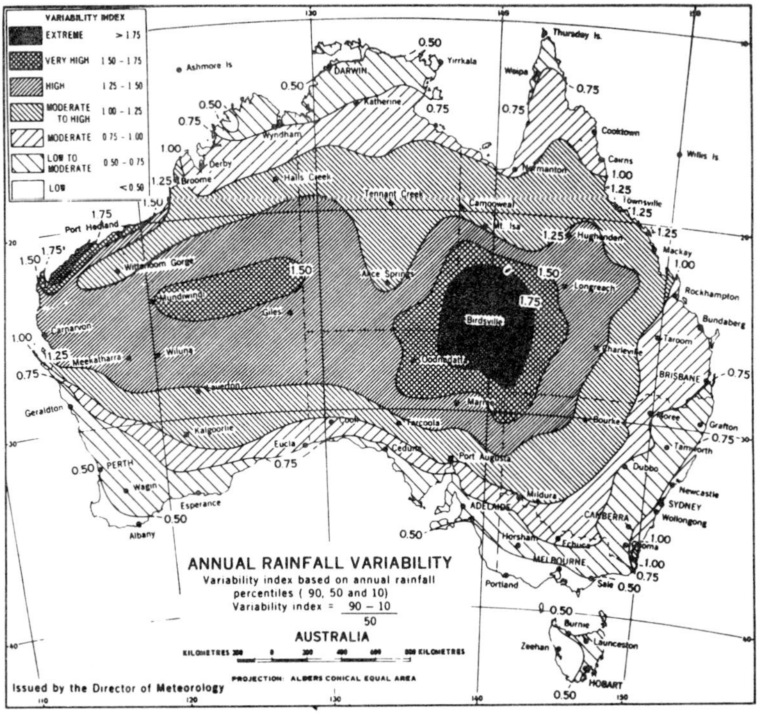

Fig. 2. A map of annual rainfall variability. The index is computed from the 90, 50 and 10 percentiles of annual rainfall, thus: Variability index = (90-10)/50.

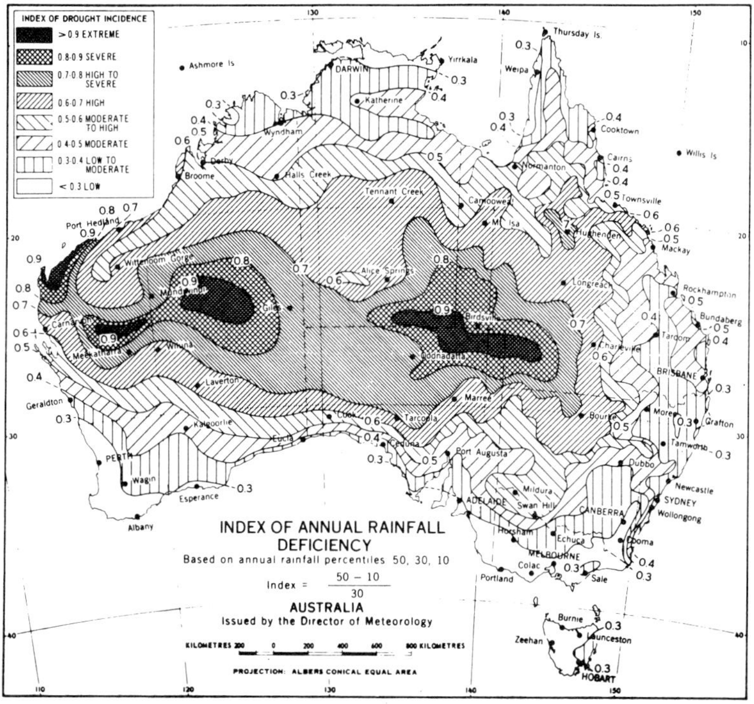

Fig. 3. A map of annual rainfall deficiency. It has been computed from the 50, 30 and 10 percentiles thus: Index = (50-10)/30.

winter rainfall is not accumulated in the soil store. Thus there are no regular plant growth rhythms and, as a consequence, no seasonal livestock grazing strategies so common in South Africa and North America. The probability of drought -- i.e. periods greater than a year without any significant periods of plant growth -- is so high for most of the rangelands that it has been suggested by Perry[3] that all management be based on drought levels but with the flexibility to take advantage of some seasonal responses of the vegetation.

(c) Relief and drainage

The arid landscapes of Australia are characteristically flat or of low gently undulating topography with relief being generally less than 50 m. It has been estimated that some 60 per cent of the arid lands can be classified as "smooth plains". The lowest point on the continent is Lake Eyre, the focus of the largest area of coordinated interior drainage with an elevation of -15 m, and the highest elevations of 1,500 m are to be found amongst the peaks of the Macdonnell Ranges in Central Australia. The increased rainfall yield from the Macdonnells and other ranges is small but the runoff from them is often of considerable importance for stock waters or in the production of forage in run-on areas for distances of tens of kilometers.

The continent may be divided into three main regions; the vast Western Plateau of Shield (which constitutes some 60 per cent of the land mass); the Interior Lowlands (25 per cent) and the Eastern Uplands (15 per cent) which, of course, fall outside the arid lands.

The Western Plateau makes up the central and western part of the arid lands and it has conferred level to rolling plateau surfaces covered by sand and gibber deserts, relieved only rarely by rocky hills. The land-scapes are of great age and very stable.

Lying between the Eastern Uplands and the Western Plateau are the Interior Lowlands; a series of great inland plains part covered with sandy desert and gibber. The Murray-Darling drainage basin forms the south eastern lobe, forked by the Cobar block, and separated by the Barrier and Grey ranges from the great arid drainage basin of Lake Eyre. This normally-dry salina is below sea level and is fed by sporadic water courses from the north-east, north-west and south.

More detailed accounts of the geological history and geomorphology are available in Mabbutt[4] and Mabbutt and Sullivan.[5]

The Murray-Darling rivers are the only water courses of any consequence to flow from the more humid uplands through the arid lands to the sea. The climatic boundary of the arid lands follows closely those of the broad drainage groupinds. Arheic drainage characterises most of the Western Plateau except for the north western shoulder whilst endorheic drainage characterises the central edges of the plateau and most of the interior lowlands. An almost imperceptible divide separates the latter from the exhorheic drainage of the Carpentaria Basin.

The low rainfall and high evapotranspiration over most of the continent results in a very low yield of fresh water. A comparison is made with a mean of the rest of the world in Table 2. This lack of surface water storages and supply in the arid lands has been overcome by water harvesting technology and has been compensated for by use of ground water.

TABLE 2
A comparison of rainfall and run-off for Australia and the rest of the world

	Australia	Average for all land areas of the world
Average annual rainfall (mm)	419	660
Average annual runoff (mm)	33	249
Runoff/rainfall	0.08	0.38

(d) Soils

The soils of the arid lands, and of Australia in general, are distinctive when compared with those of the Northern Hemisphere (Hubble).[6] Their most characteristic properties are a general low nutrient status with extremely low levels of phosphorus and nitrogen (Charley),[7] which is coupled with very poor physical attributes such as low macroporosity and surface sealing. Large areas of texture contrast soils exist.

Only about 5 per cent of the Australian continent has dominant soils which are entirely restricted to the non-arid lands; most soils in mesic Australia also occur in the arid interior or show some features of aridity. The use of the terms "desert" or "arid" soils is inappropriate for these and other reasons (Jackson).[8] A broadscale map and description of the soils of the arid lands is given in Hubble[6] and in Perry.[1]

(e) Vegetation

The flat or gently undulating landscapes of infertile soils are clothed in a flora of characteristic composition and structure. This flora consists mainly of sclerophyllous species which have diversified greatly from the ancient and more widespread stock.

Important genera are *Eucalyptus, Eremophila, Acacia* and the cosmopolitan group of halophytic shrubs *Atriplex, Maireana, Rhagodia* and *Bassia*. It is thought that sclerophylly is as much a product of the widespread low phosphate status of soils as of aridity (Beadle; Beard).[9, 10]

Of the endemic genera of the arid lands about 40 are specific to it; these genera are mainly in the families Chenopodiaceae, Asteraceae and Cruciferae. Endemic and widespread tree species are for example Acacia aneura (mulga), A. Harpophylla (brigalow) and Eucalyptus oleosa (mallee) (see Moore and Perry).[11]

The flora has been sorted into vegetation formations through climatic edaphic interactions, with fire as a component since the arrival of aboriginal man (20 - 30 x 10^3 years BC) and modified by non-indigenous herbivores in the last 150 years. The major vegetation formations are mapped in Fig. 4 and summarized in Table 3. A more extensive description may be found in Moore,[12] Moore *et al.*,[13] and Perry.[1]

It can be appreciated from Table 3 that the Acacia shrublands (or mulga lands) together with the hummock grasslands (spinifex), a vegetation type unique to Australia (Beard),[10] together cover 60 per cent of the arid lands.

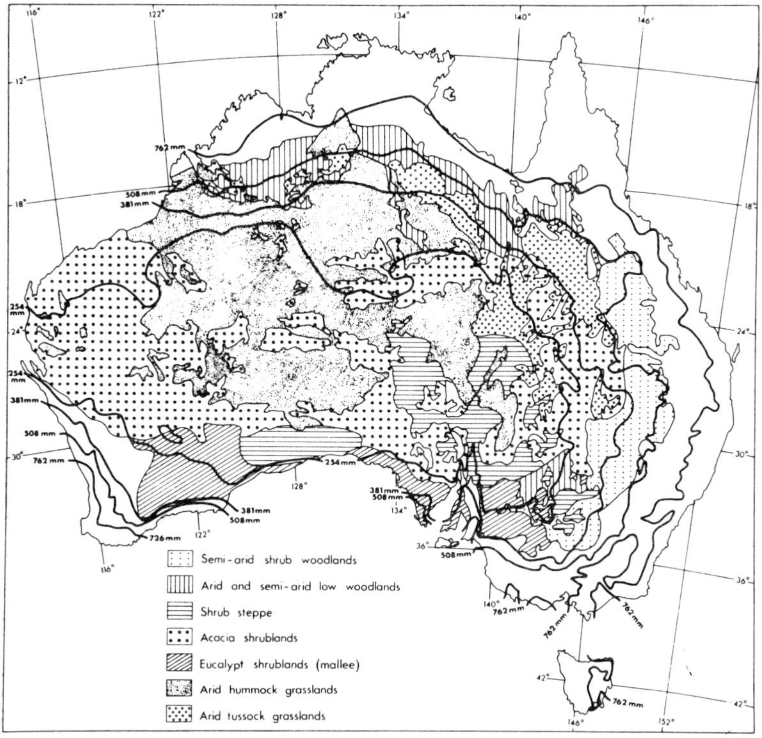

Fig. 4. A simplified vegetation map for the Australian arid lands (After Moore, 1977).

TABLE 3
The major arid land vegetation communities ranked in order of the total area covered

Vegetation	Area (x 10^3 km^2)	per cent
Acacia shrublands (mulga)	1,623	31
Hummock grasslands (spinifex)	1,613	30
Tussock grasslands (Mitchell grass)	569	11
Shrub steppe (saltbush)	434	8
Eucalypt shrublands (mallee)	416	8
Shrub woodlands	385	7
Low woodlands	272	5
Total	5,312	

The vegetation communities of most pastoral value are the shrub steppe and the tussock grasslands, both of which are largely grazed by sheep.

(f) Wildlife

Within the continent itself it is difficult, if not incorrect, to describe a peculiarly arid fauna. Many of the larger marsupials, e.g. the kangaroos, that inhabit the arid lands have very wide distribution and are largely concentrated in the arid lands because of heavy hunting pressures in the cultivated lands. Kangaroos in particular have increased in numbers since the advent of the European pastoralist who established permanent water supplies and eliminated the natural predator, the dingo. An excellent summary of the interaction of kangaroos and domestic stock has been given by Newsome.[14]

There is a large exotic component of the wildlife in the arid lands for large populations of feral camels, goats, foxes and pigs exist in various areas. These escapees from domestication have had nowhere near the devastating effect of the deliberate introduction of the European rabbit (Leight).[15] The numbers of the latter have now been biologically controlled by the introduction of myxomatosis (Rolls).[16]

Overall, the wildlife populations of the arid lands have been altered by the coming of domestic stock; the large marsupials have increased whilst many small marsupials have declined or become extinct through habitat destruction. Wildlife are generally regarded by pastoralists as competitors with, or, in the case of dingoes, foxes and pigs, predators of livestock. They are not as yet seriously regarded as a harvestable resource either for protein or hunting for recreation.

A BRIEF HISTORY OF ARID LAND USE

It is instructive to briefly consider the history of the pattern of land use as we see it now in 1978. This will help in separating out the pressures for change.

Firstly aboriginal man occupied and used the whole of the arid lands for at least the last 6,000 years. There appear to have been 150-200 demarcated tribal areas within the arid lands with about 450 people per tribe. These tribal boundaries approximated to hexagons and each contained within them a veriety of landscape types. The people were, and still are, hunter/gatherers and they used fire to manipulate the landscape, both to drive and hunt game and to encourage the pyrrhic succession of important food plants.

Then came European pastoral man. The arid and semi-arid lands were almost wholly occupied by pastoralists within the 50 year period 1850-1900, a relatively short period of time compared to the occupation of similar vast areas in the United States of America. This occupation was associated with the grazing of first cattle and then sheep and was at first in large holdings along the permanent streams. It was fuelled by a ready availability of British capital and by large profits made by the first settlers from the sale of wool, and livestock and then from the subsequent development, subdivision and resale of the land.

In the 1850s and 1860s the main expansion was into western New South Wales, central Queensland and northward from the agricultural areas in South Australia. In the 1870's and 1880's occupation extended in the drier far west of New South Wales and Queensland and the occupation of the Northern Territory and the arid areas of Western Australia was begun. In most areas occupation was at first with cattle, and these were subsequently displaced by sheep as development of watering points and fencing were made. The peak use of these lands was reached by the 1890s in the east, but not until the 1920s in the Northern Territory and the west and so it was in the east that the problems of overstocking first arose (Barnard).[17]

These problems were most severe and best illustrated by events in western New South Wales. The 1880's saw the development of high expectations, both of the land itself because of relatively good seasons and of the profits that were possible from wool-growing, sheep sales and capital gains. This was followed, in the 1890s, by a run of poor and drought seasons and falling prices, which led to widespread land degradation and financial collapse (King).[18] The problem was also compounded by the expansion of area of occupation of the European rabbit (Oryctolagus cuniculis) and its rapid increase to plague proportions which occurred at the same time. Sheep numbers, which reached about 15.5 million in 1891, fell to less than 6 million in 1900 and this was associated with extensive soil erosion and the abandonment of some 2 million hectares. This collapse and the subsequent recovery in the N.S.W. arid lands is set out in Fig. 5.

THE PRESENT PATTERN OF LAND USE IN ARID AUSTRALIA

The present pattern of use of the arid lands is depicted in Fig. 6. Mining, which plays an important role in the arid lands, occupies such small amounts of land that it cannot be shown.

In all, only 3.97 x 10^6 km^2 or 74 per cent of the arid lands are used by man on any permanent basis. The remaining 26 per cent (1.4 x 10^6 km^2), comprising largely hummock grassland, is unused and is most likely to remain so.

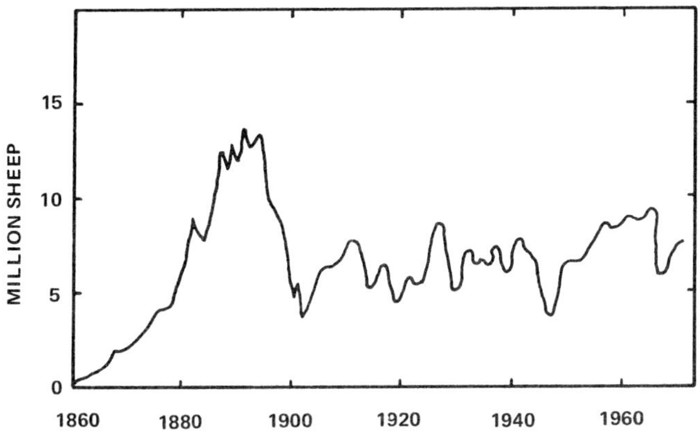

Fig. 5. The trend in sheep numbers in the arid rangelands of New South Wales.

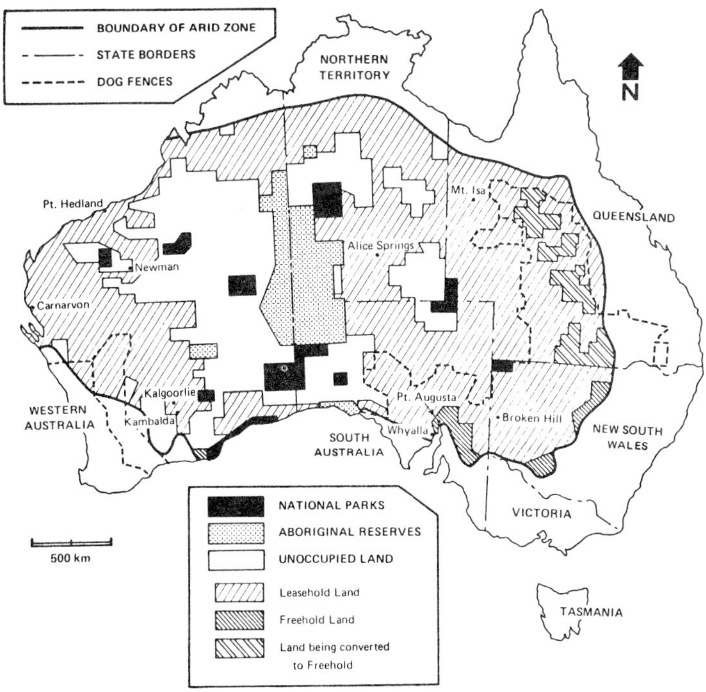

Fig. 6. The distribution of land use in arid Australia.

(a) Pastoralism

The largest land user of all is pastoralism with production being restricted to the conversion of native forage to meat and/or wool on an extensive basis. In a gross sense, the pattern of production, i.e. the extent of settlement and the herbivore type, has been stable for the last 50 years. Seasonal variation of rainfall and hence of stock numbers has always been present but it is rarely synchronised over all the rangelands. In the recent short term, because of the market and labour changes, there has been some trend from sheep into cattle.

The "dog fences" marked on Fig. 6 largely separate the sheep grazing lands from those where cattle alone are depastured. These dog fences are unique to Australia. The fences are designed to exclude the sheep predator, the dingo (Canis familiaris dingo); they are about 2 m high and there are about 11,100 km of them stretching across the landscape. They sometimes coincide with state boundaries and are funded and maintained by various methods. There could be no sheep industry in the arid lands if the populations of dingos were not suppressed.

The stock totals for the rangelands over the period 1962-1974 are presented in Table 4. The relative importance of the rangelands in terms of livestock numbers carried steadily declined over the period shown in the table. Although in absolute terms, there have been increases, the corresponding growth in livestock numbers in the non-arid areas was substantially higher.

TABLE 4
Livestock in arid and non-arid zones - selected years

	1962	1964	1966	1968	1970	1972	1974
			Sheep (millions)				
Arid lands	36.8	40.2	30.7	33.3	33.2	30.6	28.6
Non Arid lands	121.0	124.8	126.8	133.6	146.9	132.3	116.6
Total Australia	157.8	165.0	157.5	166.9	180.1	182.9	145.2
Arid lands as % of total Australia %	23.3	24.4	19.5	19.9	18.4	18.8	19.7
			Beef cattle (millions)				
Arid lands	3.9	4.1	3.7	3.9	4.3	5.0	5.9
Non Arid lands	9.1	10.1	9.5	10.8	13.6	18.3	21.3
Total Australia	13.0	14.2	13.2	14.7	17.9	23.3	27.3
Arid lands as % of total Australia %	30.0	28.9	28.0	26.5	24.0	21.5	21.7

From statistics provided by the Bureau of Agricultural economics

After a decline in sheep numbers in the early sixties, sheep numbers in the rangelands both in absolute and relative terms have remained reasonably stable up to 1974. The significant changes which have occurred have been in beef cattle numbers in non-arid areas; these were more than double the 3.3 per cent annual rate of growth recorded for the rangelands.

The relative movements in sheep and cattle numbers may be attributed partly to the market prospects for wool and beef. Over the period described, declining prospects for wool and relatively stable and rising prices for beef were associated with a decline in sheep numbers and a marked growth in cattle numbers, particularly in the early 1970s. It is also clear from Table 4 that climatic variation has been a significantly greater influence in the rangelands than in the non-arid areas. At present the arid rangelands carry about 30 x 10^6 sheep and about 6 x 10^6 cattle - representing about 20 per cent of the nation's flocks and herds.

The dollar values of these livestock and livestock products are affected directly and indirectly by export markets and movements in economic conditions overseas, both beef and wool markets being affected by different forces. However, the upsurge of mineral exports and the changed structure of Australia's export earnings generally has attenuated the variations in income from livestock products on the national economy as a whole. This, as we will discuss later, has had important effects on the social wellbeing of the pastoralists.

The Australian rangelands are organized and managed on a very different basis than in most other arid land areas. Firstly, and of most importance, all the livestock enterprises are geared to a monetary, export market-oriented system which contrasts markedly with the subsistence systems of most of the world's arid lands. Secondly, the livestock industry is organized on an extensive, sedentary basis; there is no nomadism, no herders and virtually no seasonal grazing (Perry).[19]

There are about 7,000 individual units or pastoral properties within the rangelands giving a mean management unit size of ≈ 500 km^2. In fact the size distribution is very skewed. Sheep properties tend to be small, 200 km^2 or so, and cattle properties very large, 12,000 km^2 or so, the size of the unit of land being primarily determined by the estimated number of livestock that can be carried multiplied by the carrying capacity of that particular type of country.

In all, about 96,000 people reside on rangeland properties or in associated small towns and so derive their income from pastoralism. This represents about 41 per cent of a total population of 234,000 people residing in the arid zone.

(b) Aboriginal Home Lands

In the last two decades there has been almost a revolution in the relationships of aboriginal people and their traditional land. Changes in government policy, which reflected changes in both public awareness of aboriginal problems and needs, and a growth in the political awareness of the peoples themselves, has allowed quite large areas to be held in trust by the aboriginal peoples for their own traditional uses. The aboriginal people occupy these lands, almost to the exclusion of Europeans, and they use the land as hunter/gatherers and for some cattle production. To these people the land has great religious and social significance and after 150 years, control of the land has returned to them.

(c) National Parks

The growth of areas of land set aside for recreation, preservation of flora and fauna and for wilderness areas has been a phenomenon of the last decade. There has been a corresponding growth in the awareness of the recreational opportunities provided by the arid lands resulting in large numbers of people travelling into selected areas of the arid lands and away from traditional resort areas along the coastline. The latter are also close to, or have exceeded, saturation point.

From 1970/71 to 1974/75 the number of people visiting already established tourist areas in the arid lands has increased appreciably. Over this five year period there was about a 150 per cent increase in the tourist numbers visiting the Far Western Regions of New South Wales, particularly Broken Hill (9,900 - 23,700), with a similar increase of about 90 per cent in visitors to Ayers Rock and Mt. Olga (30,200 - 58,400) in Central Australia.

It appears that this growth trend in tourist trade will continue within the arid lands. It appears not to be limited by permanent accommodation facilities, e.g. hotels, motels, etc., for most prefer camping or caravan holidays, but rather by communications and the availability of adequate information and educational services within National Parks or the mining or pastoral industry. Access to and experience of the "ethos of the bush" is sought after.

(d) Mining

Mining is an insignificant user of land in the arid lands but it employs a large number of people directly and so indirectly supports the four largest towns in the arid lands.

The mining industry makes a significant contribution to the national economy. A listing of the total value of mine production is given in Table 5. It can be seen that the value of production has increased from \$192 x 10^6 (1965) to \$1,002 x 10^6 in 1974. The relative contributions of the commodities have varied but overall mines within the arid lands contribute some 40 per cent of mine production in Australia.

TABLE 5
Value of Australian mine production* arid zone (\$ million) over a 10 year period

Commodity	1965	1974
Copper ores and concentrates	39	192
Lead ores and concentrates	82	119
Zinc ores and concentrates	31	86
Gold bullion etc.	23	35
Iron ore	10	480
Petroleum - crude oil	-	29
- natural gas	-	10
Black coal	3	4
Opals	4	25
Bismuth ores and concentrates	-	8
Salt	-	15
Total arid lands (')	192	1002
Total Australia (+)	542	2732
Total arid as a percentage of total Australia (Ø)	35%	37%

* Production of major commodities only

(') Excludes nickel

(+) Includes all commodities; major and minor: including nickel

(Ø) Understated because of the exclusion of nickel and several minor commodities n.a.p., i.e. not available for publication

As a percentage of total mines and quarries in production, those in the arid lands increased from 13 per cent of the national total in 1965 to 34 per cent in 1973/74. As a consequence of this growth the relative employment in these mines rose from 29 per cent (13,300 persons) to 35 per cent in 1973/74 (22,200 persons) of the national totals.

The above general pattern of growth has not been identical between mine products however, for each mineral has over this period had its own peculiar circumstances of production and market conditions. Future levels of production and financial return will be determined by overseas markets, local economic conditions and government policies. Thus although mining production differs from pastoral production in terms of the relative intensity of labour, capital and land use, the forces which determine the viability and stability within these two industries, and hence the towns and social infrastructures dependent on them, are much the same.

A summary of the economic aspects and the ecological aspects of the four main arid land users are set out in Tables 6 and 7.

TABLE 6

A geographical and economic comparison of arid land users in Australia

User	Area ($x10^3$ km^2)	Relative Proportion	Population	Production value ($)
Pastoralism	3,500	65%	96,000[1]	600×10^6
Aboriginal home lands	380	7%	5,000[2]	-
National parks	110	2%	-	?
Mining	-	< 1%	22,000	$1,000 \times 10^6$
Unused	1,400	20%	-	-
Total	5,300		234,000[3]	

1. This figure includes all people who work on properties and who live in towns of<1,000 people.
2. There are no exact figures available - this is best estimate available.
3. This is the total of people in the arid lands - most are urban dwellers of the four large mining towns.

TABLE 7
An ecological comparison of arid land use in Australia

User	Duration of use	Type of impact on the land resources	Result of impact	Future trends
Pastoralism	≤ 120 yrs	A sedentary consumptive use. Introduction of novel domestic and feral herbivores. Increase in herbivore densities (x 100). Suppression of wildfire.	Widespread soil erosion and plant community changes. Alteration of wildlife population sizes and distributions.	Adjustment in structure and some decline in area used.
Aboriginal home lands	6,000 yrs (?)	Nomadic consumptive use as hunter (short term)/ gatherers (long term). Regular firing of the landscape.	Pyrrhic successional stages are exaggerated giving high localized diversity.	Increase in area.
National parks	25 yrs	Sedentary localized and selective non-consumptive use of the land.	Suppression of grazing and wildfire.	Increase in area and in intensity of use.
Mining	≤ 100 yrs	Localized intensive disruption and discarding of land resources.	Complete destruction and, in part, indirectly influencing grazing, tree-felling and population growth.	Selective growth.

PRESSURES FOR CHANGE

Changes in the patterns of arid land use are occurring in Australia. It should be pointed out that all the pressure responsible for these changes originate from outside the arid zone itself and the future of the arid zone will not be shaped and determined by those who occupy and use the lands. It has been political pressures from within urban Australia that have largely been responsible for the growth in aboriginal home lands and of national parks. These changes have taken place in the last decade. The aboriginal home lands movement resulted from a change in government policy away from the previous attitudes of paternalistic assimilation. The growth in national parks was a result of the nation-wide increase in environmental awareness and a desire for urban man to reserve access to the ethos and lands of arid Australia. Pastoralism in our arid lands figured largely in the early life of our nation and, though we are now the most urbanized nation in the world (86 per cent of our population live on 3 per cent of our land) we all wish to identify with and enjoy the arid lands. This enjoyment may take the form of wilderness experience, or simply that of a brief association with a way of life that is very different from that of urban man.

The economic pressures have arisen from sources outside Australia: the production of the rangelands is linked to Australian agriculture as a whole and this is an export oriented industry. Changes in world markets and trading relationships have considerable effect on the financial viability of pastoral enterprises. There has also been considerable growth in the export of minerals from Australia and this has brought about adjustments within the domestic economy to the detriment of rural agricultural production.

However interesting are these changes involving aboriginal home lands and national parks, they involve comparatively small areas of land. It is changes in the extent and structure of pastoral land use that is of interest here. Since the beginning of this decade, the pastoral industry generally and the arid land operators in particular, have been experiencing severe economic pressures, in the form of a cost price squeeze, that has severely tested the financial viability of many operators. Costs of production (transport, labour etc.) have increased at a much greater rate than the prices paid for meat and fibre. Traditional, secure export markets for these products have all but disappeared and have been replaced by markets that are very variable in both volume and in price.

There are few options available to the pastoral operator to improve his financial status. He can try to reduce costs (nearly all paid labour has been dispensed with), or he can try to increase productivity by increasing his lease size through purchase or increasing the stock density on his own lease. Unfortunately there exist land tenure constraints to the enlargement of holdings. These constraints were applied to meet earlier political directions for closer settlement and now they constrain sensible adjustment and serve as a threat to the land for they restrict the options to just an increase in intensity of use.

Thus, as well as acute economic pressures on the individual operator that are exacerbated by societal changes, these legislative constraints ensure that this financial pressure will be transmitted directly to the land resources -- the land will have to yield more. It is the coincidence of economic pressures and drought conditions both now and in the past that has provided the greatest threat and/or degradation of the arid lands (Tribe; Perry).[20,21]

A social consequence of these periods of cost/price squeeze has been an acceleration of the drift of population from the rangelands or rural areas generally. Urbanization has been proceeding for decades but with several ebbs and flows of people and capital in response to the relative economic activity of the rural versue urban sector.

What is critical to the future of rangeland management at this time is that the rangeland livestock industries are in a severe and enforced adjustment period at a time when the rural and urban economic systems are obviously uncoupled. Urban prosperity now appears to be lease dependent on rural incomes and the products of arid and semi-arid regions than ever before. We predict that the period of enforced adjustment coupled with relative changes in economic status between rangeland/non-rangeland populations will precipitate greater social infrastructure change in the rangelands than we have seen before. It will be nigh impossible to attract and very difficult to retain people to maintain the contemporary social and service infrastructures in many regions now with declining revenues but of expectation of increased living standards on all levels of the Australian community.

The impasse is that it cannot reasonably be expected that the costs of providing a changed or upgraded infrastructure could be met by the remaining pastoralists themselves and that social expectations are unlikely to fall, fed visually as they are by occasional visits to large urban centres and the media. People will move, if they can, in order to participate in the modern scene, in enlarged educational experiences for their children and wider job opportunities for themselves. The once-important but now tiny pastoral towns may soon cease to exist.

The future of the arid lands is not in doubt for it is an integral part of the nation's economic functioning. It is certain that because of both social and economic pressures originating from both without and within the nation, great and radical changes will occur in the patterns and intensity of land use and in the numbers and social structure of the people who use these lands.

A SUMMARY OF THE AUSTRALIAN EXPERIENCE OF ARID LAND DEVELOPMENT AND MANAGEMENT

In the century or so that European man has used the arid lands, about 10 per cent of the area used has been badly degraded through overgrazing. By world standards this is a low figure (Dregne)[22] but still not one to be proud of. Mistakes have been, and are still being, made. We identify three type-causes of land degradation.

The first type-cause is the result of ignorance of the nature and productivity of the land resources on the part of governments when setting the conditions for use. An example is the deliberate "mining" of the rangelands which was apparently widespread in the early days of pastoral settlement (1860-1890) when state governments, anxious to prevent squatting and nomadism and yet retain some equity in the pastoral boom, offered leases for as little as eight years' duration. The effect on the country was disastrous (Perry).[19] In this case the lesson has been learned.

The second type-cause results from the failure of governments to recognize that the productivity of arid lands is highly variable and that as a consequence any prescriptive controls on land allocation should be equally as flexible.

As an example it is generally held that far more resources degradation has taken place unwillingly than as a wilful overuse outlined as the first type-cause. It is the well-known case of where the management decisions are forced by economic pressures, such that the ecological feedback of "condition" of the country has to be ignored. Classically the economic pressures arise through declining product prices, extensive capitalization with appreciable debts to be serviced, all of which has to be produced from range-land lease, the size of which was determined by governments in times of extraordinarily high product price. Often the situation has been aggravated by the advent of drought.

The pastoral industry in the rangelands appears generally to have been through at least three cycles of severe economic pressures. The first was in the 1890s, which gave us the greatest speculative land boom/collapse of all, the 1930s and now the 1970s. This lesson has still not been learned. Perhaps it awaits the evaluation of our five state experiment.

The last of the types is the one which most strongly reflects the inadequacies of traditional management and recruitment to management. The many examples of unwittingly degraded rangelands pose the question of why have managers failed to respond positively to declining range condition? Several points can be made which we believe to be generally true for rangelands world wide.

The first is that managers are effective decision-makers for a surprisingly short time, often only 20 years or so. Generally managers assume control in their late twenties, many from parents or relatives, and relinquish it in the early fifties to retire elsewhere. Whilst family names persist in areas, the turnover of individuals is high.

Managerial experience of, say, 20 years or so can be mismatched to the occurrence of climatic extremes, e.g. drought, such that only one severe drought or stress period might be experienced in this 20 year period. The chances of learning to cope with these extremes is thus small; crucial tactical decisions will be made based on what was done by relatives before, what the neighbours are doing now, or just not made.

This limited experience tends also to distort the individual manager's perception of his physical environment. For example, it often encourages an unreal over-optimism of rainfall -- "we haven't had an average year for 30 years" --

and yet rainfall is one of the few records of the environment that is carefully recorded. And perhaps in vain, for it is impossible to predict rain, monthly totals are poorly correlated with each other, and a severe drought and a "dry spell" are identical in the beginning.

Also, there is an associated tendency to under-estimate the degree of degradation of the country. This is most probably a combination of bias with the difficulty of predicting the non-linear path of land resource degradation. For in all but the most robust of arid ecosystems, a heavy grazing impact will exceed a threshold whereafter erosion, loss of water and nutrients act as a strong positive feedback giving an exponential decline in forage plant productivity.

Lastly, there is a tendency to overestimate the resilience of the resource -- "it always comes back, it only needs a good season". The above failings are human but at least this lesson has been acknowledged by land administrators. However, the task of developing sound range assessment techniques and the extension of these into the pastoral community is still in progress (Perry).[19]

In summary, it must be emphasized that whatever mistakes have been made, by governments or by individuals, the worst outcome has been financial ruin. No one has starved for there are no subsistence pastoralists in the arid lands of Australia.

THE FUTURE

Australians will continue to use the arid lands as long as it is economically efficient to do so. The newer uses, national parks etc., will continue to grow at least over the next decade.

The production of wool and meat from these lands is still a large and nationally important export dollar earner. However, changes in national economic and foreign market conditions will most likely force changes in the structure and functioning of the rangelands. Changes in the structure (property size) of the industry will require, in some states, adjustment of legislation. We have not fully evaluated or compared the long term consequences for the land resources of our existing legislation to arrive at the most flexible future system. This task is still before us. It is most likely that there will be an exodus of both people and capital from these lands.

New technology will have little impact on land use. There may be some advances in chemical shearing, etc., but there will be no widespread plant introduction or the grazing and harvesting of exotic animals. Similarly, no intensification of agricultural production through irrigation schemes can really be envisaged. The financial costs and returns of such ventures effectively preclude them.

Much has to be done to improve the perception of the country by both land administrator and by the individual operator. Detailed resource surveys now cover about 70 per cent of the pastorally occupied rangelands and, where possible, resource information is being used in planning.

Improvement in the short term management of our arid lands will come through the use of LANDSAT type imagery for it promises regular cost efficient monitoring of very large tracts of land. At the individual operator level work will continue in the development and deployment of range condition assessment methodologies (Perry).[19]

In many respects Australia is fortunate for she does not need to use her arid lands to directly feed her population. With a small capital investment a good financial return can be gained from the growth and export of livestock products. By and large we are able, in both short and long term, to adjust our use of these lands to the productivity of the land. These adjustments result in ebbs and flows of dollars and not lives. There can be no fail/safe way of using arid lands -- all use must be inbuilt safe/failure.

REFERENCES

1. R.A. Perry (1970). Arid shrublands and grasslands. In, R.M. Moore, Ed., Australian Grasslands. A.N.U. Press, Canberra. pp. 246-259.

2. W.J. Gibbs (1969). Meteorology and climatology. In, R.O. Slatyer and R.A. Perry, Ed., Arid Lands of Australia. A.N.U. Press, Canberra. pp. 33-54.

3. R.A. Perry (1974). Strategies available for managing multispecific communities in Australia. In, Plant Morphogenesis as the Basis for Scientific Management of Range Resources. USDA Miscellaneous Pub. 1271. pp. 124-39.

4. J.A. Mabbutt (1969). Landforms of arid Australia. In, R.O. Slatyer and R.A. Perry, Eds., Arid Lands of Australia. A.N.U. Press, Canberra. pp. 11-32.

5. J.A. Mabbutt and M.E. Sullivan (1970). Landforms and structure. In R.M. Moore, Ed., Australian Grasslands. A.N.U. Press, Canberra. pp. 27-43.

6. G.D. Hubble (1970). Soils. In, R.M. Moore, Ed., Australian Grasslands. A.N.U. Press, Canberra. pp. 44-58.

7. J.L. Charley (1972). The role of shrubs in nutrient cycling. In, Wildland Shrubs, Their Biology and Utilization. USDA Forest Service, General Technical Report INT-1. pp. 182-203.

8. E.A. Jackson (1962). Soil studies in Central Australia; Alice Springs-Hermannsburgs--Rodinga areas. CSIRO Soil Publication No. 19, Melbourne.

9. N.C.W. Beadle (1966). Soil phosphate and its role in molding segments of the Australian flora and vegetation, with special reference to xeromorphy and sclerophylly. Ecology 47:992-1007.

10. J.S. Beard (1975). The evolution of Australian desert plants. In, D.W. Goodall, Ed., Evolution of Desert Biota. University of Texas, Austin. pp. 51-64.

11. R.M. Moore and R.A. Perry (1970). Vegetation. In, R.M. Moore, Ed., Australian Grasslands. A.N.U. Press, Canberra. pp. 59-73.

12. R.M. Moore (1975). Australian arid shrublands. In, D.N. Hyder, Ed., Arid Shrublands. Soceity of Range Management, Denver, Colorado. pp. 6-11.

13. R.M. Moore, R.W. Condon and J.H. Leigh (1970). Semi-arid woodlands. In, R.M. Moore, Ed., Australian Grasslands. A.N.U. Press, Canberra. pp. 228-45.

14. A.E. Newsome (1975). An ecological comparison of the two arid zone kangaroos of Australia, and their anomalous prosperity since the introduction of ruminant stock to their environment. Quarterly Review of Biology 50:389-424.

15. J.H. Leigh (1974). Diet selection and the effects of grazing on the composition and structure of arid and semi-arid vegetation. In., A.D. Wilson, Ed., Studies of the Australian Arid Zone. II. Animal Production. CSIRO, Melbourne. pp. 102-26.

16. E.C. Roll (1969). They All Ran Wild: The Story of Pests on the Land in Australia. Angus and Robertson, Sydney.

17. A. Barnard (1969). Aspects of the economic history of the arid land pastoral industry. In, R.O. Slatyer and R.A. Perry, Eds., Arid Lands of Australia. A.N.U. Press, Canberra. pp. 209-28.

18. C.J. King (1957). An Outline of Closer Settlement in New South Wales. Government Printer, Sydney.

19. R.A. Perry (1977). The evaluation and exploitation of semi-arid lands: Australian experience. Philosophical Transactions of the Royal Society, London. Series B, 278:493-505.

20. D.E. Tribe (1977). The conservation and improvement of resources: the grazing animal. Philosophical Transactions of the Royal Society, London. Series B278:565-582.

21. R.A. Perry (1970). Productivity of arid Australia. In, H.E. Dregne, Ed., Arid Lands in Transition. American Association for the Advancement of Science, Washington, D.C. pp. 303-316.

22. H.E. Dregne (1978). Desertification: man's abuse of the land. Journal of Soil and Water Conservation 33:11-14.

APPLICATION OF GEOSCIENTIFIC KNOWLEDGE TO EVALUATION OF LAND AND WATER RESOURCES IN SOME DESERT AREAS OF EGYPT

ABDU A. SHATA
Desert Institute, El Matariya, Cairo, Egypt.

ABSTRACT

Since its inauguration in 1950, the Desert Institute has undertaken a variety of integrated surveys in areas selected for land reclamation and agricultural development. The main object of these surveys was the evaluation of available natural resources (water, soil, vegetation, animal, non-conventional energy ... etc). When carrying out such surveys certain aspects concerning the importance of the application of geoscientific ideas were elucidated. In addition to the established importance of knowledge about geomorphology, hydrogeology, ... etc., the elaboration of new ideas about morphopedology, hydrostratigraphy, hydrostructure, hydrogeophysics, ... etc, is vitally needed. Selected examples concerning this subject are taken from some desert areas, now under reclamation in Egypt.

BROAD LINES OF DEVELOPMENT

During the past two decades various desert areas of Egypt were considered for agricultural development, either by use of available groundwater as well as surface runoff water or by use of the Nile water. These areas comprise the following (Fig. 1):

The Mediterranean Littoral Zone

This zone has an area of about 20,000 km^2 and is dominated by arid and subarid climatic conditions. In this zone, the landscape is generally a low coastal plain, which is occasionally bounded on the southern side by a rocky platform. This platform, acting as a watershed, is dissected by a good number of drainage arteries, which become active during the rainy seasons (winter rain). With the exception of Wadi El Arish, which extends to the highlands of South Sinai (over a distance of about 400 km), all such drainage lines are short. The water supply comprises in this zone the shallow groundwater aquifers in the Late Tertiary as well as in the Quaternary, surface runoff and Nile water. The soils in this zone are dominated by calcareous loam and represent weathering products of the rocky platform. In this zone scattered patchy areas suitable for agricultural development are found but great efforts are needed to maintain a successful scheme (Fig. 2).

The Fringes of the Nile Delta

The young flood plains of the Nile Delta, which are intensely cultivated, are bounded on the east and west by elevated gravelly plains. These merge northward into lacustrine muddy plains, which are occasionally covered by dune sand accumulations. This part of Egypt has dominantly arid climatic conditions and the surface is almost degraded with regard to soil and vegetation. On the eastern side, the gravelly plains are crossed from west to east by the Ismailia Canal, and on the western side, the plains are crossed from southeast to northwest by El Nasr Canal. These two canals form the main supply of surface water for irrigation in the fringes of the Nile Delta. Groundwater stored in the deltaic gravels (thickness about 200m) is also used for supplementary irrigation (Figs. 3 and 4). In this part of Egypt wide tracts are under reclamation. In spite of the sound achievements made, portions of the already cultivated areas are, at present, attacked by drainage water seeping from the adjacent elevated plains (Fig. 5).

The Sinai Peninsula

The Sinai Peninsula (62,000 km^2), is one of the areas in Egypt having relatively high agricultural possibilities. It is essentially a rocky platform, dominated by arid climatic conditions, but it extends northward into the temperate zone of the Mediterranean Region (rain fall may rise to 250 mm). In Sinai five geomorphic units are known and comprise the southern elevated horst block, the central calcareous plateau, the northern folded ridges, the eastern graben depression, the western graben depression and the northern coastal plain. This configuration determines the drainage basins of Sinai and determines consequently the land and water resources of each.

In Sinai seven drainage basins of this type are represented, the most important of which is Wadi El Arish Basin, having an area of about 22,000 km^2. In this basin, the nonconsolidated deposits (soils), cover an area of about 6000 km^2 and the water resources are ample. Such resources comprise the groundwater aquifers of the Nubian Sandstone and the surface runoff of Wadi El Arish itself (Figs. 6, 7, and 8).

The Western Desert Depressions

These comprise the series of natural excavations (not including the Qattara Depression), which form a disconnected belt, extending from the near reaches of the Nile at Abu Simbel to Siwa El Ghaghbub at the Egypt-Libyan Border. This belt, known as the New Valley, has an area of about 20,000 km^2 and is dominated by extremely arid climatic conditions. The bottom portions of such depressions are generally near sea level and take almost the shape of a "bajada." These are occupied by lacustrine deposits (Quaternary) and are at present subjected to severe salinization processes. The landscape, in almost all the depressions, is unique and is characterized by the following (Fig. 9).

(i) a depression proper (Bajada); closed, subclosed or open.

(ii) a steep escarpment; forming the edge of a calcareous plateau from one side or more (Cuesta).

(iii) an elevated plain; extending from the other side or sides (pavement or sarir).

Within the depression areas sand dune accumulations are found in great amounts and attack the little cultivated patches (oases). The main source of water is the Nubian Sandstone aquifer which shows a very complex stratigraphy and structure (Figs. 10 to 14). The water exists under artesian conditions and contributes locally to the supply of the carbonate aquifers overlying it (Siwa). The soils in the depression areas are of lacustrine and alluvial origin and are mixed, in varying amounts, with the aeolean sand.

The High Dam Lake Fringes

These occupy a small portion of the desert peneplain of South Egypt, which extends into the Sudan. The southern portion of the New Valley is, therefore, a continuation of the High Dam Lake fringes and the total area is of the order of 15,000 km^2. This portion of Egypt is dominated by extremely arid conditions and the surface is subjected to very strong deflation processes. The desert peneplain shows a variety of landforms, of which we mention:

(i) dead watershed areas, either underlain by Basement Rocks or by sedimentary carbonate rocks;

(ii) inland drainage lines, directed either to the Nile (main base level or erosion), or to the inland hollows (bajada or playa) (Figs. 15, 16, 17);

(iii) a great number of almost circular and shallow hollows, presumably associated with Late Tertiary and Quaternary groundwater activities;

(iv) a great number of small isolated hills (butts), underlain either by volcanic rocks or by altered sedimentary rocks;

(v) the occasional occurrence of petrified forests.

In this portion of Egypt, the soil profiles are generally shallow, except in the "old playas," where certain amounts of salinization processes are witnessed. The main source of water is the Nubian Sandstone (both artesian and water-table conditions) and the surface water of the High Dam Lake, which contributes also to the supply of the Nubian Sandstone aquifer (Fig. 18).

The Red Sea - Gulf of Suez Foreshore Areas

This portion of Egypt has extremely arid climatic conditions and the areas suitable for agricultural development are very limited. Although the land resources are major, very wide tracts are subjected to salinization processes. Only in West Sinai, where the deltaic portions of some wadis escaped salinization, can the existing land and water resources permit the establishment of disconnected cultivated plots (Wadi Sudr).

LESSONS LEARNED

Since its inauguration in 1950, the Desert Institute has taken the responsibility of carrying out a variety of integrated surveys over much of such areas. Emphasis has been given to the evaluation of water, land, plant and animal resources. Because the water and land resources are considered the back bone of any agricultural development in similar areas within arid regions of the world, the staff of the Institute has introduced some new approaches to its geoscientific investigations. Such approaches have proven valuable in differentiating relatively stable from relatively unstable landscapes and in predicting the changes that will be induced by changes in the water balance and in the use of land.

As an example of the application of such approaches to water resources, reference will be made to the major artesian aquifers in the Nubian Sandstone in Egypt as well as in the adjacent countries of Northeast Africa. Although advances have been made in our knowledge, since the beginning of this century, about water storage and recharge in this aquifer, various problems are still met with. Better understanding of such problems can be made through the following approaches of geosciences.

Hydrophysiography

This plays an important part in determining the present recharging of this aquifer (if there is any). Generally speaking, it is hesitantly agreed that this aquifer is recharged from the rainy upland areas located to the southwest in Central Africa, from the rainy areas and also from the Nile itself in areas located to the south in Sudan, from the rainy upland areas in Cyrenaica located to the northwest and eventually from the rainy upland areas in the East Mediterranean region located to the northeast. This pattern is rather complicated, and the present investigations being made by the Desert Institute aim at determining the quantities of water obtained from each direction. Hydrophysiography assists also in determining the discharging areas. These are either developed in the natural excavations (New Valley, Qattara Depression...etc.) or in the River Nile itself. Estimation of the amount of groundwater discharged in such areas is in progress (for the Qattara Depression it is of the order of 3.0 MCM/day).

Hydrostratigraphy

The Nubian Sandstone aquifer is not composed of a massive sandstone layer. It is composed of a complex and heterogeneous succession of sandstones, conglomerates and greywackes, with interbeds of clay and siltstone. All these show conspicuous lateral change of facies and exerts; therefore, there is considerable heterogeneity in the nature of the aquifer. Again, the Nubian Sandstone, as a formation of epicontinental origin, is practically devoid of fossils and, therefore, it is hazardous to make any reliable correlations between an outcrop or a rock cutting from a borehole, in one locality, with others from the same formation in another locality. This condition makes it difficult for the hydrogeologist to define, with a certain degree of accuracy, the different water bearing zones of that aquifer. However, adequate zonation can be made by measuring a good number of sections, drilling of a good number of test holes (coring) and then studying the samples obtained in different ways including, essentially, paleontology and radioisotopes.

Hydrostructure

This aspect is particularly emphasized when outlining the configuration of the surface of the Basement Complex, upon which the Nubian Sandstone aquifer rests directly. Regionally speaking, this surface slopes gently in the northward direction at the rate of 10 to 20 m/km. This picture is far from simple, as there are several undulations, corresponding to swell axes, dominantly oriented in the NE-SW direction, and to synclinal basins of the same orientation as well as other orientations (NW-SE). Again, the Nubian Sandstone as well as the Basement Complex, is dissected by faults striking in different directions, of which faults with E-W and N-S strikes are common.

Such complexities have a direct bearing upon the thicknesses and lithologies of the aquifer and determine, therefore the directions of flow (the thickness varies from a feather edge in South Egypt and Sudan to 3000 m in North Egypt). We can imagine, therefore, that the hydraulic properties of the Nubian Sandstone aquifer are subjected to wide variation; even on the local level.

Hydrology

Several attempts were made to standardize the hydrological parameters of the Nubian Sandstone aquifer by use of both analogue as well as digital computers, but the results obtained are far from being satisfactory. This means that plenty of work is needed, both on the regional and on the local levels. Having this in mind we might as well refer to some determinations made:

- Porosity coefficient = 5% to 25% (calculated values)
- Permeability = 1.42 to 5.75 m/day
- Transmissivity = 360 to 665 m^2/day
- Piezoconductivity = 1.4×10^6 m^2/day
- The direction of flow is regionally from SW to NE but local complications are known.

Hydrogeochemistry

The importance of this aspect is demonstrated when determining the quality of the water in relation to the water bearing formation and when attempting to define the relationship between the water stored in the Nubian Sandstone since the past wet periods (varying between 30,000 years, according to C14 measurements and 1,850,000 years according to Helium-Argon measurements) and the water added to it from recent intake areas. With regard to this last point various attempts are now made by the Desert Institute to get results from the High Dam Lake and from Siwa Oasis.

Taking another example of the application of the new approaches developed by the Desert Institute to the geoscientific investigations of land resources, reference will be made to the vast area considered for agricultural development in the western fringes of the High Dam Lake and its extension into the southern portion of the New Valley. In this vast area sands predominate very markedly and take the form of large slightly undulating plains. These are bounded on the west and north by an escarpment showing an alternation of limestone, gypseous marl and phosphatic limestone. These form the parent rock material for the soil deposits accumulated in that area. The climate in the area is extremely arid though there are indications that it was preceded by wet conditions. The nature of the present climate is responsible for the destruction of whatever good soil cover may have been formed during the past wet periods and for the production of the desert dust and drift sand deposits common to the area. Locally, however, alluvial and lake-like deposits are preserved in the low lying areas (either in the old drainage lines or bajadas). These take the form of terraces, clay hammocks or mud flats.

On a morphopedologic basis, the western fringes of the High Dam Lake and the southern portion of the New Valley are differentiated into the following units, each having its type of soil cover: barren calcareous plateau surface; barren steep slope; talus slope covered with the weathering products of a series of beds composed of limestone, phosphate and gypseous marl; bouldery surface; rough rocky sandstone surface; pre-Nubian sandstone surface composed of weathered Basement; old playa surface dominated by mud flats; peneplained Nubian Sandstone surface; series of Nile Terraces.

From the field surveys and sophisticated laboratory investigations, much has been learned about the potentials of both the water and the land resources in each area. Two points must be emphasized:

1. In almost all such areas water and land resources are available, but for agricultural development there is need to consider the ecological balance and the economics of development.

2. Cooperation between research institutions and development organizations must be strengthened in order to help in solving immediate problems that may be faced during development and to avoid future problems such as the deterioration of the main groundwater aquifers and the salinization of the land.

Fig. 1

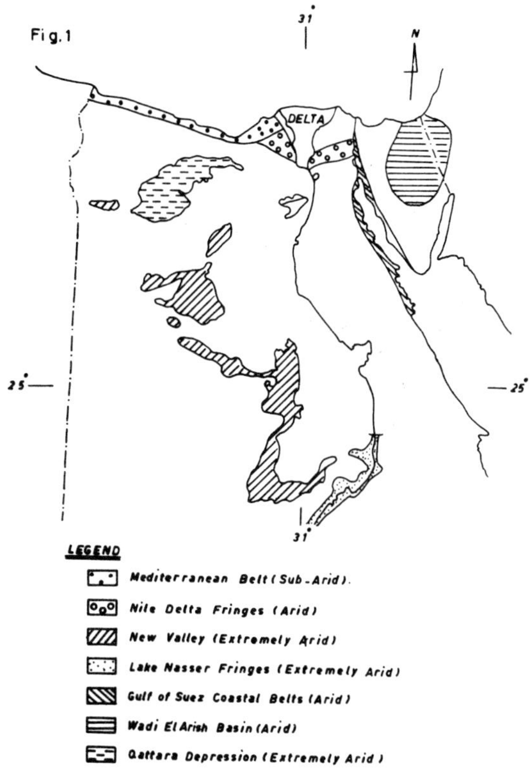

Fig. 2

MAIN AQUIFERS
MEDITERRANEAN COASTAL ZONE EGYPT

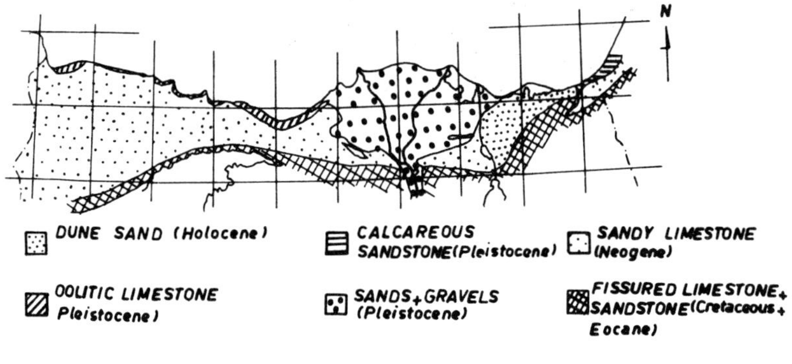

Fig. 3

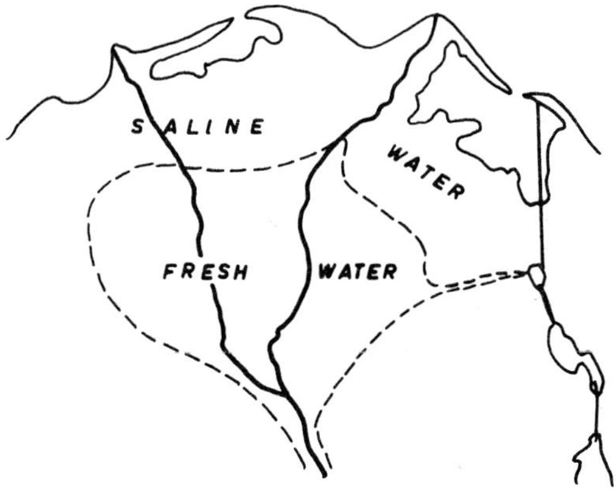

NILE DELTA
FRESH WATER BODY

Fig. 4.

LAND RECLAMATION PROJECTS AROUND
THE NILE DELTA

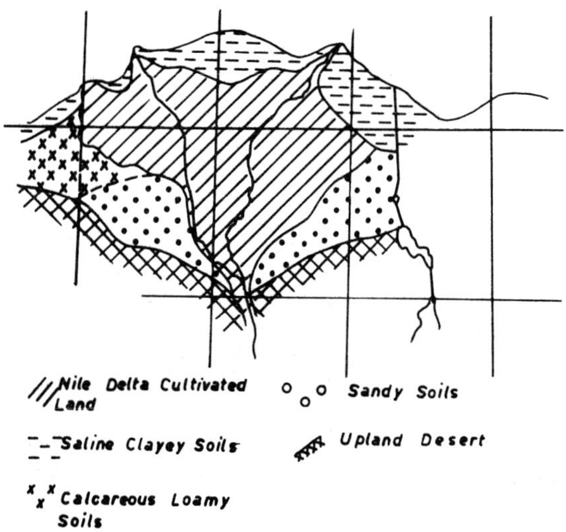

Fig. 5
LANDFORMS
IN
WEST DELTA REGION

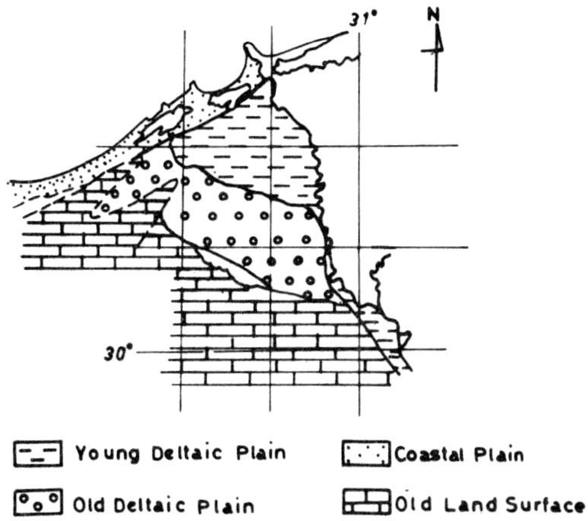

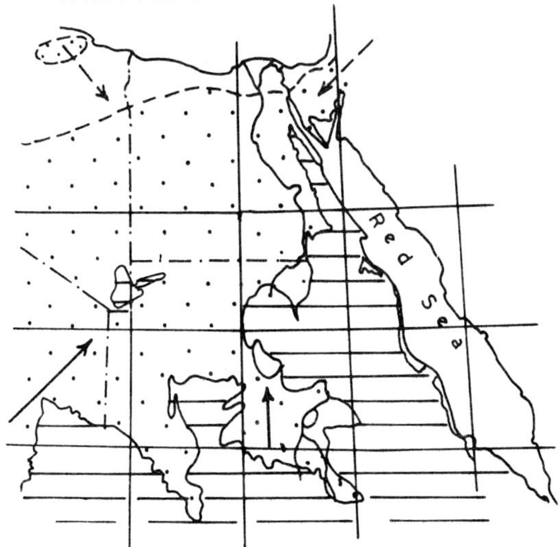

Fig(6) Distribution of Nubian Sandstone Aquifers in Northeast Africa

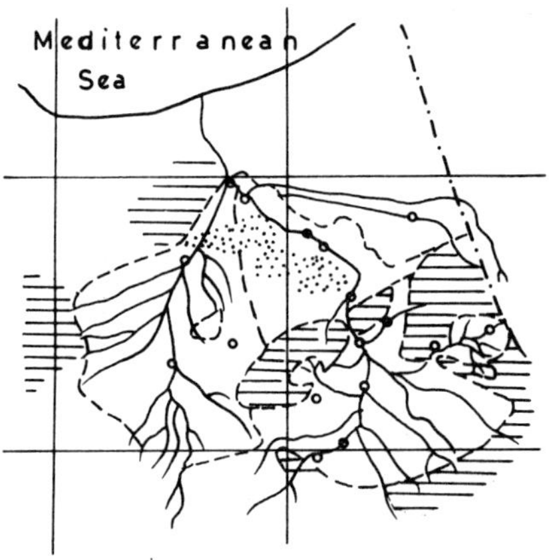

Fig.(7) Hydrographic Basins And
Watershed Area of Northeast Sinai

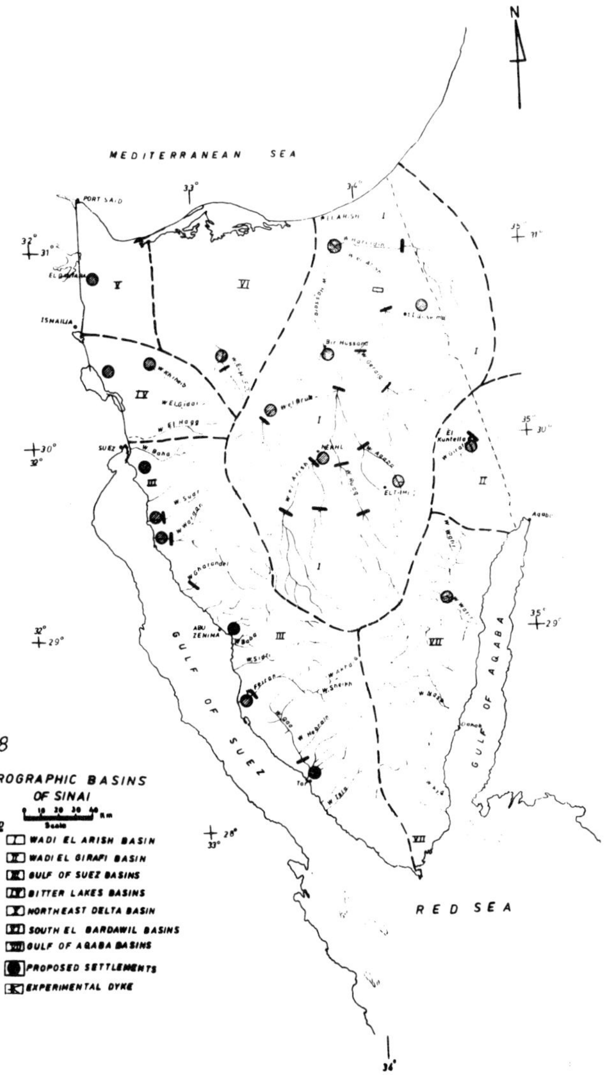

Fig. 8

HYDROGRAPHIC BASINS OF SINAI

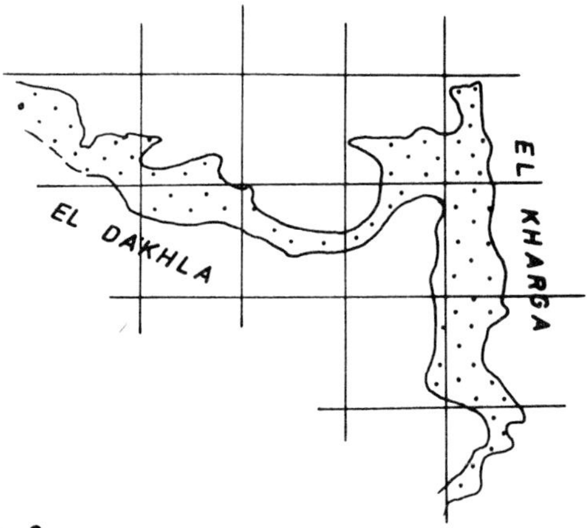

Fig. 9

LOCATION OF EL KHARGA AND EL DAKHLA DEPRESSIONS

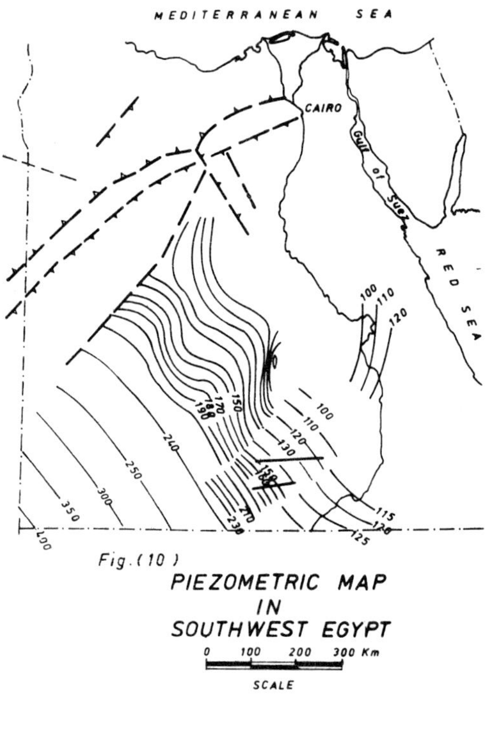

Fig.(10)

PIEZOMETRIC MAP IN SOUTHWEST EGYPT

0 100 200 300 Km

SCALE

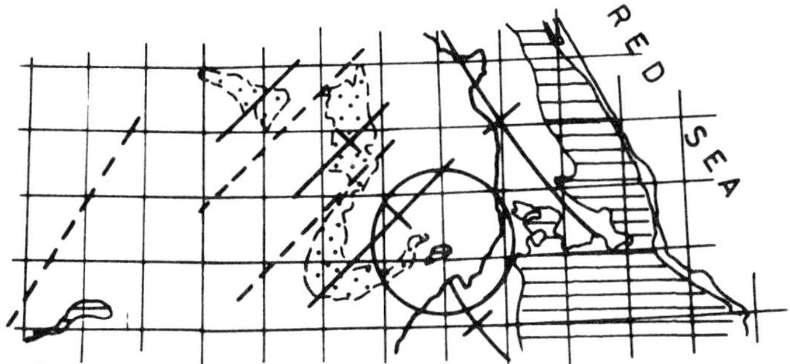

Fig.(11) Structural Trends in South Egypt

Swell Axes   Synclinal Basins

Fig. 12

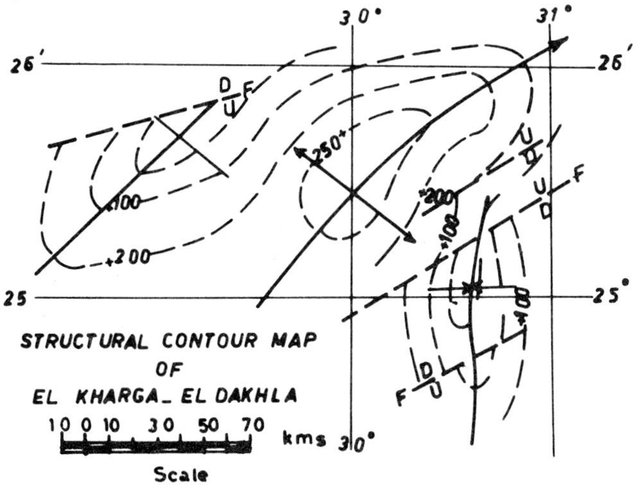

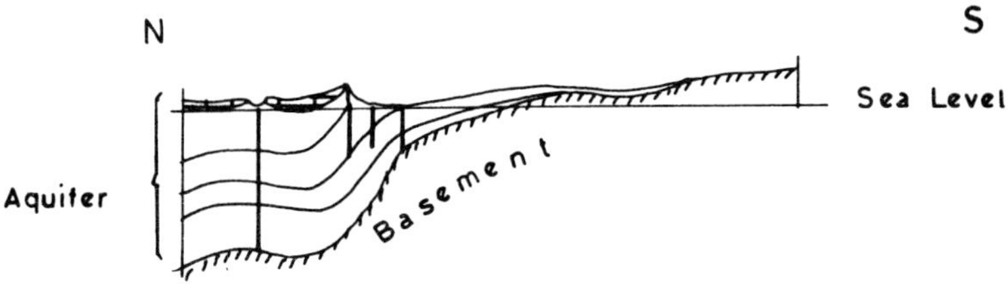

Fig. (13) Hydrogeological Section in Southwest Egypt

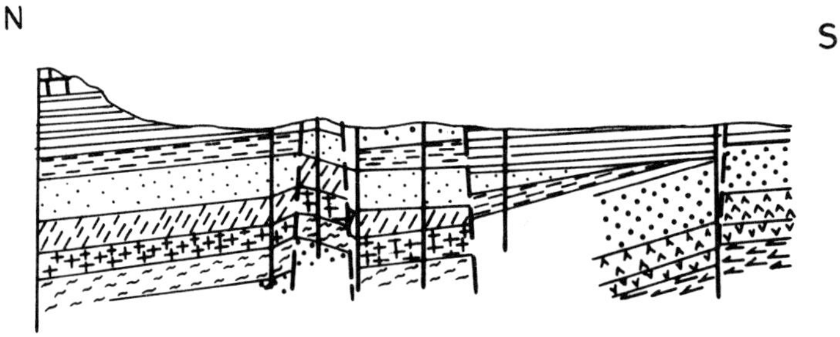

Fig. (14) Complex Structure of Nubian Sandstone Aquifer in El kharga Oasis.

Fig. 15

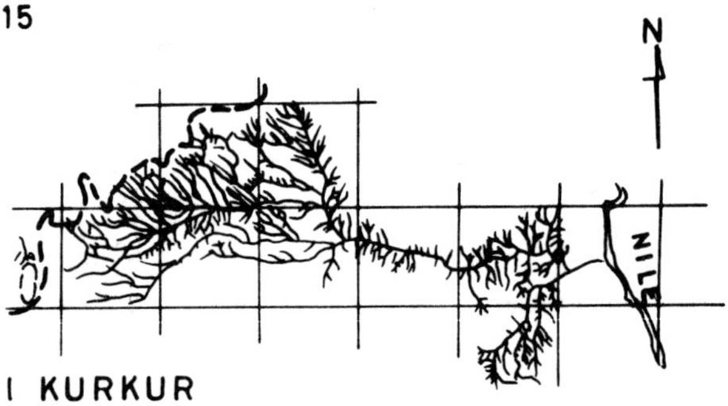

WADI KURKUR
Coplex pattern in An Extremely Arid Ragion

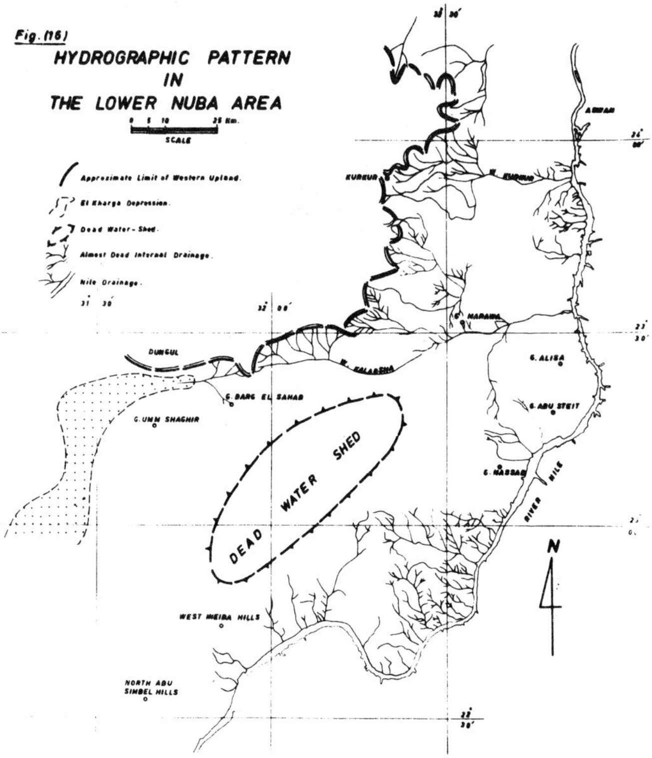

Fig. (16)
HYDROGRAPHIC PATTERN IN THE LOWER NUBA AREA

Fig(17) MORPHO_ PEDOLOGIC FEATURES OF THE LOWER NUBA AREA

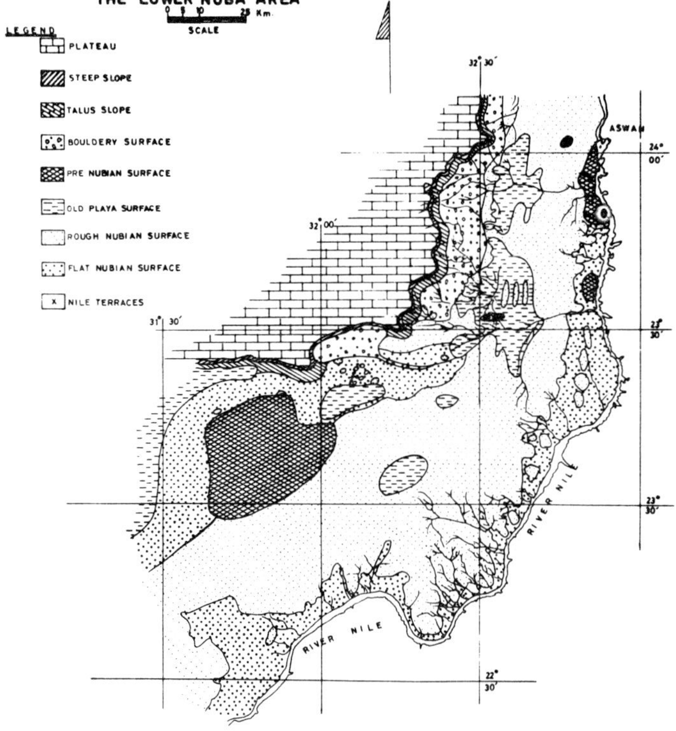

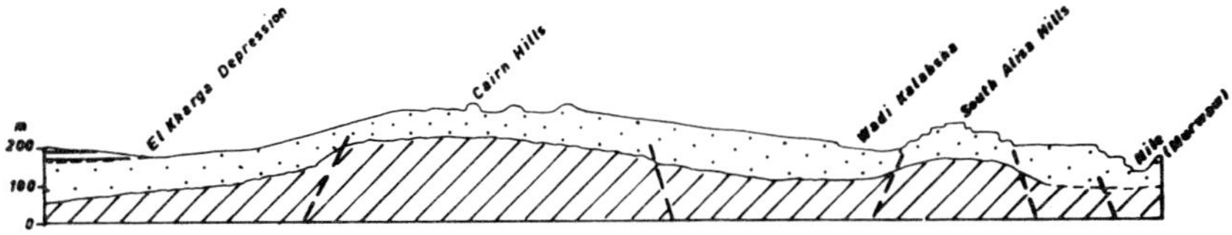

Fig. (18)
GEOLOGICAL SECTION BETWEEN THE NILE AND EL KHARGA

WATER RESOURCES AND LAND RECLAMATION IN EGYPT

M. MOUSA SEDDIK, Chairman
SALAH SHAFEI, External Branches Manager
The General Company for Research and Groundwater (REGWA), Cairo, Egypt

ABSTRACT

The paper surveys water resources in Egypt, including Nile water, groundwater, and rainfall. A short survey is given about its development and utilization and its relation to land reclamation.

Nile water resources were developed to 55.5 milliard cu.m/year after construction of the Aswan High Dam. Equatorial Nile projects will save water to irrigate one million feddans.

Groundwater extracted in the Nile Valley is about 2.9 milliard cu.m/year, of which one milliard is used for irrigation.

The groundwater available in desert areas which could be utilized in development is about 2.5 milliard cu.m/year, sufficient for irrigating 500,000 feddans.

The groundwater available in the western coastal zone is about 2.5 million cu.m/year, sufficient to irrigate 5,000 feddans. Rainfall may add 135,000 feddans to the cultivated area when means to utilize it are improved.

The future cultivated area may reach 10,785,000 feddans, distributed as follows:

8.8	*million feddans:*	*Utilizing Nile water*
1.2	*million feddans:*	*Utilizing drainage water and minimizing evaporation from reservoirs*
0.65	*million feddans:*	*Utilizing groundwater*
0.135	*million feddans:*	*Utilizing rainwater*
10,785	*million feddans*	*TOTAL*

Arid zones are, from the Agricultural point of view, those that require artificial means for concentrating rain or flood waters on the cultivable soil to ensure a crop. It is generally considered to be limited to land with an average yearly rainfall of less than 250 mm. Egypt is among the most arid countries, as the average annual rainfall ranges from 150 mm at the coastal areas of the Mediterranean sea to almost nil at Aswan.

WATER RESOURCES

The River Nile

The Nile, 6700 kms long, is the main source of irrigation in Egypt. The Egyptians long ago realized the importance of the Nile in their life, and perfected the art of irrigation 5000 years ago, the art of diverting streams to land by weirs and canals, and flooding by basin irrigation and perennial irrigation.

History tells us that the Pharaohs were the first to control the Nile flood by shoring up the river's banks and constructing dams. Ancient Egyptian engineers, 4000 years ago, designed and executed the first masonry work to control the river Nile, "the Lahun," and to provide an automatic overflow basin to guard against dangerous floods. Their object was also to ensure a high summer supply to aid navigation and help irrigate gardens and summer cultivation, as mentioned in the letter of the scribe Pibesa in the reign of Ramses II.

Since the beginning of the 19th century, Egyptians looked forward to the development of agricultural expansion by introducing summer cash crops, converting land under basin irrigation to perennial irrigation. They were aware that the Nile flow divides the year into two water seasons, of which one is abundant while the other is of low level. Since that date and up till the middle of the 20th century, Egyptians were able to control river flow in both seasons and tame it to fulfill agricultural needs. Series of barrages were built at various reaches to raise water levels upstream and enable irrigation water to flow freely in branch canals and streams. Annual storage reservoirs were also built to store part of Nile water during flood seasons, to be used during the low level seasons. During that period, detailed hydrological studies for the Nile flow were carried out and revealed three facts: (1) principal projects erected along the river channel are closely related to each other, (2) it is difficult to meet agricultural development needs in Egypt by constructing more annual storage dams which are filled and emptied during the year; (3) further agricultural development requires long term storage reservoirs. These realizations led to a comprehensive scheme for long term storage of silt free water from the upper Nile and Equatorial lakes, and other related projects.

Early after the 1952 revolution, it was decided to construct the Aswan High Dam for long term storage, to save the Nile flod water, which is 80% of total Nile flow, lost every year into the Mediterranean Sea. Construction of the Aswan HIgh Dam started in 1960, the first stage was terminated in May, 1964, and the second was completed in July, 1970. The Dam was put in use upon the termination of the first stage in 1964.

For further development of Nile resources, other proposed projects include: (1) the equatorial Nile projects for long term storage in the great lakes; (2) conservation of Bahr El Jebel water; (3) minimizing evaporation losses in the Nile reservoirs.

Groundwater

Egypt has been familiar with the development of groundwater resources since ancient times. Groundwater occurs under various conditions: (1) water table conditions in unconfined aquifers, as is the case in the Nile Delta and Upper Egypt; (2) artesian and semi-artesian conditions in confined aquifers, as is the case in the depressions of the Western Desert and New Valley areas; (3) perched water tables, as is the case of Fuka basin in the Western Desert coastal area. Historic evidence from ancient temples indicates that groundwater was utilized in Egypt since the Pharaonic period, for both drinking and agriculture.

The modern history of groundwater exploitation began in 1905 in the Western Desert oases where a British comapny drilled 50 shallow wells, 50-200 meters deep, at Mahariq and Sherka area in Kharga oasis. Due to several difficulties, the company had to abandon the project after nine years. Development interest was renewed in 1955 when the Government started drilling 20 deep wells, up to 650 meters, in Kharga and Dakhla oases. In 1959 the Egyptina General Desert Development Authority was established to carry out a research and execution program to develop the desert areas.

Rainfall

The main area in Egypt which depends on rainfall is the northern Mediterranean coastal area. The average annual rainfall is 150 mm. This average decreases as we go to the south, is 50 mm at Wadi Natrun and almost nil at Aswan. Rainfall takes place during November, December, January, February and March. Since ancient times, this small amount of rainfall has been utilized in soil conservation and water spreading projects for pasture improvement and barley cultivation, groundwater replenishment, and filling the hand-dug cisterns which are scattered along the area and used for drinking purposes.

LAND RECLAMATION IN EGYPT

Horizontal expansion of cultivable land in Egypt may be summarized as follows:

Utilizing Nile Water

Before 1900

Efforts to control and make better use of Nile water can be traced back to the 19th century, when the Delta Barrage was constructed to raise the water level during summer period, so that the Delta main canals could be fed by irrigation water needed for summer crops. By 1886, the area under perennial irrigation was 3.0 million feddans, which consumed the entire natural river flow in the low flow period, and during which, the flow was put under full control by closing the river outlet to the sea by earth dams every year to make sure that no summer flow was wasted. During flood time, the remaining area under basin irrigation, which yielded one crop, was about 2.0 million feddans.

1900-1920 Development

The Aswan Dam was built (1898-1902) to provide 980 million m^3 of stored water during the flood period. By 1902, the Delta Barrage was strengthened, Assiut and Zifta Barrages were completed and by 1908, Isna Barrage was constructed. These barrages were built to raise the river water level to feed the upstream irrigation canals. During 1907-1912, Aswan Dam was heightened to increase storage capacity to 2.5 milliard m^3.

By 1920, the area under perennial irrigation in Egypt was 4.0 million feddans, while that under basin irrigation was 1.6 million feddans. The total yearly water requirement for Egypt at that time was 40 milliard m^3.

1920-1964 Development

Rapid increases in population necessitated urgent study to ensure an additional supply of water sufficient for horizontal expansion, in order to provide sustenance for the growing population.

The second heightening of Aswan Dam was completed in 1933 to increase storage capacity to 5.2 milliard m^3, and the construction of Gebel El Awlia Dam was completed in 1937, to provide Egypt with 2.5 milliard m^3 at Aswan to meet summer water requirements for the cultivated area. For better river control, and to save water for Aswan Dam reservoir filling, the Naga-Hammadi Barrage (1930), strengthening of Assiut Barrage (1938), Mohamed Ali "New Delta" Barrage (1940), strengthening of Esna Barrage (1947), and the Edfina Barrage (1951) were completed.

The area under cultivation in Egypt at that time was 6.15 million feddans, of which one million were still under basin irrigation. The annual water requirements in Egypt were 48 milliard m^3.

1964-1972 Development

Immediately after the completion of Aswan High Dam's first stage, its reservoir was used for additional flood water storage. The reservoir capacity is 164 milliard m^3/year, from which Egypt's share is 55.5 milliard m^3 (the remainder being Sudan's share). Egypt's share should allow it to reclaim 1.3 million feddans of new arable land, in addition to the conversion of one million feddan from basin to perennial irrigation, raising the total cultivated area to 7.8 million feddans.

The Equatorial Nile Projects might add another 1.0 million feddans. Good water management may also add 0.8 million feddans. Using drainage water suitable for irrigation may add another 0.4 million feddans. That could raise the planned cultivated area in Egypt to 10.0 million feddans, utilizing Nile water.

Utilizing Groundwater

Nile Valley

The total amount of groundwater extracted in the Nile Valley areas is about 1.6 milliards m^3/year in Lower Egypt and about 1.3 milliards m^3/year in Upper Egypt, distributed among the governorates as follows:

Governorate	Annual extraction
Cairo	238,404,000
Guiza	66,836,154
Qalubia	277,143,625
Sharkia	142,429,426
Dakhalia	137,661,005
Gharbiya	26,976,687
Menufiya	59,735,882
Tahrir Province	612,757,856
Beheira	11,033,216
Beni Suef	12,259,200
Minya	258,322,520
Assiut	101,733,040
Sohag	602,250,650
Qena	352,118,000
Aswan	996,450

Total in Lower Egypt about 1.6 milliard m^3
Total in Upper Egypt about 1.3 milliard m^3

Thus, the total annual extraction in the Nile Valley is about 2.9 milliard m^3. Nearly 65% of this amount is used in civic needs and the rest in irrigation. That means the amount of groundwater used in irrigation in the Nile Valley of Egypt is one milliard m^3/year.

Desert Areas

Before 1955

Information regarding shallow wells of the western desert oasis was formerly available only from records used for taxation purposes. The source of reliable information concerning the locations of many of these wells, and also the extent of cultivated land was the 1:10,000 topographical maps revised in 1925-1930 by the Geological Survey of Egypt. In 1939 to 1955, 8 deep wells were drilled at Kharga oasis and 7 deep wells were dug at Dakhla oasis using cable tool drilling, by the Ministry of Public Works.

Information concerning the deep and shallow wells in Kharga and Dakhla oases may be summarized as follows:

Deep Wells

Name of well	Date completed	Depth in meters	Yields (m^3/day)
Kharga oasis			
Mahariq	1947	487.7	600
Borg	1940	503.3	425
Qasr	1939	507.2	1741
Faruqiya	1939	470.3	2615
Bustan	1950	457.5	1720
Ginah 2	1952	342.5	3400
Ginah 1	1942	483.4	-
Gomhuria	1955	542.5	5500
Dakhla oasis			
Budkhulu	1941	240.1	8705
Qasr 2	1947	353.8	6400
Qasr 1	1943	221.1	4910
Gedida	1940	259.2	7482
Qalamun	1940	249.2	1670
Faruqiya	1940	233.0	8683
Ismant	1951	335.5	8200

Shallow Wells

	1941		1953	
District	No. of wells	Discharge m^3/day	No. of wells	Discharge m^3/day
Kharga oasis	354	123200	412	105800
Dakhla oasis			805	253500

Note: Data from taxation records.

About 5800 feddans of land were cultivated at Kharga Oasis and 10,000 feddans were cultivated at Dakhla Oasis in 1955.

The situation in other oases was limited to private efforts, 200 feddans were cultivated at Farafra, 1000 feddans at Baharya and 2000 feddans were at Siwa oasis irrigated from shallow wells.

After 1955

In 1953, a three year agricultural expansion program was approved by the National Production Council. The program included drilling of 20 deep wells at Kharga and Dakhla oases to increase the land irrigated by a total of 26,000 feddans. In 1955, rotary drilling was introduced into the oases and the drilling of 21 deep wells started in August at a depth of up to 650 meters in Kharga and Dakhla oases.

In 1959 The Egyptian General Desert Development Authority was established to carry out an intensive program for research, investigation, and exploitation of water resources in Egyptian deserts. The following research program was carried out in the Western desert depressions, the so-called "New Valley Project."

	Feddans
Aerial magnetic survey	53,000,000
Surface geology survey	10,200,000
Electric resistivity	7,700,000
Gravimetric survey	5,000,000
Soil survey from aerial photos	53,000,000
Reconnaissance soil survey	16,000,000
Semi-detailed soil survey	680,000
Detailed soil survey	200,000

The authority also drilled 46 wells and piezometers for geological and hydrogeological studies; 348 productive wells were drilled to irrigate the reclaimed lands.

Areas under Cultivation in Feddans (1976)

Crops	Kharga		Dakhla		Total	
	Old land	New land	Old land	New land	Old land	New land
Perennial fruit trees	3,420	6,500	9,890	11,850	13,310	18,350
Winter crops	9,920		21,740		31,660	
Summer crops	5,850		13,030		18,880	

This research program led to an estimate of the quantity of groundwater which could be extracted from the reservoir basin and the land which could be reclaimed utilizing this water.

Area	Available Groundwater million m^3	Possible irrigated area 1000's Feddans
Siwa Oases	125	25
Bahariya Oases	250	50
Farafra Oases	700	140
Dakhla Oases	500	100
Kharga Oases	250	50
South Kharga Dep.	675	135
Total	2500	500

Rainfall

Average annual rainfall at the coastal area is 150 mm. Although this quantity is not sufficient for economic crops, there is a network of wadis which collect runoff from their catchment areas and concentrate it on the alluvial areas at the end of the wadis. There are 218 wadis along the western coastal zone, with catchment areas varying from less than 1 km^2 to 240 km^2. The runoff collected is utilized in soil conservation and water spreading projects and also in groundwater replenishment and filling the ancient cisterns scattered over the area. The number of cisterns in operation now are 486 with total capacity of 215,000 m^3.

The study of this area led to an estimated of: (a) 25 million m^3/year of groundwater available, sufficient for irrigation 5000 feddans of olives, almonds and fig trees; (b) rainfall and runoff sufficient for 135,000 feddans yearly.

CONCLUSION

Water resources in Egypt provide enough water to cultivate an area of approximately 10,785,000 feddans, as follows:

- 6,500,000 feddans: cultivated till 1952.
- 1,300,000 feddans: from Aswan High Dam.
- 1,000,000 feddans: by water from equatorial Nile projects and Bahr El-Jabal water conservation project.
- 400,000 feddans: by drainage water suitable for irrigation.
- 800,000 feddans: by good water management
- 150,000 feddans: by groundwater in the Nile Valley.
- 500,000 feddans: by groundwater in desert depressions.
- 135,000 feddans: by rainfall

10,785,000 Feddans

DEVELOPMENT OF WATER RESOURCES AND ITS INFLUENCE ON THE ECOLOGY OF SAUDI ARABIA

H. H. HAJRAH
King Abdul Aziz University, Jeddah

ABSTRACT

The paper represents a comprehensive comparative study of the efforts elaborated by the concerned departments in Saudi Arabia for the Development of water resources. The ecological features of the vast Arabian peninsula are summarized. It is possible from an ecological point of view to divide Saudi Arabia into: a) the Western province including the south western and western high lands, and sea shore, b) the eastern province including the central Najd plateau and eastern shore on the Gulf. Results of a historical review of cultivated lands, their water systems and the effects of recent discharging of ground water for drinking purposes on cultivated lands especially in Wadi Fatma and Kholaise are summarized and recommendations are advocated for the benefit of land cultivation.

INTRODUCTION

Water for drinking and house-hold purposes is a vital need in comparison with the other industrial or irrigation uses. In the last twenty years several plans have been proposed for development of water resources. The problems faced by the planners were the rather restricted amounts of ground water and also the low recharge rate of the aquifers by rain.

ECOLOGICAL FEATURES OF SAUDI ARABIA

Saudi Arabia is situated in Arid and Semi-Arid zones with the series of Sarawat mountains extending almost parallel to the Red Sea coast and known as Asir and Al-Hijaz mountains. The southern parts are higher than the northern. Accordingly, Saudi Arabia is distinguished into two main provinces, the western and the eastern.

The western province includes mountinous high lands followed by plains extending to the Red Sea shore known as the Tihama plains, the width of which continuously decreases to the north. To the south these plains are dissected by many valleys which pour their water to the sea. The Western province depends upon rains as the main source of water which are kept within the restricted aquifers in volcanic rocks.

The eastern province includes the Najd plateau, the central regions and the eastern coast. It is mainly sedimentary rocks which slope constantly to the Arab Gulf. More than twenty formations have been found to contain water. These aquifers, at various depths, have varying amounts and quality of water. Table (1) summarizes the main characteristics of the main aquifers in the province. The depth of aquifers ranges from 10 to 2000 m and in certain localities there are flowing springs.

The following ecological features are the main factors affecting development of water resources in Saudi Arabia. Table (2) shows the climatic means for eleven stations chosen to cover most of the areas of the Kingdom for the period 1966-1976.

Rainfall: Most of the rains fall over the south west areas, particularly over high ground and mountains which are affected by monsoon rains or tropical rains in summer. Rainfall decreases over the coastal areas toward the north. The eastern and middle areas around Dahran and Riyadh, for example, have recorded more rainfall than the western coastal areas, mainly in the transitional seasons, spring and autumn. The mean annual rainfall at most of the eleven stations does not exceed four inches except in the Hijaz and the Assir mountains (Kh. Mushaits and Taif).

Temperature: Mean annual temperature is high except in the mountain areas. There are considerable differences between temperatures in summer and winter and the diurnal temperature range is large except in the mountain areas, giving the features of desertic climate. The highest temperature recorded during this period is 49°C in Riyadh, though its elevation is 626 m above sea level. The lowest temperature recorded for this period is - 7°C in Hail, 995 m above sea level.

Evaporation: The main factors affecting evaporation are surface winds, relative humidity, temperature, and the nature of the land surface.

Due to the lack of records for evaporation at these stations, water vapor pressure was taken as an indicator of

Table (1) The main ground water aquifers

Formation	Situation	Depth	Quality	Production	Utilization
Al-Saag	Al-Kassim	150 - 2400	V. good	Flowing	Irrigation
Al-Waheed	Wadi-Al-Dawaser	100 - 1100	Excellent	Flowing	Partially for irrigation
Tabook	Al-Kassim & Tabook	100 - 250	Different	Pumping	Irrigation
Al-Manjoor	Riyadh	1300 - 2200	Good	Flowing	Drinking for Riyadh and Al-Dawaser cities
Al-Bayad	Al-Dawaser & Riyadh	370 - 500	Excellent	Pumping	
Al-Wasce	Riyadh & Eastern	250 - 500	Good	Flowing	Drinking for Riyadh and injection of oil wells
Umm Raddah	Southern & Eastern	100 - 500	Different	Flowing	Partially developed
Al-Dammam	Eastern & Northern	50 - 120	Poor	Different	Irrigation at Al-Katif and drinking at Naeeriya
Al-Neogen	Al-Ahsaa & Eastern	00 - 567	Different	Flowing & pumping	Al-Ahsaa for irrigation

Table (2) Annual mean of precipitation, temperature (maximum & minimum), atmospheric vapor pressure, direction and speed of winds for the period (1966-1976) at 11 stations distributed all over the Kingdom

Station	Annual mean for precipitation (mm)	Annual mean for Temp. (°C)	Maximum temp. during this period	Minimum temp. during this period	Average Atmo. vapor press. (Millibar)	Direction and speed of wind (knott)	
1 Jeddah	63.0	27.9	48.2	+11.4	23.1	SE	48
2 Gizan	30.3	30.2	41.0	+15.8	28.5	SE	54
3 Tabuk	57.7	21.7	47.0	- 6.0	8.3	SW	60
4 Hail	122.9	22.0	46.0	- 7.0	7.6	W	56
5 Riydah	132.7	25.2	49.0	- 1.0	7.8	NW	58
6 Kassim	120.8	23.5	45.3	- 3.2	8.7	SW	64
7 Taif	156.9	22.1	39.2	- 1.0	9.1	N	66
8 Khamis Mshit	235.0	18.9	33.8	+ 0.2	9.7	NW	45
9 Wajh	21.0	24.3	44.0	+ 5.0	21.2	SW	50
10 Zahran	78.4	25.7	48.0	+ 1.4	18.0	NW	48
11 Madina	44.8	28.1	47.1	+ 1.0	7.8	S	46

evaporation. Vapor pressure in the coastal areas is high and hence evaporation is low in these areas, while vapor pressure is low in inland areas and thus evaporation rates are high. It is reasonable to conclude that a great amount of the rainfall is lost by evaporation in the inland areas.

Winds: Table (2) shows the maximum speed and direction of surface winds which were recorded in the period (1966-76) in each of the eleven stations. The prevailing winds in the Kingdom are as follows: NE-N-NW in the north and west areas of the Kingdom; N-NE in the middle and eastern parts of the Kingdom, and W-SW in the western parts of the Kingdom.

Wind velocities increase in the transitional seasons, spring and autumn, and also in winter over the north-western and eastern coasts and also the middle part. Most of these strong winds are S-SW winds which blow before the passage of the cold fronts with frontal depressions. These winds cause sand storms or rising sands. They are also accompanied by a strong rise in temperature and occur mainly in spring and early autumn, and are known as "Samoom" in Saudi Arabia as well as in Kuwait and the Gulf areas.

WATER RESOURCES

Saudi Arabia has scarce water resources of poor quality. In this vast area of desert no running river exists. Surface and ground water are the main sources of water together with water produced through desalination processes.

Surface water: Rains are the primary source of water in the Arab Peninsula. Thus they have been the object of all development efforts through construction of various size dams and reservoirs for drinking and household purposes, especially those near towns. Lagoons have been dug for watering cattle, e.g., the more than 13 Nebieda lagoons between Medina and Iraq. Rain water has been transported through covered and uncovered channels.

Springs and surface wells have been known in the Arab Peninsula from earliest times. At the Alahsaa Oases, springs are still used for the irrigation of more than 7000 hectares and the Al-Kharg springs irrigate about 1000 hectares. Surface wells, or "artificial springs", are used particularly in the coastal and near-coastal zones, e.g., Ein Zeibaida, which has been the source of drinking water for Mecca for more than 1000 years. Other surface wells are the Ein El-Zarkaa (Medina), Eyon Wadi Fatma (Jeddah), Ein Al-Mothannah, Al-Waht and Al-

Wahit (Al-Taif). Recently artesian wells have been dug at Jizan, Yanboo Al-Bahr, Rabigh, Tabook, Al-Llaa, and Yanboo Al-Nakhl. The water of these springs was transported through covered channels known as "Al Khoyoof." Many springs have vanished or exhausted and their traces are still observed. Ground Water: Development of ground water started as early as 1373 H when the first well was drilled at Al Kassim near Breida which was suffering of water shortage. Actual development of ground water started at 1383 H where some millions of Saudi Rials were spent to evaluate the potentialities of the ground water all over Saudi Arabia. The detailed studies need more time due to the vast area of the Kingdom. Table (1) summarises the important aquifers and their depths. Desalinated sea water: This is becoming an important source of drinking water. It is planned to complete more than 28 desalination plants along the red sea shore, from Farasan island in the south to Hagl at the northern extremity of Gulf of Aqaba, and Arabian Gulf shore. The first plant started in 1969 and the total productivity of all plants when completed at about 1983 will be more than 500 million gallons per day.

Soil is an important factor affecting the development of water resources. Sandy soils are characterized by rapid loss of water due to evaporation. Thin soils with shallow solid base facilitate their drifting by the action of wind and/or rain. Salty soils contaminate surface water with salts and also consume large amounts of water for their washing. Accordingly, the nature of soil will accelerate or retard the developing of water resources.

The rapid social development in certain parts of the country requires immediate implementation of water development plans. The main source of water in the western and southern provinces is rain. On the other hand, desalination will be the solution for the coastal parts. Ground water will be the main source of water in the central and eastern provinces.

WATER CONSUMPTION

The rate of consumption of water is very high and this is due to the absence of rules controlling its use, and the use of traditional methods instead of the new technologies of drip or sprinkle irrigation.

There is a continuous decrease in the number of springs due to over pumping for household and drinking purposes in the large cities. Eyon Wadi Fatma (between Mecca and Jeddah) which were originally 360 springs dropped down to 30 springs in 1945 and to 26 springs in 1967. However, the total productivity of the area has increased due to the development of deep wells which are mainly used for supplying water to Jeddah (Tables 3 & 4)

Table (3): Quantities of water produced for Jeddah

Source	No. of wells	No. of Producing wells	Depth (meters)	Production*	Water received* (Jeddah)	Remarks
Wadi Fatma	49	42	48-50	15	10	Differences used by villages for drinking
Wadi Kholais	41	20	35	10	8	
Desalination	-	-	-	10	10	

* in millions of gallons

There is a continuous increase in the consumption rate of water for household and drinking purposes. While it was only 3 G/D in 1941, it increased to 32 G/D in 1963 and it is constantly increasing with the increase of water supplied to the town. The consumption of water for household and drinking purposes results in the decrease of available water for agriculture and the decrease of productivity of land and farms as clearly shown in Wadi Fatma. Dams and reservoirs were constructed in the southern province for storing the relatively heavy precipitation. This was followed by establishing a network of channels for irrigation. The regions characterized by thick sedimentary deposits contain huge quantities of ground water which made it possible to cultivate vast areas. Desalination of Red Sea and Gulf water is the future unlimited source of fresh water. Re-using of water and controlled agriculture techniques have attracted special attention during the last few years.

Table (4): Historical development of water resources and inhabitants of Jeddah

Year	Source	Quantity	Inhabitants
1926	Kindasa (+ Tanks & wells)	135 Ton (+40) "	Less than 25,000
1941	Wells	350 "	25,000
1946	Wadi Fatma	650,000 Gallon.	More than 25,000
1953	Wadi Fatma	5,000,000 "	" " "
1963	Wadi Fatma	11,000,000 "	60,000
1970	Wadi Fatma	11,000,000 "	More than 500,000
	Wadi Kholais	6,000,000 "	
	Desalination	5,000,000 "	
1978	Wadi Fatma	10,000,000 "	About 1,000,000
	Wadi Kholais	8,000,000 "	
	Desalination	10,000,000 "	

IRRIGATION TECHNOLOGY FOR DESERT LAND RECLAMATION

E.V. RICHARDSON
Professor of Civil Engineering

V.A. KOELZER
Professor of Civil Engineering

Colorado State University, Ft. Collins, Colorado

ABSTRACT

The arid soils of the desert can be most productive when irrigation water is available. However, because they may contain large quantities of salts or be underlain with salt layers, contain no organic matter and ofen have low moisture holding capacity, they require careful water management. Sprinkler, drip and bubble irrigation methods have specific advantages over traditional surface irrigation methods for the development of arid soils. The initial cost of development using sprinkler, drip or bubble may even be less costly than assumed. The selection of drip, sprinkler or bubble irrigation depends on many factors such as type of crop, cost of energy and water availability. The paper will briefly describe these methods and provide an example of a sprinkler development of sand dunes.

INTRODUCTION

The irrigation of desert lands provides both an opportunity for agricultural development and economic gain and potential for economic disaster. Aridisols of the desert can be very productive with irrigation. However, those lands outside of alluvial valleys may contain large quantities of salts, have no organic matter, may be low in fertility and often have low moisture holding capacity. These factors require very careful water management. Sprinkler, drip and bubble irrigation systems provide an effective method of water management and can usually provide better water management than the traditional basin, furrow and border surface irrigation systems.

This paper is confined to a discussion of center pivot sprinklers, drip and bubble irrigation systems. There are many different sprinkler systems in use today, such as buried solid set, side roll, tow-a-line, big gun, hand move, traveling, or boom. By far the largest single sprinkler system in the development of new lands in the United States is the center pivot system. It has unique capabilities that recommend its use for field crops in desert land development.

Drip and a variation of drip called the bubbler have special qualifications for certain applications, in particular, for orchards and some vegetable crops, in water scarce areas and in the use of highly saline waters. Because there is considerable interest in these two systems for desert land development; discussion of these methods is included in this paper.

The paper cannot discuss all aspects of these systems, but can only briefly describe them. For more information the reader is referred to the literature cited. However, the bibliography is not all inclusive. The paper will describe some of the general advantages of the three systems in contrast to surface irrigation and then will describe each system in more detail. In general, the paper deals with the experiences and investigations of the authors in developing new croplands in Nebraska and Colorado. In the process of our work we have had to investigate various sprinkler systems, drip and bubbler rather than the traditional basin, furrow and border surface irrigation because the former systems provide more effective water management.

GENERAL

Water Management is applying the correct amount of water to the soil to meet plant water requirements, evaporation from the soil and any additional water needed for salinity control without excess runoff or deep percolation.

The water requirements of plants is not constant but changes with plant growth with the largest requirements generally occurring during flowering. Water available for plant use is that water held in the soil between field capacity and the wilting point. Field capacity is the amount of water soil will hold against drainage by gravity and wilting point is the amount of water in the soil when plants begin to wilt.

Saturation is the amount of water in a soil when all air space is occupied by water. Plants' ability to extract water from the soil decreases with a decrease in the amount of water available in the soil and with an increase in concentration of soluble salts. Soils at saturation for any length of time adversely affect plant growth.

Water holding capacity and infiltration rate (rate water moves down and into the soil) of soils depends on its texture and structure. Coarse textured soils -- sandy soils -- have low water holding capacity and high infiltration rates, whereas fine textured soils -- clay and silty clay soils -- have high water holding capacity and low infiltration rates.

Salinity management is required on all irrigated lands but is especially important with desert soils. Salinity management requires efficient irrigation and is necessary not only to maintain the productivity of the lands being irrigated but also to prevent deterioration of the groundwater aquifer or surface water supplies by the return flow from excess irrigation.

All irrigation waters contain a certain concentration of salts. Salt concentrations may range from one to two hundred parts per million (ppm) of total dissolved solids to as large as 2000 to 3000 ppm. Successful surface irrigated agriculture has been maintained in the Arkansas River Valley of Colorado, USA, using surface irrigation techniques with concentrations of salts of 2000 ppm. However, careful water management practices have to be used.

Traditional salinity management practice has been to put excessive amounts of water on the field to leach the salt from the soil. Excessive water is watering greatly in excess of the water required by the plants for transportation and the water lost from the soil surface by evaporation: or, the evapotranspiration (ET) requirement of the crop. This ET varies with crops and there are various equations for calculating the ET of various crops such as the Penman,[1] Blaney-Criddle,[2] and Jensen-Haise[3] equations.

Water greatly in excess of the ET of the crop leaches the salts from the root system. However, the excess water returns to the aquifer contaminating it or returns to the river via surface and subsurface drains. These return flows either to the aquifer or to the surface water system create problems downstream.

Research has shown (Agricultural Research, USDA)[4] that crops utilize mostly the water from the top portion of the root zone and that plants can tolerate larger salt concentrations near the bottom of the root zone. Thus, it is not necessary to apply excess water to leach the salts from the profile. With proper water management whereby excess salts are moved out of the upper part of the root zone, good yields are maintained, and deep percolation and salt water contamination is prevented. The applicability of the research has been demonstrated in field studies in Colorado (Agricultural Engineering Department),[5] Arizona and other areas.

Not only does good water management decrease the hazards of salinity but it saves water that can be used elsewhere, decreases the capital costs of the irrigation system (smaller ditches, laterals or pumping plants and decrease drainage requirements) and leaves valuable nutrients in the root zone.

All good irrigators, regardless of the irrigation methods, practice good water management and profit from the increase in crop production. Water management increases yields. Good fertilization practices increase yields. Good agronomic practices increase yields and good seeds increase yields. But one of these practices does not increase yields as much as all of them would in total. In fact, there is a multiplying effect of water management in combination with the other practices. Yields often triple and quadruple when water management is used along with the others.

Sprinkler, drip, and bubbler provide an effective method of providing good water management. That is not to say that good water management cannot be obtained with surface irrigation, but only that with sprinkler, drip and bubble and other controlled closed conduit systems that bring the water from source to the consuming plant, water control is easier to accomplish. These methods can be utilized on land that is too rough for surface irrigation (center pivot sprinklers can be used on any land that can be farmed). In Nebraska, where over 16,000 center pivot sprinkler systems are installed, the only requirement with regard to land slope is that farm machinery can traverse it.

Sprinkler, drip and bubble, because they apply the correct quantity of water to the crop at the time it is required can be used on soils ranging from light sandy soils with large infiltration rates and small holding capacity to heavy soils with small infiltration rates and large holding capacity.

In sandy soils, surface irrigation methods, even with lining of ditches, small runs, etc. will use excess water and cannot replenish the soil moisture content often or fast enough to prevent crop loss. This is often also the case for hand set, wheel roll and drag sprinkler systems. In the sandhills of Nebraska and Colorado, USA, crops of 100 to 200 bushels of corn are raised on sand with less than two percent organic matter. All nutrients are supplied, often by the center pivot sprinkler.

In heavy soils surface irrigation may require that water be held on the land for excessive times in order to replenish soil moisture. This may result in excessive tail water runoff with furrow irrigation or excessive ponding with basin irrigation. However, level basins with or without furrows, or tailwater return system and other improved surface irrigation techniques that are being developed provide methods to utilize surface irrigation on these soils. With any soils except for the very rough or the very sandy good water management can be obtained with surface irrigation. However, training farmers, water distribution managers and government officials to practice good water management (both on-farm and in the distribution system) with surface irrigation is more difficult than with center pivot sprinklers, drip or bubble.

CENTER PIVOT SPRINKLER SYSTEM

A center pivot sprinkler system consists of a machine that has a series of sprinklers (either impact type or spray nozzles) mounted on a pipe (pipe diameter being 4, 6, or 8 inches in diameter or combination of diameters depending on size of area to be irrigated) that is in turn mounted or carried by a row of five or more mobile towers. The towers move in a circle around a center pivot where water is pumped into the pipe. The source of water may be a well located near the pivot or from a canal with ditch or pipe to the pivot. The movement of the towers around the pivot is controlled by the outermost tower and an alignment device that uses simple micro switches to move the other towers to keep up with the outer tower. The rate of advance is variable--usually from a complete circle in 18 to 24 hours to a complete circle in three or four days. Water application rate is directly proportional to the rate of travel. Thus, the moisture needs (ET) of the crop can be exactly met by varying the rate of advance. The towers are driven by either electric motors or water drive and the pumps may be electric, diesel or gas driven. Pipe pressure at the pivot ranges from 60 to 100 pounds per square inch (psi) for high pressure systems and 30 to 70 psi for low pressure systems. The systems as developed are simple to operate, easy to assemble, relatively fool-proof, and need few spare parts. In fact, plants to manufacture the machines using some readily available components would be relatively easy to construct in any country and do not require a large capital outlay.

Because center pivots cover a circular area they miss the corner and thus only cover 79 percent of the area in any square. If land is scarce then the loss of 21 percent of the land from production could be important. By nesting the circles, the use of cornering devices or other sprinkler systems, the land loss can be decreased. However, if land is not expensive then it is usually better to leave the corners fallow or utilize them for storage, homes, etc.

If one center pivot is used in isolation, crop production is less at the outer edge because of the large loss of soil water to the surrounding land and higher evaporation from the land and transpiration from the plants resulting from the dry desert air. That is, the inner part of the circle benefits from the oasis effect. A big gun sprinkler on the end tower enlarges the circle and helps decrease crop loss. To take advantage of the oasis effect it is well to install a number of center pivots on a block of land and confine the loss of production to the outer edge of a large area rather than around a single center pivot.

In a recent land development (1977) in eastern Colorado ten center pivots were installed on 1900 acres. In total 1475 acres were irrigated. The land was hilly, in one area the slope was so steep that tractors could not go up the slope and had to operate downhill. The soil was sand (less than two percent organic) and all nutrients had to be supplied using commercial fertilizer. The installation started in September and the first corn crop planted the following April. The crop averaged 137 bushels or 4.1 tons per acre.

Cost of installing the center pivots were as follows:

Pumps and motor	$ 99,744
Wells	74,343
C.P. Sprinkler	308,260
P.P	24,182
Concrete pads	2,367
Power (line, transformer etc)	26,350
	$ 535,246

or $363/acre. In addition to the above costs, $26,600 was spent in land development--filling in some gullies, eliminating a small ephemeral lake so that the towers and farm machinery could operate.

Operating costs, exclusive of labor, insurance, taxes, etc. were as follows:

Seed	$ 14,140
Herbicides, pesticides, and fertilizer	54,159
Utilities (electric)	46,024
Repair & Maintenance	7,150
	$ 121,473

Overall the capital and operating cost can be recovered with an average yield of 130 bushels per acre at a price of $2.00/ bushel.

The above costs are for installation where irrigation equipment, labor and equipment to install the machines and the total infrastructure is all set up. In desert lands of Egypt near Cairo a recent estimate of the installation of a single machine pilot project for 130 acres using a tubewell was $1270/acre. These costs would decrease as the infrastructure to install and service the machines developed. A pilot project would of course help develop the needed infrastructure needed to construct and operate the system and develop the methodology for farming with center pivots. For example, the towers tend to cause ruts in heavy soils. Sands normally are no problem. It is easier to operate farm machines in straight lines (across tower tracks) than in circles. The pilot project would develop techniques to farm where rutting may be a problem. Also, other questions that a pilot project would answer are costs, water efficiency method of applying fertilizer, maintenance and so forth.

TRICKLE DRIP

Trickle or drip irrigation systems are a network of closed conduits that supply small amounts of filtered water periodically to the soil at each plant. The system operates at low pressure (5 to 14 lbs per square inch). Only the soil in the root zone of each plant is supplied with moisture. The network of conduits (usually plastic pipe of 1/2 to 3/4 inch in diameter) is laid on the ground along the crop row. The crop is usually fruit trees and grapes, but vegetables and in Hawaii, sugar cane, have been irrigated by trickle systems. From these lines outlets called emitters are placed to provide the water to the plant area. These emitters are needed to dissipate the pressure of the pipe network and provide a trickle of water to the plant.

The soil moisture in the root zone of the plant is kept in the desired range by intermittent operation of the system. This intermittent operation can be either manually or automatic. Water scheduling can be accomplished by calculations whereby on and off times are established or by soil moisture sensors that determine when a given water deficiency should be replenished.

The advantages of the system as given by Keller and Karmeli[6] and K. Shoji[7] are efficient use of water, because only a small surface area of the soil is wetted and evaporation is minimum. Weed growth, and loss of water by weeds, is inhibited. Although weeds grow around the emitters, in orchards, there are large areas of bare ground that trickle irrigation keeps dry. Trickle, as with the other systems mentioned, allows for good water management and the control of salinity. Also, because it can provide a continuously high soil moisture content trickle is advantageous when water with large concentrations of salt must be used. Trickle also moves the large concentrations of salt to the outer or fringe area of the root zone which keeps the salt concentration in the soil in the major part of the zone at tolerable levels. Trickle in contrast to sprinkler irrigation eliminates leaf burn and fruit spotting because of salt accumulations on the plant. Of course, fertilizer can be applied through the system and it is only supplied to the root zone of the plant being irrigated, thus decreasing the amount of fertilizer required.

The major disadvantage to trickle is clogging of the emitters either by sediment in the water, algae growth or chemical precipitates. To prevent clogging by sediments filters must be used, chemicals are added to prevent algae growth and possibly acids or other chemicals are added to remove or prevent precipitates. Rodents sometimes damage pipes. All of the above increase cost and in particular, filtering increases energy requirements and the complexity of operating the system. In general, trickle will be an efficient, effective and economical system of irrigation for high value crops, scarce water supply, sediment-free water, and high labor costs.

BUBBLE IRRIGATION SYSTEMS

The bubble irrigation is a closed-conduit system similar to trickle but operates at much lower pressure (0.7 psi or less vs. 5 to 14 psi for trickle) and with no emitter. Pressures available from open ditches are often sufficient. Bubble does not require filtered water but if surface water is extremely dirty a method of flushing the main lines and laterals may be needed. This can be simply a valve at the end of the line so that most of the pipe flow can be periodically discharged. Generally the main line and laterals are buried (about 2 ft) to protect the pipe and to aid in cultivation.

The bubble system consists of a network of low pressure pipes of 2 to 4 inches in diameter. Inexpensive, thin-walled, corrugated plastic pipe is sufficient. Small hoses (1/4 to 1/2 inch) go from the plastic pipe to the plant to be irrigated. Water flows full in the pipe by gravity and delivers the water to the plant by way of small hoses which take the place of emitters in the trickle system. The flow rate to the plant is controlled by the elevation of the end of the small hose either by fastening the end to a stake or the tree.

The proper elevation of the hose end is determined by trial and error with the system in operation until each plant is getting the proper amount of water to replenish the soil moisture as determined by calculations. Simple calculations of head loss and surveying in a preliminary elevation line speeds up the trial and error process.

Bubble irrigation is well described by Rawlins and is an attractive alternative to trickle. It has the same advantages and does not require the filtering and associated high energy requirements. It does, however, require fairly uniform land. The land can be sloping but because the system operates by gravity large changes in grade cannot be tolerated. Costs for materials for the bubble irrigation system is estimated by Rawlins (personal communications) as around $370 per acre. Materials for the same system in Polyvinyl chloride pipe required for trickle would be $690 per acre.

SUMMARY

Successful irrigation of desert lands requires good water management which is difficult to obtain with surface irrigation systems. In particular, when developing new lands center pivot irrigation systems provide an excellent method of obtaining good water management and large crop yields at relatively low overall costs for field crops such as beans, corn, sugar beet and cotton. The system is particularly well suited for rough ground and is a good method for soil types from sand to light soil.

For fruit trees, and some vegetables, a low head, closed conduit irrigation system called bubble has many advantages such as low cost, simplicity of operation, low energy requirements and water savings. It does require relatively uniform land surfaces although it can be engineered for fairly steep slopes if they do not have large adverse breaks in grade. Where energy may not be a problem and there is rough land, a system called trickle or drip may be a good alternative to the bubble.

REFERENCES

1. H.L. Penman (1948). Natural evaporation from open water, bare soil and grass. Proc. Royal Soc. London A193:120-146.

2. H.F. Blaney and W.P. Criddle (1947). A method of estimating water requirements in irrigated areas from climatological data. USDA Soil Conservation Service Memo.

3. M.E. Jensen and H.R. Haise (1963). Estimating evapotranspiration from solar radiation. J. Irr. & Drain. Dir. ASCE 89:NIR4.

4. Agricultural Research, 1974. Attacking salinity on irrigated land. USDA. Vol. 23, No. 6.

5. Agricultural Engineering Department, 1976. Irrigation Field Day Report. Colorado State University, Ft. Collins.

6. J. Keller and D. Karmeli (1975). Trickle irrigation design. Rain Bird Sprinkler Co., Glendora, California.

7. K. Shoji (1977). Drip irrigation. Scientific American (November).

8. S.L. Rawlins (1978). Uniform irrigation with low-head bubbler system. U.S. Salinity Laboratory Pub. No. 602.

See also:

N.C. Brody (1974). The Nature and Property of Soils. McMillan Pub. Co., New York. 8th ed.

J. Doorenbos and W.O. Pruitt (1977). "Crop water requirements." FAO Irr. & Drain. Paper 24, Rome.

European Commission on Agriculture (1973). "Trickle irrigation." FAO Irr. & Drain. Paper 14, Rome.

Lockwood Corporation (1977). "Center-pivot design manual." Gering, Nebraska.

A.W. McCulloch, J. Keller, R.M. Sherman and R.C. Mueller (1976). "Irrigation handbook." American Division of Lockwood Corporation, Woodland, California.

W.E. Splinter (1976). "Center-pivot irrigation." Scientific American (June).

SAHARA FARMING IRRIGATION PROBLEMS AND REMEDY RECOMMENDATIONS

HASSAN F. MIRREH
Ball Agricultural Systems, Wadi Dawasir

ABSTRACT

Continental climate characterised by severe weather conditions prevails in the Sahara, resulting in high potential evapotranspiration demands during the summer. The low water supply of Sahara soils with limited crop rooting patterns cause crops to become under water stress during periods of high potential evapotranspiration. Water stress developing in crop plants at certain critical stages may seriously impair the quality and grain yield of crops. Wheat plants under water stress at times of pollination and seed-fill developed barren heads and/or shrivel seeds. Irrigation methods currently practiced in the Saharas require modification to make them adaptable to Sahara conditions. Techniques to upgrade the water storage capacity of Sahara soils need to be investigated.

INTRODUCTION

Discovery of large underground water reservoirs suitable for irrigation in the Saharas are continuously drawing attention to the Saharas as a great potential source for the production of crops in the future (Libya, Saudi Arabia, Sudan, Egypt). There is an urgent need for adequate investigation of the Sahara farming problems so that efforts spent to develop the Saharas could fruitfully contribute to the present world population explosion, and the acute food shortage.

Continental climates and conditions for retarded chemical weathering prevail in the Sahara.[1] Consequently, extreme weather conditions occur (Table 1), and the soils of the Sahara tend to be immature[2] and often consist of residual rock fragments and unaltered, or partly altered, parent minerals which all together make-up coarse textured soils (Table 2). Prevalence of eroding winds and absence of land cover in the Saharas cause wind transport and deposition of soil materials

Table 1: Climatic Data at Three Different Sahara Locations

		Month	Jan.	Feb.	Mar.	Apr.	May	June	July	Aug.	Sept.	Oct.	Nov.	Dec.
SAUDI ARABIA	AL SULAYYIL	Av. Daily Temp.	13.9	17.5	21.7	26.5	31.7	32.5	34.8	34.5	30.3	24.0	19.5	16.5
		Max. Temp.	27.6	35.6	37.2	39.2	43.7	45.0	45.0	45.0	44.0	37.0	35.0	29.9
		Min. Temp.	2.3	3.0	5.8	14.0	18.4	18.4	22.6	21.1	14.1	10.0	6.1	3.6
		R/Humid. %	47.0	39	27	26	18	11	10	9	10	13	26	35
		Evapor. (cm)	22.8	29.3	44.5	43.1	53.8	55.2	69.7	55.2	49.8	38.5	31.0	23.1
LIBYA	KUFRA	Av. Daily Temp.	13.0	15.3	19.1	24.1	27.3	30.7	30.4	30.4	28.0	34.3	18.5	14.5
		Max. Temp.	30.6	37.2	40.6	44.2	44.8	46.2	43.9	44.1	42.0	40.6	37.2	32.6
		Min. Temp.	0	0.8	1.0	2.0	9.2	16.7	17.3	17.3	13.3	3.9	1.8	-1.1
		R/Humid. %	47.0	39.0	33.0	28.0	26.0	25	26	26	30.0	34	43	45
		Evapor. (cm)	13.2	15.2	20.0	23.8	27.8	33.1	35.0	32.1	33.8	19.1	14.9	12.5
	TAZERBU	Av. Daily Temp.	12.4	14.9	13.2	23.4	26.7	29.8	29.1	29.2	27.5	23.3	17.3	13.6
		Max. Temp.	30.2	38.0	41.5	44.0	46.5	46.2	44.2	44.3	44.3	40.5	35.4	32.2
		Min. Temp.	-2.5	-5.0	0	6.0	9.0	11.6	10.6	15.6	11.0	5.0	0.4	-1.0
		R/Humid. %	50	43	38	34	29	28	32	33	35	40	47	46
		Evapor. (cm)	12.5	13.7	20.1	23.0	30.2	32.8	30.1	31.2	23.0	18.6	14.1	10.9

Table 2: Particle Analysis of Sample Soil Profiles of Some Sahara Soils

	Profile No.	Depth (cm)	Particle Analysis %Sand	%Silt	%Clay	%Gravel
LIBYAN SAHARA (Sarir)	M37	0-10	96	1.5	1.5	1.0
		10-50	91	3.4	3.4	1.7
		50-150	96.6	1.4	1.4	0.6
	07	0-5	90.4	5.6	2.0	2.0
		5-40	88	6.4	2.4	3.2
		40-150	87	5.2	1.4	5.6
SAUDI SAHARA (Wadi Dawasir)	3	0-38	97	1	2	
		38-64	86	5	9	
		64-102	96	16	15	
		102-165	36	43	21	
	4	0-13	97	1	2	
		13-45.7	91	2	7	
		45.7-71	86	4	10	
		71-107	89	5	6	

Table 3: Soil-Water Relationships of Sample Soil Profiles of Some Sahara Soils

Profile	Depth (cm)	F.C. %	P.W.P. %
M37	0-10	7.4	3.0
	10-50	3.0	1.6
	50-150	6.2	3.4
07	0-5	6.2	3.4
	5-40	4.3	2.3
	40-150	12.2	6.6

to varying degrees depending on terrain and wind speed.[3] Continued transport and deposition of soil materials in this manner often results in shallow soils left behind at the sites from where soil materials had been blown-off and soil profiles containing alternating layers of coarse and fine sand developing at the sites of aeolian deposits.[4] Both of these soil types tend to limit crop root penetration and development.[5] Limited crop root growth patterns when coupled with the low water holding capacity of the Sahara soils (Table 3) and the extreme weather conditions of the summer cause crops to suffer from water stresses and create considerable irrigation problems. This paper will discuss the various crop watering problems encountered in the Saharas.

MEASUREMENTS

Growth and development of centre pivot irrigated wheat fields in the Saharas of Libya and Saudi Arabia were carefully studied. Abnormal growth symptoms were inspected; the locations delineated and notes on plants, root excavations, root zone soil moisture content and soil profile studies were recorded and photographed when pertinent. Daily weather data were recorded and utilized to compute crop consumptive use then compared with available soil water at various probable crop root depths. Irrigation machines performance and rate of water delivery were recorded. At harvest time, wheat yield data and root length measurements were studied in numerous one square meter plots randomly distributed over normal and abnormal wheat plant sites.

OBSERVATIONS

Water stresses occurring at certain critical stages in the crop growth cycle can seriously impair the quality and grain yield of crops. Wheat fields at Kufra and Sarir projects in the Libyan Sahara were observed to contain large areas where normally growing plants until the time of heading, developed wholly or partly barren heads afterwards. Further studies indicated that symptomatic plants of this nature dominate among shallow rooted wheat plants.[5] This was even more pronounced among plants in weedy areas and/or areas lying at the lee-ward outskirts of center pivot irrigated wheat fields. Symptomatic plants were observed to cover 30-50%

of the wheat fields. The loss in the quality and grain yield of wheat caused by barren wheat heads and/or shrivel seeds cannot be over emphasized.

The pattern of distribution of symptomatic wheat plants in the field suggested that the cause may be water shortage.[5] The system of irrigation employed in the Libyan Sahara projects is self-propelled centre pivot irrigation machines supplied with water by submersible pumps at discharge rates in the range of 75-80 ℓ/sec.[6] The frequency and the amount of irrigation water delivered by these machines could be monitored through the adjustment of the machine speed setting.[5] Table 4 contains information regarding the amount of irrigation water delivered at each of the machine speed settings most commonly employed together with calculated effective irrigation water reaching the soil after allowing for 20% sprinkler evaporation losses.

The low water storage capacity of the Sahara soils and the existence of root inhibiting layers in the soil when combined with the prevalence of high potential evapotranspiration in March-April-May are conducive to the development of acute water stresses in wheat plants;[5] unless very short irrigation intervals are practiced. Winter planted wheat in the Libyan Sahara comes to pollination and seed-fill during the period March-April-May. Optimum water supply to wheat plants at the pollination and seed-fill stages is very critical. Water stresses developing in wheat plants during these stages interfere with pollination and impair wheat seed-fill. Mirreh et al[5] observed that wheat plants that suffered water stresses at the time of pollination and seed-fill developed barren wheat heads and/or shrivel seeds that seriously reduced the quality and grain yield of wheat in the Libyan Sahara. They explained that the low water supply of the Sahara soils and the shallow roots of wheat plants growing the areas containing root inhibiting layers could not cope with the high potential evapotranspiration demand that was prevalent at the time of pollination and seed-fill. They reasoned that the acute water stresses developed in the wheat plants at these stages interfered with adequate pollination and seed-fill, thereby causing barren wheat heads and/or shrivel seeds. This may also be true for other cereals and seed crops that require water for similar physiological functions at the stages of pollination and seed-fill.

Table 5 shows the calculated amounts of soil available water for crop uptake between irrigations at various soil

Table 4: Irrigation machine speed settings, gross amount of irrigation water (mm) corresponding to each speed setting, and calculated effective irrigation water reaching the soil

Speed setting	Depth of Irrigation Water (mm)	
	Gross	Effective
17	4.7	3.6
34	9.3	7.7
42	11.5	9.2
57	12.4	10.0

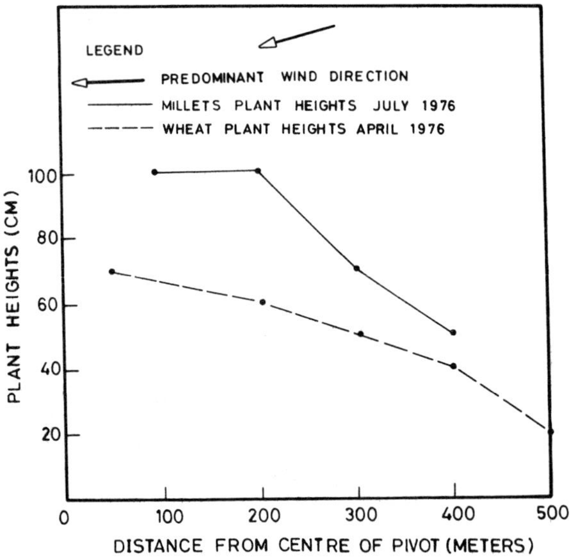

Fig. 1: Wheat and Millets Plant Heights Along the Radius of Centre Pivot Irrigated Crop Fields.

Table 5: Wheat Consumptive Use, Root Zone Soil Water Storage Capacity, and Corresponding Adequate Irrigation Frequencies

Month	Nov.	Dec.	Jan.	Feb.	Mar.	Apr.	May
Water Req. (mm)	3	3-4	4-5	5-7	7-9	9-11	11-13
Av. W.R.D.* (at 10-15cm)	3-4.5	3.4-5	3.4-5	3.4-5	3.4-5	3.4-5	3.4-5
Irri. Freq. (days)	3/3	3/4	3/5	3/7	3/9	3/11	3/13
Av. W.R.D. (at 15-20cm)	-	4.5-6	4.5-6	4.5-6	4.5-6	4.5-6	4.5-6
Irri. Freq. (days)	-	-	4.5/5	4.5/7	4.5/9	4.5/11	4.5/13
Av. W.R.D. (at 20-25cm)	-	-	6-7.5	6-7.5	6-7.5	6-7.5	6-7.5
Irri. Freq. (days)	-	-	6/5	6/7	6/9	6/11	6/13
Av. W.R.D. (at 25-30cm)	-	-	7.5-9	7.5-9	7.5-9	7.5-9	7.5-9
Irri. Freq. (days)	-	-	7.5/5	7.5/7	7.5/9	7.5/11	7.5/13
Av. W.R.D. (at 30-35cm)	-	-	9-10.5	9-10.5	9-10.5	9-10.5	9-10.5
Irri. Freq. (days)	-	-	9/5	9/7	9/9	9/11	9/13
Av. W.R.D. (at 35-40cm)	-	-	-	10.5-12	10.5-12	10.5-12	10.5-12
Irri. Freq. (days)	-	-	-	10.5/7	10.5/9	10.5/11	10.5/13
Av. W.R.D. (at 40-50cm)	-	-	-	12-15	12-15	12-15	12-15
Irri. Freq. (days)	-	-	-	12/7	12/9	12/11	12/13

* Av. W.R.D. : Available Water Root Depth

depths together with calculated wheat consumptive use during the growing season which lasts from November until May in the Libyan Sahara. Also included in this table are calculated irrigation frequencies considered adequate for each predicted crop rooting pattern. Reference to Tables 4 and 5 indicate that even the fastest centre pivot machine speed setting cannot supply the shallow rooted crops with the adequate irrigation frequencies during the months February to May. Consequently, all crop plants which have root depths 20 cm or less are apt to be under water stress during these months. Soil water content tests between irrigations at symptomatic plant sites indicated that the shallow root zone soils were very dry (Table 6).

Table 6: Root-Zone Soil Moisture Content Between Irrigations at Symptomatic Plant Sites (April 1977)

Site No.	Symptomatic Plants/m^2	% Water wt/wt
1	239	2
2	80	4
3	146	4
4	193	3
5	163	3
6	127	5
7	140	3
8	100	3

Wheat yield data from plots at symptomatic and non-symptomatic plant sites together with measured root depths of the harvested crop plants are shown in Table 7. The information in Table 7 indicates that under shallow crop rooting conditions the total and grain yield of wheat are very low. It becomes evident that unless wheat irrigation under Sahara conditions is adequately managed, the risk of cropping only low amounts of wheat straw may be high.

Tables 8-10 give the calculated amounts of available soil water for crop uptake between irrigations together with calculated consumptive use and irrigation frequencies at various

Table 7: Total and Grain Yield of Wheat at Symptomatic and Nonsymptomatic Plant Sites Together with Corresponding Wheat Plant Root Depths

N.S. Plant Sites*	Root Depth (cm)	Yield			Symptomatic Plant Sites	Root Depth (cm)	Yield		
		Total g	Heads g	Grain g			Total g	Heads g	Grain g
1	> 50	2110	790	570	1	15	940	360	120
2	> 50	2550	960	700	2	19	925	420	135
3	> 50	2010	760	560	3	15	990	335	180
4	> 50	1960	740	520	4	12	810	200	70
5	> 50	2790	1000	750	5	9	750	200	80
6	> 50	3000	1150	860	6	13	1500	290	85
7	> 50	1560	795	565	7	14	1050	220	60
8	> 50	2600	840	620	8	15	890	240	75
9	> 50	2260	1080	780	9	17	1050	190	40
10	> 50	1860	920	680	10	13	1060	200	50
11	> 50	3090	1090	780	11	18	1010	200	50
12	> 50	2590	920	680	12	14	832	312	68
13	> 50	3250	1140	840	13	12	915	395	53
14	> 50	2600	850	620	14	10	960	219	50

* N.S. : Nonsymptomatic

t-Test	10.99**	18.18**
5%	2.16	2.16
1%	3.01	3.01

predicted crop rooting patterns for alfalfa, wheat and assorted vegetable crops grown or in plan to be grown under Wadi Dawsir Sahara conditions Saudi Arabia. Thorough study of the profiles of the soils in the areas to be cropped together with understanding of the rooting patterns of these crops are essential for correct monitoring of the irrigation water and the planning of adequate irrigation frequencies to satisfy their varied watering demand. Failure to observe these directions have caused anomalous physiological development of all the wheat varieties planted and retarded the establishment of seeded alfalfa fields.

Extreme differences in growth had been observed to exist within center pivot irrigated crop canopies in the Libyan Sahara. Growth and yield of alfalfa, wheat, millets, and maize were observed to be significantly less towards the periphery as compared to the growth and yield of these crops towards the center of the pivot circles (Fig. 1). Open pan evaporation, temperature, and relative humidity studies along the radius of the centre pivot circles revealed increasing rates of pan evaporation, increasing temperatures and decreasing % relative humidity from the centre of the pivot circles outwards (Figs. 2,3). These observations indicate the existence of variation in potential evapotranspiration within center pivot crop canopies and could explain the variation in growth and yield observed along the diameter of the center pivot irrigated crop circles.

Present tendencies in sprinkler irrigation design are to achieve high uniform water distribution coefficients. Such designs, though well adapted to temperate climate conditions, may not necessarily be well suited under Sahara irrigation conditions. This may be true particularly when the sprinkler precipitation is designed to satisfy the water requirements of the crops growing towards the center of crop canopies thereby resulting in non-uniform crop growth patterns. Designs of sprinkler irrigation systems to satisfy the relatively high water requirements of the crops growing towards the periphery of the Sahara crop fields may result in soluble fertilizer and irrigation water losses towards the center of the fields where the crop water requirements are low.

It becomes evident from the information in Tables 5, 8, 9 and 10 that deep rooted crops are less susceptible to water stresses under severe weather conditions than those that are shallow rooted. Consequently, efforts should be spent to enhance deep crop rooting patterns. Such practices increase

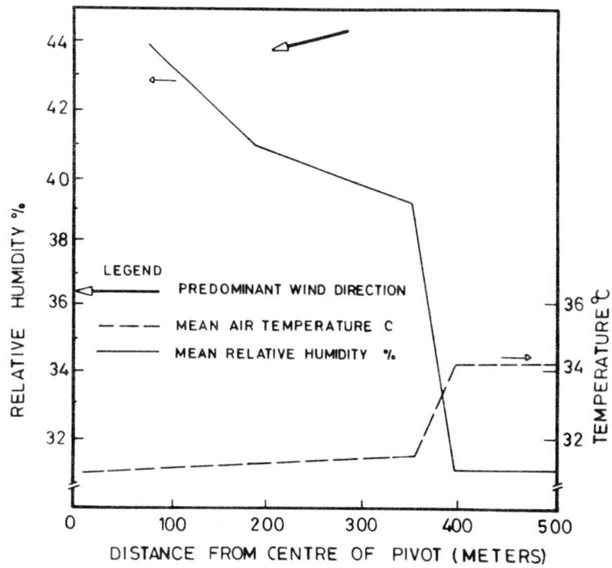

Fig. 2: Mean Air Temperature C, and Mean % Relative Humidity Measured 20 cm Above Ground Along Lee-ward Radius of a Centre Pivot Irrigated Maize Field July, 1976.

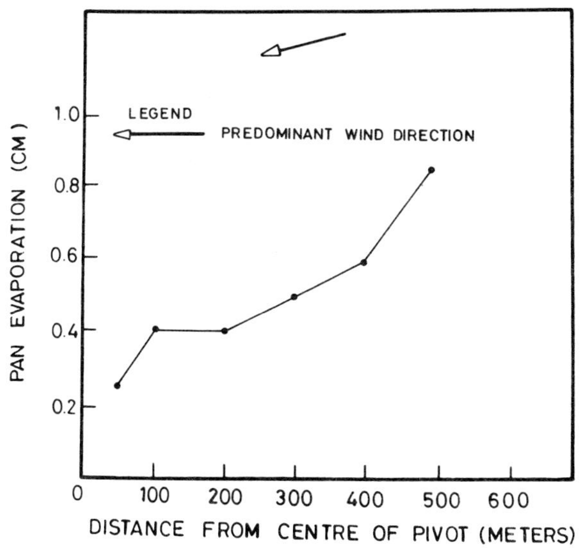

Fig. 3: Mean Daily Pan Evaporation Rates Along the Radius of a Centre Pivot Irrigated Wheat Field April, 1977.

Table 8: Alfalfa Water Requirements and Irrigation Frequency Schedules (Days) as Influenced by Available Soil Moisture and Crop Root Depth.

Month / Cu.	Jan. 4mm	Feb. 5mm	Mar. 5mm	Apr. 6mm	May 8mm	June 11mm	July 14mm	Aug. 13mm	Sept. 9mm	Oct. 6mm	Nov. 5mm	Dec. 4mm
Av. W.R.D.*	3-6	3-6	3-6	3-6	3-6	3-6	3-6	3-6	3-6	3-6	3-6	3-6
10-20 cm	4.5mm	4.5mm	4.5mm	4.5mm	4.5mm	4.5mm	4.5mm	4.5mm	4.5mm	4.5mm	4.5mm	4.5mm
Irri. Fre. Days	4.5/4	4.5/5	4.5/5	4.5/6	4.5/8	4.5/11	4.5/14	4.5/13	4.5/9	4.5/6	4.5/5	4.5/3
Av. W.R.D.	6-9	6-9	6-9	6-9	6-9	6-9	6-9	6-9	6-9	6-9	6-9	6-9
20-30 cm	7.5mm	7.5mm	7.5mm	7.5mm	7.5mm	7.5mm	7.5mm	7.5mm	7.5mm	7.5mm	7.5mm	7.5mm
Irri. Fre. Days	7.5/4	7.5/5	7.5/5	7.5/6	7.5/8	7.5/11	7.5/14	7.5/13	7.5/9	7.5/6	7.5/5	7.5/3
Av. W.R.D.	9-12	9-12	9-12	9-12	9-12	9-12	9-12	9-12	9-12	9-12	9-12	9-12
30-40 cm	10.5mm	10.5mm	10.5mm	10.5mm	10.5mm	10.5mm	10.5mm	10.5mm	10.5mm	10.5mm	10.5mm	10.5mm
Irri. Fre. Days	10.5/4	10.5/5	10.5/5	10.5/6	10.5/8	10.5/11	10.5/14	10.5/13	10.5/9	10.5/6	10.5/5	10.5/3
Av. W.R.D.	12-15	12-15	12-15	12-15	12-15	12-15	12-15	12-15	12-15	12-15	12-15	12-15
40-50 cm	13.5mm	13.5mm	13.5mm	13.5mm	13.5mm	13.5mm	13.5mm	13.5mm	13.5mm	13.5mm	13.5mm	13.5mm
Irri. Fre. Days	13.5/4	13.5/5	13.5/5	13.5/6	13.5/8	13.5/11	13.5/14	13.5/13	13.5/9	13.5/6	13.5/5	13.5/3
Av. W.R.D.	15-18	15-18	15-18	15-18	15-18	15-18	15-18	15-18	15-18	15-18	15-18	15-18
50-60 cm	16.5mm	16.5mm	16.5mm	16.5mm	16.5mm	16.5mm	16.5mm	16.5mm	16.5mm	16.5mm	16.5mm	16.5mm
Irr. Fre. Days	16.5/4	16.5/5	16.5/5	16.5/6	16.5/8	16.5/11	16.5/14	16.5/13	16.5/9	16.5/6	16.5/5	16.5/3
Av. W.R.D.	18-21	18-21	18-21	18-21	18-21	18-21	18-21	18-21	18-21	18-21	18-21	18-21
60-70 cm	19.5mm	19.5mm	19.5mm	19.5mm	19.5mm	19.5mm	19.5mm	19.5mm	19.5mm	19.5mm	19.5mm	19.5mm
Irri. Fre. Days	19.5/4	19.5/5	19.5/5	19.5/6	19.5/8	19.5/11	19.5/14	19.5/13	19.5/9	19.5/6	19.5/5	19.5/3
Av. W.R.D.	21-24	21-24	21-24	21-24	21-24	21-24	21-24	21-24	21-24	21-24	21-24	21-24
70-80 cm	22.5mm	22.5mm	22.5mm	22.5mm	22.5mm	22.5mm	22.5mm	22.5mm	22.5mm	22.5mm	22.5mm	22.5mm
Irri. Fre. Days	22.5/4	22.5/5	22.5/5	22.5/6	22.5/8	22.5/11	22.5/14	22.5/13	22.5/9	22.5/6	22.5/5	22.5/3

* Av. W.R.D.: Available Water Root Depth

Table 9: Wheat Water Requirements and Irrigation Frequency Schedule (Days) as Influenced by Available Soil Moisture and Crop Root Depth

Month Cu	Jan. 4mm	Feb. 6mm	Mar. 8mm	Apr. 10mm	May 12mm	Nov. 3mm	Dec. 3.5mm
Av. W.R.D.* 10-20 cm Irri. Fre. Days	3-6 4.5mm 4.5/4	3-6 4.5mm 4.5/6	3-6 4.5mm 4.5/8	3-6 4.5mm 4.5/10	3-6 4.5mm 4.5/12	3-6 4.5mm 4.5/3	3-6 4.5mm 4.5/3.5
Av. W.R.D. 20-30 cm Irri. Fre. Days	6-9 7.5mm 7.5/4	6-9 7.5mm 7.5/6	6-9 7.5mm 7.5/8	6-9 7.5mm 7.5/10	6-9 7.5mm 7.5/12	6-9 7.5mm 7.5/3	6-9 7.5mm 7.5/3.5
Av. W.R.D. 30-40 cm Irri. Fre. Days	9-12 10.5mm 10.5/4	9-12 10.5mm 10.5/6	9-12 10.5mm 10.5/8	9-12 10.5mm 10.5/10	9-12 10.5mm 10.5/12	--	9-12 10.5mm 10.5/3.5
Av. W.R.D. 40-50 cm Irri. Fre. Days	12-15 13.5mm 13.5/4	12-15 13.5mm 13.5/6	12-15 13.5mm 13.5/8	12-15 13.5mm 13.5/10	12-15 13.5mm 13.5/12	--	--
Av. W.R.D. 50-60 cm Irri. Fre. Days	15-18 16.5mm 16.5/4	15-18 16.5mm 16.5/6	15-18 16.5mm 16.5/8	15-18 16.5mm 16.5/10	15-18 16.5mm 16.5/12	--	--
Av. W.R.D. 60-70 cm Irri. Fre. Days	18-21 19.5mm 19.5/4	18-21 19.5mm 19.5/6	18-21 19.5mm 19.5/8	18-21 19.5mm 19.5/10	18-21 19.5mm 19.5/12	--	--

* Av. W.R.D.: Available Water Root Depth

Table 10: Assorted Vegetables Water Requirements and Irrigation Frequency Schedule (Days) as Influenced by Available Soil Moisture and Crops Root Depth

Month Cu.	Feb. 4mm	Mar. 5mm	Apr. 6mm	May 8mm	June 9mm	Aug. 7mm	Sept. 6mm	Oct. 5mm	Nov. 5mm	Dec. 4mm
Av. W.R.D.* 10-20 cm Irri. Fre. Days	3-6 4.5mm 4.5/4	3-6 4.5mm 4.5/5	3-6 4.5mm 4.5/6	3-6 4.5mm 4.5/8	3-6 4.5mm 4.5/9	3-6 4.5mm 4.5/7	3-6 4.5mm 4.5/6	3-6 4.5mm 4.5/5	3-6 4.5mm 4.5/5	3-6 4.5mm 4.5/4
Av. W.R.D. 20-30 cm Irri. Fre. Days		6-9 7.5mm 7.5/5	6-9 7.5mm 7.5/6	6-9 7.5mm 7.5/8	6-9 7.5mm 7.5/9		6-9 7.5mm 7.5/6	6-9 7.5mm 7.5/5	6-9 7.5mm 7.5/5	6-9 7.5mm 7.5/4
Av. W.R.D. 30-40 cm Irri. Fre. Days		9-12 10.5mm 10.5/5	9-12 10.5mm 10.5/6	9-12 10.5mm 10.5/8	9-12 10.5mm 10.5/9		9-12 10.5mm 10.5/6	9-12 10.5mm 10.5/5	9-12 10.5mm 10.5/5	9-12 10.5mm 10.5/4
Av. W.R.D. 40-50 cm Irri. Fre. Days			12-15 13.5mm 13.5/6	12-15 13.5mm 13.5/8	12-15 13.5mm 13.5/9			12-15 13.5mm 13.5/5	12-15 13.5mm 13.5/5	12-15 13.5mm 13.5/4

* Av. W.R.D. : Available Water Root Depth

the volume of soil that could be foraged by roots, thereby increasing the efficiency of applied fertilizer and irrigation water use.[4] This could be accomplished through adequate identification and elimination of root growth inhibiting layers in the soil profile.

Techniques designed to upgrade the water storage capacity of the Sahara soils should be investigated.[7] Classical techniques which employ various types of organic matter may not be very useful in this regard since all types of organic matter may very rapidly burn under the high temperature conditions prevalent in the Sahara.[8,9] However, this does not preclude the use of relatively stable synthetic soil conditioners, e.g. Hygromull, Agrosil. These soil conditioners have shown promise when tried at a number of sites in the Kingdom and in other few countries.[10]

Other techniques that could be tried to improve the efficiency of applied fertilizer and irrigation water use are buried layers made of asphalt or plastic or carpet like layers of organic or inorganic ammendments.[11,12,13] These techniques have been successfully used in a number of countries. However, such techniques should be practiced with care since they may cause some salinity problems under saline soil and/or saline irrigation water conditions.

REFERENCES

1. M. El Gabaly and H. Kadr (1962). Clay mineral studies of some Egyptian desert and Nile alluvial soils. J. Soil Sci. (UAR). 100-105.
2. H. Jenny (1941). Factors of Soil Formation. McGraw-Hill.
3. A. Zein El Abdin and M.M. Abdalla (1958). The soils of the Egyptian desert. III. The improvement of some physical and chemical properties of wind borne band sediments under farm management. Bull. 166. Cairo University Press.
4. H.F. Mirreh (1978). Sahara farming, soil fertility problems and fertilizer recommendation. (In press).
5. H.F. Mirreh, M.A. Hussein and A.O. Attir (1977). Barren wheat heads and shrivel seeds at Sarir. First Wheat Seminar, Agric. Res. Center, Tripoli, Libya.
6. H.F. Mirreh (1976). Ball centre pivot irrigation machines. Follow-up Report. Sarir Production Project, Libya.
7. S. Klimes (1956). Effects exerted on the water economy of sandy soils by substances containing organic or inorganic colloids applied in layers. Cong. Inter. Sci. Paris, Vol. I:1:36.
8. A.H. El Damaty and M. Mobarak (1962). Studies on virgin sandy soils at Tahreer Province of the U.A.R. Part I and Part II. J. Soil Sci. (UAR). 2:195-221.
9. S.A.Z. Mahmoud, S.M. Taha, A.H. El Damaty and M.S. Mobarak (1968). Effect of green manuring on the fertility of sandy soils at the Tahreer Province, U.A.R. J. Soil Sci. (UAR) 8:113-127.
10. A. Sayegh (1978). Agricultural Research Centre, Ministry of Agriculture and Water, Riyadh, Saudi Arabia (Personal communication).
11. S. Egerszegi (1959). Economical and lasting effect of organic fertilizers in sandy soil. Acta Agronomic Academiac Scientiarum Hungaricac lx Faciculi 3:3:319-340.

12. S. Egerszegi (1964). Plant physiological principles of efficient sand amelioration. Agrokemia Es Talajtan, 13 supplementum, 209-218.

13. S.A.S. El Bagouri and M.A. Abdel Salam. Long term studies on amelioration techniques of sandy soils. Desert Inst. Bull. (Egypt). 22(2):351-368.

14. F. Makled (1965). Sheet layers in iycmeter experiment in sand soil. 6th Cong. Irrigation Practice Seminar for Middle East and South Asia. Amman, Jordan.

THE DEVELOPMENT OF EFFECTIVE IRRIGATION SYSTEMS FOR LONG-TERM AGRICULTURAL PRODUCTIVITY

GAYLORD V. SKOGERBOE
Professor, Department of Agricultural and Chemical Engineering, Colorado State University, Fort Collins

ABSTRACT

The "heart" of an irrigation system is the farm subsystem. The purpose of an irrigation system is to grow food and this "action" takes place in the root zone. The purpose of the water delivery subsystem and water removal subsystem is to support this "action". Therefore, the proper design of an irrigation system requires, first of all, that the farm subsystem be adequately designed so that the water delivery subsystem can be designed to provide the quantities of water at the times required by the plants. The most important constraint in the design procedure is the necessity for assuring adequate drainage through the root zone in order to maintain a root zone salt balance to insure continued long-term agricultural productivity.

The approach which has been applied to irrigated agricultural development in the past is characterized by separating the development of water resources from the management aspects of water resource utilization. The "management" approach attempts to achieve water development objectives by applying a variety of measures after studying the entire system. Unfortunately, in most cases, water management and the many disciplines required to produce efficient management are relegated to the post analysis of engineering works, which aggravates not only the implementation of technology, but really constrains or makes extremely difficult the implementation of a host of services requiring strong institutional measures. A combination of technological changes and institutional modifications will usually be required to effectively manage irrigated agriculture for long-term productivity.

INTRODUCTION

Whether the goal is minimizing diversions to new croplands because of limited water supplies, reducing future diversions to irrigated agriculture to provide water supplies for new demands, minimizing water quality degradation in receiving streams resulting from irrigated agriculture, or maximizing agricultural production on existing croplands, the solutions are identical: improved water management practices.

Irrigation is one of the most important agricultural practices developed by man, with irrigation being practiced in some form since the earliest recorded history of agriculture. The economic base for many ancient civilizations was provided by irrigation. Irrigation farming not only increases productivity, but it also provides flexibility which allows shifting from the relatively few dryland crops to many other crops which may be in greater demand. Irrigation contributes to strengthening other facets of a region's economy in that it usually creates more employment opportunities than rainfed agriculture through intensive and diversified cultivation, the stimulation of important agri-businesses and public institutions, the provision of products for export, and the creation of a healthy domestic market (Skogerboe).[1]

History is replete with declining civilizations due to declining agricultural productivity, which has usually been the result of improper soil and water management practices. This potential danger is present today at many places throughout the world. This comes at a time of dangerous food shortages in a hungry world, when the goal is to increase per acre crop yields.

To maintain agricultural productivity in irrigated agriculture--and we must do more than that today--salts applied onto the croplands, which are dissolved in the irrigation water supplies, must be moved below the root zone in order not to retard plant growth. Therefore, it is mandatory that water supplied to a crop must exceed the actual water requirement of the plants to include evapotranspiration needs, leaching requirements, seepage losses and in most cases other transit or ditch losses which in most countries are substantial.

As is known, usually the quantity of irrigation water diverted from a river far exceeds the cropland water requirement. Data from many irrigation regions indicate that seepage losses from canals and laterals throughout the irrigation distribution systems are extremely high. Added to this problem is the excessive application of water on farm fields, which results in surface runoff from the lower end of the field (tail-water runoff) and/or large quantities of water moving below the root zone (deep percolation). The combination of seepage and deep percolation losses cause groundwater levels to rise (waterlogging). In many irrigated regions, the groundwater levels have reached the vicinity of the root zone which frequently results in the upward movement of groundwater due to capillary action. When upward moving water reaches the soil surface and evaporates, the salts contained in the moisture are left behind on the ground surface. This process of salinization has not only resulted in declining agricultural production, but has caused many lands to become essentially barren.

The quality of water draining from irrigated areas is materially degraded from that of the irrigation water applied as we have discovered along the Colorado and other rivers.

Agriculturists have viewed this phenomenon as a natural consequence of the several complex processes and strangely enough even today little attention has been given to the possibility that progress could be made toward controlling or alleviating this degradation of water quality.

Historically, some degree of salt concentration due to irrigation has been usually accepted as the price for irrigation development. In some areas, however, there has been so much laxity that quality degradation has become a serious matter. As pressures on water resources become greater, due to increasing populations and the necessity to produce food in increased quantity and improved quality, there is a mounting concern for proper control of serious water quality deterioration and soil salinization. The need then for more precise information as a basis for wise policy action is a matter of critical importance.

The major problems resulting from irrigation are due to the basic fact that plants are large consumers of water resources. Growing plants extract water from the supply and leave salts behind, resulting in a concentration of the dissolved mineral salts which are present in all natural water resources. In addition to having a greater concentration of salts in the return flow resulting from evapotranspiration, irrigation also adds to the salt load by leaching natural salts arising from weathered minerals occurring in the soil profile, or deposited below. Irrigation return flows provide the vehicle for conveying the concentrated salts and other pollutants to a receiving stream or groundwater reservoir. It is necessary then to examine the waterlogging and salinity problems resulting from this process and to develop and implement measures to control or alleviate the detrimental effects.

Impaired crop production resulting from salinity is a major problem in many areas of the world. In fact the more one observes problems in other countries the more one is convinced that this is indeed an international problem. Unfortunately, the portions of the world now facing the greatest population pressures are the same areas which have the least amount of additional land available for agriculture and in some cases where waterlogging and salinization are rampant. In such areas, increased food production must come from more intensive farming with consequent increased yields and, as the economist reminds us, the modernization of agriculture requires that output continues to rise with decreasing per unit input costs. On the one hand, there is a great need to increase the productivity of such lands, but on the other, agricultural production is continually being damaged due to rising groundwater tables (waterlogging) and increased salinity in the soils and groundwater supplies.

MAJOR WATER PROBLEMS

The purpose of irrigation is to increase crop production, but the real difficulty lies in sustaining increased crop production over a long time period. History is replete with hydraulic societies which flourished and then floundered, or became extinct. Why?

Maintaining or increasing agricultural productivity in an irrigated area requires, first of all, that a salt balance be achieved in the root zone, and secondly, that not too much water is applied such that the groundwater levels rise until the water table is near the ground surface, thereby resulting in waterlogging and increased salinity levels in the root zone. Thus, a balance must be reached in order that sufficient water is applied to the croplands to leach salts from the root zone, but not so much water that groundwater levels nearly reach the ground surface. The history of hydraulic societies has been primarily one of applying too much water. And over-irrigation continues today in most portions of the world.

In order that water does not limit crop yields, the proper amount of water must be applied to the cropland at the proper times. The timing and quantity of required irrigation water is a function of primarily climate, soils, and the stage of crop growth.

In many regions and countries, increasing urbanization and industrialization requires the reallocation of agricultural water uses in order to meet these new water demands. In order to accomplish this water reallocation, improved water management practices must be instituted in the agricultural sector, thereby reducing agricultural water diversion requirements.

The real detriment in over-irrigation is that groundwater levels rise, thereby resulting in waterlogging of the soils and a continual decline in agricultural productivity until the land may finally become barren. In addition, over-irrigation results in water quality degradation of irrigation return flows and consequent detriments to downstream water users.

DEFINITION OF AN IRRIGATION SYSTEM

An irrigation system can be subdivided into three major subsystems: (a) the water delivery subsystem; (b) the farm subsystem; and (c) the water removal subsystem. The water delivery subsystem can be further subdivided into two components; namely, (a) the transport of water and pollutants from the headwaters of the watershed to the cross-section along the river where water is diverted to irrigate croplands, and (b) the transport of water and pollutants from the river diversion works to the individual farm. The farm subsystem begins at the point where water is delivered to the farm and continues to the point where surface water is removed from the farm. Also, the farm subsystem is defined vertically as beginning at the ground surface and terminating at the bottom of the root zone. The water removal subsystem consists of (a) the surface runoff from the tail end of the farm, and (b) water moving below the root zone (Skogerboe and Law).[3]

PLANNING FOR EFFECTIVE IRRIGATION WATER MANAGEMENT

The resource base for irrigated agriculture has not substantially changed since its inception thousands of years ago. Over the last century there has been little incentive for any major innovation to improve efficiency in the use of water which now fortunately is considered a scarce resource. The provision of irrigation water since ancient times has been considered a governmental or collective responsibility, and the direct charges made for water have usually not been high enough to encourage proper budgeting of water or other necessary innovations. This custom of undercharging for water has continued to this very day; a few regions charge the farmer for the real cost of water. However, there are a few examples of extremely water-scarce areas in the world where considerable ingenuity has been applied in effectively utilizing water supplies.

Aggravating this situation of so-called "cheap water" is the fact that the development of irrigated agriculture in most places, even in the last few decades, has focused almost entirely upon the construction of water delivery subsystems. The preoccupation with the installation of "hardware" results from a naive single-discipline approach to water management (Wiener, 1972)[3] and let us be reminded that one discipline cannot begin to solve the complex physical, economic, political, and sociological problems involved. Probably the greatest deterrent to improved water management in most irrigation systems today is the inordinate focus on the water delivery subsystem and the almost complete neglect of other problems. How long can we afford to neglect other problems of the system such as the need for improved soil-plant-water management techniques, improvements in cultural practices, farm machinery, agrarian structural reforms, roads, marketing systems, advisory services and input supply systems, administration of institutions, water laws, cooperatives and water users associations, and many other factors all of which must fit together in all their interdependencies and complementarities to form a most complex system. In reality, though, especially in irrigation agriculture, we come face-to-face with a wide gap which frequently exists between "hardware development" and the development of all the other requisites for increased agricultural production.

The approach which has been applied to irrigated agricultural development in the past is characterized by separating the development of water resources from the management aspects of water resource utilization. Therefore, the record shows development being emphasized greatly while management is most often neglected. This orthodox approach has been used almost exclusively in the western United States with reasonable success. However, as the water resources become more fully utilized, the necessity for meeting new water demands (along with physical, socio-economic, and political problems of water quality degradation) require that we reject much of the conventional wisdom of the past. It should also be obvious that developing countries today neither have the time nor resources many western irrigated regions had to utilize in their development. Pressures created by rapidly increasing population alone will force them at some point to reevaluate their approaches.

In contrast to the mere development of water resources approach, the "management" approach attempts to achieve water development objectives by applying a variety of measures after studying the *entire system*, thereby attempting to modify the total system to meet new and changing demands as well as estimated future demands. Therefore, instead of constructing new engineering works to meet new demands, the focus should be on water resource management, with construction works being considered only as a tool when necessary to meet water management objectives (Wiener).[2] Unfortunately, in most cases, water management and the many disciplines required to produce efficient management are relegated to the post analysis of engineering works (e.g., much of the future emphasis will be geared towards improving existing irrigation systems), which aggravates not only the implementation of technology, but really constrains or makes extremely difficult the implementation of a host of services requiring strong institutional measures.

DESIGN OF IRRIGATION SYSTEM

The "heart" of an irrigation system is the farm subsystem. The purpose of an irrigation system is to grow food and this "action" takes place in the root zone. The purpose of the water delivery subsystem and water removal subsystem is to *support* this "action." Therefore, the proper design of an irrigation system requires, first of all, that the farm subsystem be adequately designed so that the water delivery subsystem can be designed to provide the quantities of water at the times required by the plants. The most important constraint in the design procedure is the necessity for assuring adequate drainage through the root zone in order to maintain a root zone salt balance to insure continued long-term agricultural productivity (Skogerboe and Walker).[4]

Farm Subsystem

The first important variables in designing the farm subsystem are climate, soils, and crops. The interrelationships between these variables dictate the capability of the land resource for producing food and fiber. We need to understand the interactions and relationships between water and management factors such as optimum rates of fertilizer applications, proper pest control measures, seeding and tillage practices, and other improved cultural practices. Much of this research is adaptive and protective, which due to its rather low level of sophistication, is sometimes neglected.

The next important variable will be water and its physical availability. Frequently, in arid areas of the world, water is the most limiting factor. However, the capability of the available water supplies (precipitation, surface runoff, and groundwater) for plant production is highly dependent upon the efficiency with which the water is used, which in turn is a function of both economic and institutional factors. Besides physical limitations, the questions of economics in supplying water must be answered to insure that costs are commensurate with planning goals.

Once the general scope of an irrigation project has been determined, then the more detailed design procedures can follow. The critical factor at this point becomes the infiltration characteristics of the soil. Unfortunately, infiltration is a complex phenomenon and the intake function of a particular field will vary during the irrigation season, as well as varying from season to season. There are a number of laboratory and field methods available for determining the intake characteristics of a soil.

Using climatic data, the potential evapotranspiration of the various crops can be calculated. These computations will provide the information regarding water consumption with time, provided sufficient moisture is made available in the root zone.

The next important step is designing the farm irrigation layout so that sufficient moisture will be available in the root zone when required by the plants. The root zone is capable of storing moisture for future plant use. Again, soil characteristics determine the amount of storage, as well as the capability of the plant to extract the moisture from this "reservoir." At the same time, the leaching requirement for maintaining a salt balance in the root zone must be kept in mind. Consequently, the farm irrigation layout must be capable of supplying not only the plant water requirement, but also the leaching requirement.

The proper design of the farm irrigation layout is crucial for: (1) uniformly distributing the necessary moisture throughout the field; and (2) minimizing deep percolation losses so as not to aggravate problems in the water removal subsystem (e.g., waterlogging and salinity).

Generally, the development of irrigation projects has not entailed the design of farm irrigation layouts suited to the individual characteristics of each field. Instead, only the general method of irrigation may be adopted (e.g., basin or furrow irrigation). The farmer is usually left to his own means in irrigating his fields, without having the benefit of technical (or economic) assistance. The situation is further aggravated because, along with adopting a general method of irrigation, an average irrigation efficiency for this method is "pulled out of the air." If this so-called average water use efficiency was even close to being correct it would be most fortunate--let alone taking into account the variability of this "magic" number of any specific field throughout the irrigation season, as well as the variability from field-to-field. Usually, the application of this average efficiency results in large quantities of deep percolation during the early part of the irrigation season, which in turn contributes to waterlogging of the soils and consequent poor crop yields (or eventual failure of the irrigation enterprise).

Water Delivery Subsystem

The design of the individual farm irrigation layouts should dictate the design of the water delivery subsystem. The irrigation layout design, if properly accomplished, will show the necessary quantities and timing of water deliveries at the farm inlets. The water delivery network must be designed to meet the farm water requirements. Except for alluvial channels conveying large sediment loads, the design of the conveyance works is rather "mechanical".

One of the essential facilities for successfully operating an irrigation conveyance network is adequate and numerous flow measurement devices. To begin with, since each farm has a particular water requirement, then the only means by which the proper amount of water can be delivered is by measuring the water at the farm inlet. After all, the farmer cannot be expected to use good water management practices if he doesn't even know the quantity of water being managed. besides each farm inlet, a flow measurement structure should be provided at all division points in the water delivery subsystem.

The real problem in the water delivery subsystem is the institutional framework controlling the operation of this portion of the irrigation system. Generally, the operation of the conveyance facilities has not been related to the requirements for sustaining a long-term productive agriculture. In particular, institutional factors have acted as constraints to improved water management or increased agricultural production.

The primary requirement for sustaining an irrigation system is an institutional framework that is compatible with the design requirements for the water delivery subsystem, which in turn has been dictated by the proper design of the farm irrigation layouts, as well as any constraints imposed by the water removal subsystem. Thus, even if all three components of the irrigation system have been properly designed, the lack of an adequate institutional framework for operating the system in accordance with the design criteria will likely lead to either failure of the system, or at least having agricultural production levels below (or far below) expectations. Many developing country institutions for irrigation systems were developed by outsiders for other than the modernization of agriculture. In some countries, the objective was a simple system which would almost operate itself. Some of these institutions were regulatory rather than service oriented and little attention was ever given beyond the canal outlet.

Water Removal Subsystem

The principal function of the water removal subsystem is to allow proper drainage below the root zone so that adequate leaching of salts from the root zone will occur. The most satisfactory mechanism for insuring adequate drainage is proper operation of the water delivery subsystem. By so doing, a drainage problem will probably not occur, or at least the necessity for drainage works will be minimized. This is much better than allowing the problem to occur, then constructing drainage facilities to correct the damage. Unfortunately, the usual solution consists of constructing additional facilities. Frequently, project reports will state: "drainage facilities will be designed after the project has been in operation for a number of years in order to more precisely ascertain drainage requirements." This is the same naive single-discipline thinking referred to in the previous section, which is the rule rather than the exception. Perhaps the statement is correct that the history of irrigation systems teaches one that man often does not learn from history.

Another important consideration in the water removal subsystem is water quality. If canal seepage and cropland deep percolation losses result in water quality degradation of the underlying groundwater supplies, then the use of these supplies may be impaired. Also, the return flows to the river may limit the usefulness of the river water to downstream users. Numerous examples of this situation can be cited throughout the world. Again, this is the rule rather than the exception.

ADMINISTRATION OF IRRIGATION SYSTEM

The proper functioning of an irrigation system is highly dependent upon an institutional framework which is compatible with the design criteria used in developing the system, as well as providing flexibility in achieving improved water management as the need arises. Satisfying on-farm water management objectives cannot be achieved without controlling water deliveries. Therefore, the administration of the irrigation system requires that satisfactory legal mechanisms exist that control water deliveries.

In developing new irrigation systems, the storage and conveyance facilities are first constructed. Then, a portion of the lands to be served by these facilities are placed under irrigation. Water is plentiful and frequently farmers are allowed too much water. This practice results in farmers developing "poor" irrigation practices. This situation points out the necessity to provide farmers adequate technical assistance in on-farm water management at the very beginning of the project.

The failure of new irrigation developments to meet estimated production goals has largely resulted from the lack of follow-up in providing essential agricultural inputs and services to farmers. Again, the lack of training in farm water management and other farm practices by those responsible for the planning, design and construction of the irrigation development leads to the shortcomings in output. Also, the lack of cooperation between disciplines is certainly a detriment in any country, but is particularly noticeable in the developing countries. Interdisciplinary teamwork is a noble concept but it is seldom internalized and implemented effectively. Perhaps the most expedient solution is to expand the training of those personnel involved in the planning, design and construction of irrigation enterprises. Such training would then hopefully provide feedback into the planning and design of future irrigation developments.

The provision of adequately trained personnel for the operation and maintenance of irrigation systems may be understood, but is sometimes difficult to accomplish. Much focus must be given to training not only the engineers and technicians, but those who work directly and indirectly with the farmer. Although agricultural experiment stations may exist, there is usually a severe limitation in accomplishing on-farm improvements because of insufficient adequately trained farm-level advisors capable of transferring applied research results directly to the farmers. In many countries, the components of research and advisory services have never been brought together in a carefully planned relationship.

We are all aware of the need for improved delivery systems for technology once it has been made available by research. How best can this transfer of technology in water management be made to the end user, the farmer? Here we need the help of the applied social sciences such as economics, sociology, anthropology, political science and even the social psychologist. Some extension system is required which will take the proven findings of research, along with utilizing the knowledge of the social sciences "package," in such a way as to convince the end user to move from the trial to the final adoption stage.

CONCLUSIONS

Successful irrigation enterprises need more attention given to "on-farm water management," rather than the continued emphasis on the "hardware" development of dams and canals. Rather than correcting problems of inefficient water management by constructing drainage facilities, the farm subsystem must be designed and managed in such a manner that the requirements for drainage facilities will either be eliminated, or at least minimized.

There is a tremendous need to expand the training of those personnel involved in the planning, design and construction of irrigation enterprises so that they are keenly aware of the importance and role of on-farm management in the long-term productivity of irrigated agriculture. In addition, much more attention must be given to providing sufficient numbers of adequately trained farm-level advisors who can effectively interact with farmers in achieving improved irrigation practices and increased agricultural productivity.

REFERENCES

1. G.V. Skogerboe (1974). Agricultural systems, In, Human Ecology, North Holland Publishing Co., Amsterdam. Chap. VII, pp. 127-145.

2. A. Wiener (1972). The Role of Water in Development. McGraw-Hill Book Co., New York.

3. G.V. Skogerboe and J.P. Law, Jr. (1971). Research needs for irrigation return flow quality control. Report No. 13030--11/71. U.S. Environmental Protection Agency, Washington, D.C.

4. G.V. Skogerboe and W.R. Walker (1978). Impact of irrigation on the quality of groundwater and river flows. In, Impact of Salinity on Water Resource Use and Development. Marcel Dekker, Inc., New York (In press).

WASTEWATER AS A SOURCE OF NUTRIENTS AND WATER FOR HYDROPONIC CROP PRODUCTION IN ARID REGIONS

A. WALLACE
University of California, Los Angeles

S.M. SOUFI
American University of Beirut,

A.M. EL GAZZAR
University of Alexandria, Egypt

ABSTRACT

Reclaimed water from a sewage plant at Calabasas, California was a satisfactory source of nutrients, except for iron for growing crop plants in hydroponic culture under pilot plant commercial conditions. Commercial culture of a number of glasshouse crop plants could be done without toxicity from salt or other materials. The crops were grown at energy savings compared with standard crop culture as it was not necessary to use manufactured fertilizers. Water which was otherwise wastewater was utilized in the process. To avoid adding additional soluble nutrients to the sewage water, iron was supplied either as foliar spray or as insoluble iron oxide applied to plant roots before and after transplanting. The latter procedure was most satisfactory. After the water had been used in the hydroponic tanks, much of the copper, zinc, manganese, iron, nickel, lead, and some of the nitrogen, phosphorus, and potassium had been stripped from the water. In arid regions of the world an on-line hydroponic culture system could be a very important aspect of a water recycling program. Economic benefit and water purification could result.

INTRODUCTION

There is an urgent need for water recycling in urban areas in many parts of the world (Kasperson and Kasperson).[1] There are four good reasons for recycling of urban wastewaters in underdeveloped arid regions: (a) Scarce water resources can be stretched; (b) Food production can be increased; (c) Jobs can be provided through the procedures used; (d) Fewer environmental problems will be encountered.

There are also four good reasons why hydroponic crop production as described in this report can be advantageously coupled with an urban wastewater recycling program.

(a) Wastewater from urban areas contains nutrients which, with the proper flow rate, can supply all the fertilizer elements needed by crop plants grown in hydroponic culture in reasonably good proportions except for iron, which can easily be supplied without adding to the salt burden of the water.

(b) Crop plants deplete or strip nutrients from the water which makes the water more suitable for the next stage of water renovation, which could even be additional hydroponic culture. Other types of aquaculture could follow in turn (Gordon).[2]

(c) The economics of the hydroponics stage or stages could pay for other stages of the water renovation process.

(d) Valuable food products could be produced especially in off seasons. An alternate to food products could be floricultural crops which could be even more valuable than food crops.

A generalized scheme for the use of hydroponic plant growth procedures in wastewater treatment processes is shown in Figure 1. The wastewater can be used for irrigation at various stages or it can have other domestic and urban uses as shown in the diagram. One of the research goals has been to produce water, through a series of hydroponics uses, which is of acceptable chemical quality for urban use. Other procedures can be used to make the water of acceptable microbiological quality.

PREVIOUS WORK

A plastic greenhouse 9.2 m^2 was constructed at the Tapia plant of the Las Virgenes Municipal Water District in Los Angeles County, California, for the initial pilot plant study. This district serves a suburban area interspersed with a few "clean" industrial plants. The pilot greenhouse, approximating one-fifth of a commercial vegetable greenhouse commonly found in California, was equipped with space heaters, evaporative cooling, and the necessary controls for maintaining minimal environmental conditions. The house contained four hydroponic growth beds (tanks) 6.2m long, 0.7m wide, and 0.23m deep. Secondary treated wastewater which had been chlorinated and subsequently dechlorinated with sulfur dioxide was piped in under pressure. The macronutrient concentration of the effluent on the average was 17 µg/ml NO_3-N, 32 µg/ml P, 13 µg/ml K,

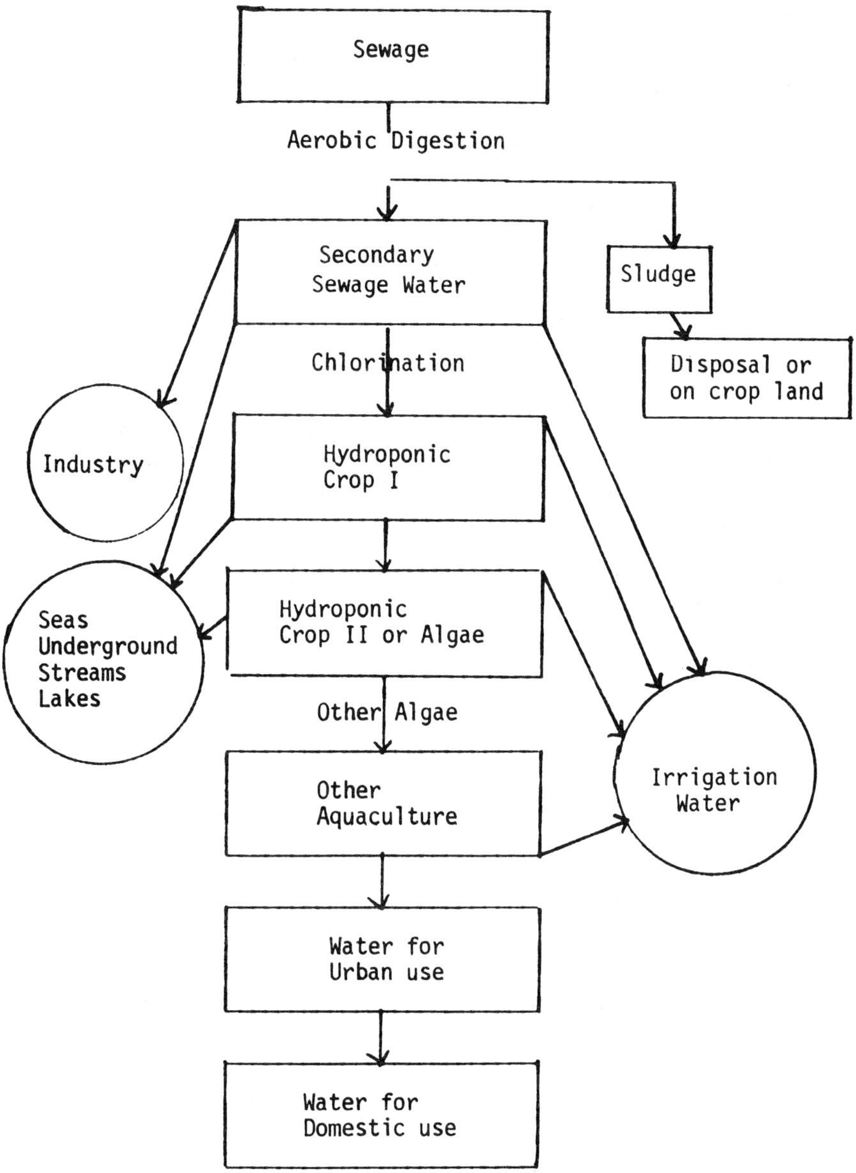

Fig. 1. Flow Diagram: Use of urban wastewater for hydroponic crop production.

66 μg/ml Ca, 33 μg/ml Mg, and 11 μg/ml S. The effluent additionally contained a number of trace elements in the concentration range normally found in domestic effluent waters including 0.04 g/ml Cu, 0.1 μg/ml Zn, 0.06 μg/ml Ni and 0.007 μg/ml Cd. The standard flow rate used in this operation was approximately 27,000 liters per day through each hydroponic tank under gravity flow.

The effluent was used directly as it came from the renovation plant, except for the addition of iron to correct an iron deficiency that developed because of the high pH (about 7.6) of the wastewater which rendered the existing iron unavailable. Preliminary experiments in containers had confirmed that iron deficiency would be a problem in the unamended water, but that it could be corrected with foliar spray, acidification of the solution bathing roots, or by solid phase iron added to the nutrient solution. In this pilot operation the iron deficiency was controlled by a weekly application of a slurry of commercially available insoluble iron oxide applied directly to the roots. The trace elements associated with the effluent seem to be strongly absorbed to the roots of the plants (Patel *et al.*),[3] but without apparent injury to the plants as long as iron was supplied. Although there were high concentrations of these heavy metals on and/or in the roots there was little translocation to the aerial portions and even less to the fruit. For example, zinc concentrations in the plants were high, but within a normal range.

The growth of tomatoes, cucumbers and chrysanthemums in the pilot hydroponic greenhouse was comparable to commercial production. Long term crops of tomatoes and cucumbers did show more than the usual stress at the end of the season. The lettuce also grew well but was not grown on a scale sufficient to evaluate commercial production.

PRESENT STATUS OF HYDROPONIC CROP PRODUCTION WITH SEWAGE WATER

The present status of the research on use of sewage water in hydroponic crop culture in the U.S.A. (Patel and Wallace);[4] Berry *et al.*;[5] Fromhagen and Mah;[6] Wallace *et al.*;[7] Wallace and Soufi[8]) is as follows:

(a) Commercial equivalent production of cucumbers, tomatoes and chrysanthemums has been achieved in the (9.2 x 9.2 meter) greenhouse.

(b) The secondary stage sewage water supplies all the nutrients needed except iron which has been most satisfactorily supplied by the use of solid phase Fe_2O_3 applied near the roots. This procedure then adds no nutrients to the water.
(c) Trace metals seemed to be stripped more effectively from the water than was the N, P and K which means that water can be even more effectively used in other processes after the hydroponic crop stage.
(d) Acidification of water does not seem to be a satisfactory means of making iron available to the plants even though it does make the iron available. Other elements seem to go into imbalance with this procedure.
(e) Intensive cropping can deplete nutrient elements in reclaimed sewage water to levels of some domestic waters.
(f) The virus aspects of use of reclaimed sewage water for hydroponic crop production has been studied in some detailed work at UCLA (Fromhagen and Mah).[6] Several different types of applied viruses were found associated with plant roots, but none were transported from roots to shoots or to fruits of plants. This is in keeping with studies of large molecules in general and suggests that the viruses are on root surfaces or free space of plant roots. If adequate processing and treatment of water is undertaken and if water does not contact edible parts, no problems are envisioned. Edible root crops will not, or should not, be grown in the system.

In further studies, reclaimed secondary-treated municipal wastewater from the Tapia reclamation plant of the Las Virgenes Municipal Water District in Los Angeles County, California was used to grow commercial horticultural crops. The partitioning of trace elements among the wastewater, roots, stems, and fruit was determined. The distribution within the plant in relation to the roots was such that calcium, magnesium, potassium, sodium and boron were concentrated in the fruit while phosphorus and most of the trace elements were excluded to some degree from the fruit. For example, there was two times as much calcium in fruit as in the root, but 800 times less Mn in the fruit than in the root. In plants grown in this particular wastewater, cobalt, nickel, cadmium, chromium, and tin were not detected in the fruit, but were present in significant concentrations on and in the roots.

Flow rate studies were run at 4, 10 and 20 liter/min/bed of cucumbers over a 10-week period. The highest flow rate gave the highest yield of both plant material and fruit, but the two lower flow rates were not constantly lower. Tissue analysis did not indicate any nutrient deficiencies even at the lowest flow rate. It appears that even at the lower flow rate the mineral nutrients were not limiting; however, other physical parameters such as oxygen availability could become limiting before the mineral nutrients.

Previous greenhouse pot cultures had indicated that iron could be made available in wastewater with pH adjustment. This was tested in the pilot-size greenhouse at Tapia by lowering the pH of one of the beds from 7.6 to 6.2 with H_2SO_4. The iron deficiency symptoms were eliminated, but the growth was not comparable to the controls; some abnormal symptoms remained. The plants had a general loose form showing symptoms similar to those of phosphate deficiency but in addition had interveinal necrotic areas on the old and mature leaves similar to those attributed to heavy metal toxicity by Hewitt .[9]

Although the pilot project demonstrated that commercial horticultural crops could be grown using municipal waste water as the source of nutrients and water, the limits of many important cultural parameters have not yet been determined.

In order to make the technique fully practical on a commercial scale while simultaneously improving the water quality the following recommendations are suggested pending further study:

(a) Do not use the green leafy parts of vegetables for food; rather use only fruit or similar structures.

(b) If heavy metal removal from water is of concern, grow short-cycle crops such as cucumbers where plants and roots will be frequently discarded.

(c) Boron is present in most wastewater at chronic toxicity levels, and efforts should be made to reduce it through improvement of soaps and detergents for the urban population.

(d) Determine the contribution of particulate matter to the trace element burden of the effluent water and its relationship to the efficiency of trace element uptake by hydroponic crops; procedures may then be developed for decreasing the burden.

(e) Determine the relationship of flow rate to nutrients removed by hydroponic crops.

(f) Determine the degree of trace element binding by the organic fraction of effluent water -- a fraction that may not be utilized by plants.

RESEARCH QUESTIONS THAT YET NEED ANSWERS

Research is yet needed in this area to be fully able to produce high quality crops and still produce useful water for urban uses. Some of these problems are listed.

(a) How to prevent occasional phytotoxic substances from appearing in the water or to minimize this effect.
(b) How to maximize nutrient-element stripping from the wastewater without sacrificing yield and quality of crop.
(c) How to screen all pathogens from the wastewater prior to water use for the hydroponic crops.
(d) How to adjust flow rates of wastewater to get maximum benefit from quantities of nutrients present.
(e) How to develop crop sequencing to take advantage of changing ratios and quantities of nutrients.
(f) How to insure adequate levels of oxygen in water for growth of specific crops.

WATER QUALITY STANDARDS

Each jurisdiction will need to set water quality standards for its area. Chemical as well as biological standards are needed. Biological recycling of water involving hydroponics could provide an opportunity for upgrading of standards in many nations. Standards for procedure are also needed. For example, in California when wastewaters are used for crop irrigation, legislation prohibits spraying of water onto edible parts of plants. Root contact with such water is satisfactory if above ground parts are to be consumed.

ACKNOWLEDGMENTS

This pilot study was supported, in part, by a grant from the Rockefeller Foundation to the Institute of Evolutionary and Environmental Biology (Malcolm S. Gordon, Principal Investigator) and the Laboratory of Nuclear Medicine and Radiation Biology (DOE Contract 04-1 Gen 0012) at the University of California, Los Angeles. The National Research Council of Lebanon provided a research grant for study in that country (M.S. Soufi, Principal Investigator).

REFERENCES

1. R.E. Kasperson and J.X. Kasperson, Eds. (1977). Water Re-use and the Cities. University Press of New England, Hanover, New Hampshire.

2. M.S. Gordon (1977). Biological recycling of dissolved nutrients in treated domestic wastewaters using hydroponic and aquacultural methods. In, Wastewater Renovation and Reuse. F.M. D'Itri, Ed.

3. P.M. Patel, A. Wallace and R.T. Mueller (1976). Some effects of copper, cobalt, cadmium, zinc, nickel, and chromium on growth and mineral element concentration in chrysanthemum. J. Amer. Soc. Hort. Sci. 101(5):553-556.

4. P.M. Patel and A. Wallace (1976). Correction of iron deficiency in tomatoes grown hydroponically in sewage water. Comm. Soil Sci. Plant Anal. 7(1):65-75.

5. W. Berry and A. Wallace (1977). Hydroponic crop culture as a simultaneous method for both treating and recycling secondary treated municipal wastewater. Hort. Sci. 12 (cover page).

6. L. Fromhagen and E. Mah (1977). Water and product quality. In, Wastewater Renovation and Reuse, F.M. D'Itri, Ed.

7. A. Wallace, P.M. Patel, W.L. Berry and O.R. Lunt (1978). Reclaimed sewage water: a hydroponic growth medium for plants. Resource Recovery and Conserv. 3:191-199.

8. A. Wallace and M.S. Soufi (1977). Water quality studies with hydroponic crop culture and reclaimed sewage water (Unpublished data).

9. E.J. Hewitt (1966). Sand and Water Culture Methods Used in the Study of Plant Nutrition. Commonwealth Agric. Bureau, Farnham Royal, Bucks., England.

ENERGY FLOW FOR AGRICULTURAL ECOSYSTEMS IN DESERT ENVIRONMENTS

GERALD W. THOMAS
President, New Mexico State University, Las Cruces

ABSTRACT

A thorough analysis of energy flow patterns should provide a better basis on which to plan desert land use and development. Such an analysis must take into consideration two major energy systems:

(1) First, the capture of solar energy by vegetation through the process of photosynethesis (food energy), the movement of this energy through ecosystems, and the ultimate utilization of a small fraction of this photosynthetic energy by man as a good or fiber product.

(2) Secondly, the flow of "cultural energy" (energy subsidies) required to "run" the food and fiber ecosystems. This latter source of energy includes manpower, horsepower, hydroelectric power, large amounts of fossil fuels, and certain other energy subsidies.

Understanding the complicated interrelationships between these two energy flow patterns is not only important to energy conservation and efficient utilization, but, indeed it may be the key to man's survival as a viable organism on the planet Earth.

This paper considers the efficiency of photosynthesis on deserts for both cultivated and rangeland ecosystems and how man influences the vegetation capture capability and ultimate utilization of this "food energy." It also presents data on "cultural energy" requirements for agriculture in typical developed and developing nations.

For the next two decades at least, it appears that the availability and cost of "cultural energy" (particularly fossil fuels) will be the most limiting factor in food and fiber production. Therefore, "appropriate technology" systems designed to adjust to available energy sources is one important key to proper development.

DUAL ASPECTS OF ENERGY ANALYSIS

The role of energy in the world's food and fiber ecosystem can be analyzed by examining the two major energy flow patterns: (1) First, the capture of solar energy by vegetation through the process of photosynthesis (food energy), the movement of this energy through ecosystems, and the ultimate utilization of a small fraction of this photosynthetic energy by man as a food and fiber product; (2) Secondly, the flow of "cultural energy" (energy subsidies) required to "run" the food and fiber ecosystems. This latter source of energy includes manpower, horsepower, hydroelectric power, large amounts of fossil fuels, and certain other energy subsidies.

Understanding the complicated interrelationships between these two energy flow patterns is not only important to energy conservation and efficient utilization, but, indeed it may be the key to man's survival as a viable organism on the planet Earth. Energy conservation and use is closely related to water efficiency and nutrient cycling patterns.

All life is supported either directly or indirectly by the solar energy captured primarily by vegetation in the process of "photosynthesis". This chemical reaction involving carbon dioxide, water and sunlight energy to produce food and release oxygen and water (through transpiration) is the most important chemical reaction in the world. Even the fossil fuels (coal and petroleum) resulted from over 400 million years of photosynthetic activity. One estimate states that, on all of the land areas on earth, some 16 billion tons of carbon each year are fixed by photosynthesis. How can man influence this total capture capability? And, once the carbohydrates are formed, how can man influence the distribution process? These are questions that relate directly to both the "food" energy cycle and the "fuel" energy problem.

In his primitive condition, man was merely one organism in a complex natural ecosystem. He fought for his life, collected wild plants for food, occasionally captured and ate other animals, and was beset with numerous diseases and the multiple problems of most competitive life forms. Under these conditions, the carrying capacity for the earth has been estimated as 10 million, a population smaller than that of London or Tokyo today.[1] As man learned to harvest surplus food, domesticate animals, and cultivate crops, the carrying capacity of the land (for man) increased.

The first major increase in carrying capacity resulted from the diversion of "surplus" photosynthetic energy directly to man from other biological organisms. In other words, man became more competitive in the natural, but still somewhat "wild" environment. The next important development, in terms of increased food production, was the utilization of techniques for planting and crop cultivation. The third major breakthrough in carrying capacity came about when man found that he could "subsidize" the system with fossil fuels and thus increase the effective harvest of food and fiber.

Consequently, through the diversion of surplus photosynthetic energy, through cultivation, through fossil fuel subsidies, and by other technological innovations, total food and fiber production on the earth has been increased many-fold. How many people, then, can the earth support? One representative to the World Food Conference in Rome stated that the world could now easily support 40-50 billion people.[2] This high estimate was shocking to most of us who have been worrying about the 6-7 billion projected by the United Nations for the year 2000.

Whether we look at energy resources, land use, water or other requirements for food and fiber production, all nations must become more concerned about population growth and associated technological development.

ENERGY FLOW ON ARID AND SEMI-ARID RANGE LANDS

A complete analysis of energy flow patterns on range lands must take into consideration the capture of energy by photosynthesis, the dissipation pattern of energy flow, and the inputs of outside or "cultural" energy to harvest and utilize the vegetation. Figure 1 presents a concept of energy flow on uncultivated lands.[3] All the photosynthetic energy captured by the "producer organisms" is eventually dissipated or consumed by animals (including man) and the organisms of decay.

The first step in the process is energy capture by photosynthesis. The statement has been made that, on the average, about one percent of the sunlight energy falling on the earth is captured by vegetation. The IBP studies show that one percent is much too high an estimate for arid and semi-arid areas.

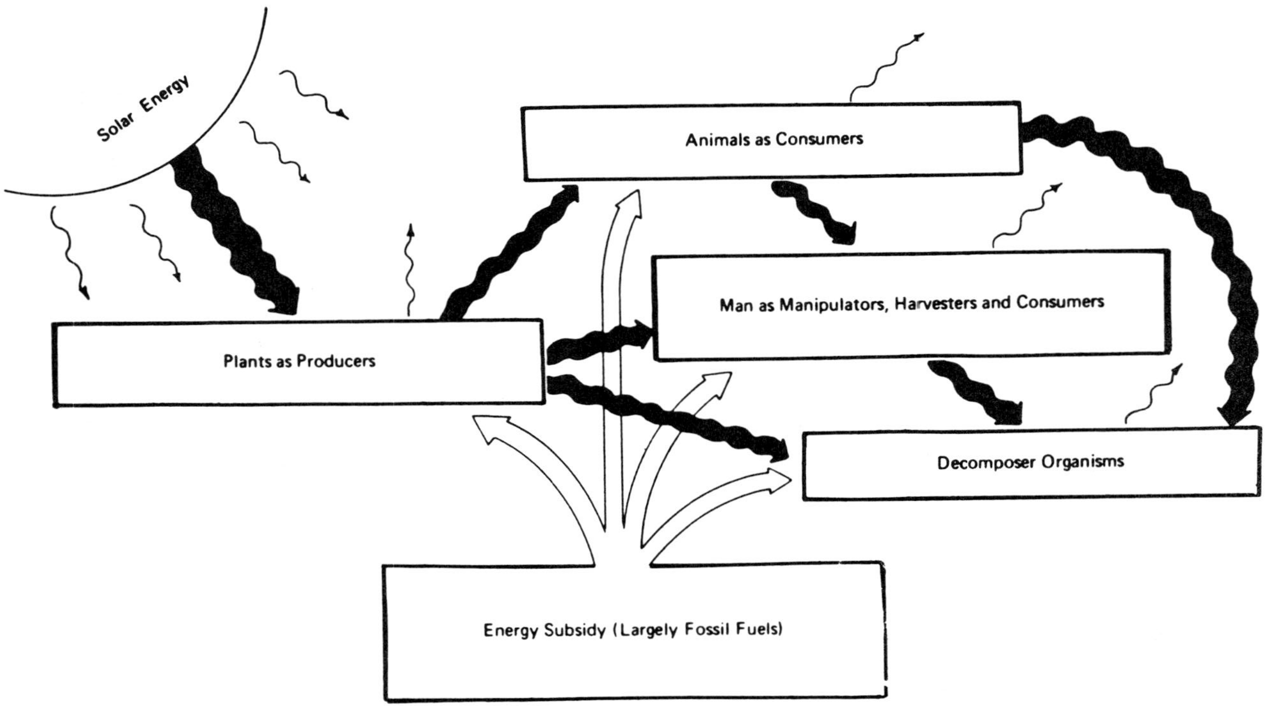

Fig. 1. A schematic diagram of solar energy capture and utilization on uncultivated lands. Fossil fuel energy subsidies are relatively small compared with intensive crop production.

Van Dyne reports that only about 47 percent of the solar input on an area is of wavelengths useful for photosynthesis.[4] At the Pawnee Short-grass Prairie Site in Colorado, USA, only 0.3 percent of the usable radiation was finally captured by vegetation growth.[4] This contrasts with 0.5 percent capture for mixed prairie under higher rainfall conditions in Canada. For the desert biome near Las Cruces, New Mexico, the efficiency of utilization of solar energy ranged from 0.10 percent down to 0.03 percent over a three-year period (Pieper *et al.*).[5] In these studies, efficiency of energy capture was less in grazed pastures than on those protected from livestock grazing.

Once the energy is captured by the vegetation (producer organisms), the dissipation process begins. Respiration and growth take place, and primary consumers feed on the vegetation. Here again, some interesting results are being developed by the IBP studies that change, rather substantially, some of the old assumptions. For example, termites, in both the desert and grassland biomes, appear to be far more important in vegetation harvest than one might assume. In our studies in New Mexico, termites consumed 10 times as much biomass energy as livestock, and these termites are not nearly as visible or as active as some termite colonies in Africa, South America or Asia.

In addition to termites, there are large amounts of energy diverted to other forms of insects and microorganisms at all stages in the food chain. In many of the United States' range ecosystems, the amount of green herbage consumed and detached by grasshoppers appears to approach or exceed the amount ingested by livestock (Van Dyne).[4] And, of course, in certain years, range caterpillars or locusts consume virtually all of the available vegetation energy as they move through an area.[6]

Van Dyne[4] states that a characteristic of U.S. range ecosystems "which seems surprising is the relatively small proportion of net primary production that enters the grazing food chain. Thus, when one summarizes the energy flow through the Shortgrass Prairie system, of the total solar energy input in terms of the five-month growing season, cattle capture only 0.0003 percent -- but this is a very important and tasty percentage!"

With primary production levels near Las Cruces ranging from 500-744 (Kcal/M^2), diversion of vegetation to cattle was only 1.9 - 3.8 percent. Utilization by livestock on the grassland biome was somewhat higher, but the fact remains that the utilizable energy on most of these rangelands is less than most of us had assumed. Energy going to rodents and rabbits is high -- in some years reaching half of the amount utilized by livestock. How can we divert more solar energy to human food through livestock production? This question also deserves careful study and analysis.

Further reductions in energy are shown in Figure 2, which presents a concept of energy partitioning and disposition by ruminants (Hyder).[7] Cook[8] reported that Utah ranchers could sell, in a good cow-calf operation, about 10 percent of the energy consumed by the animal. A study in California showed that 1/1000 of the radiant energy falling on an area was available to cattle, and only 1/40,000 of the original energy reaches the food product, meat (Love).[9] How can energy be saved in this process? An examination of grazing habits of livestock, the digestion process (including consideration of additional "super bug" experiments in the rumen), slaughter and packing house techniques, grading standards, and consumer habits will readily show vast opportunities for improvement. An outstanding paper on the subject of energy conservation by livestock was presented by B.P. Carbon at the February 1974 AAAS meeting in San Francisco.[10]

In an excellent article entitled, "Increasing Red Meat from Rangelands," Box[11] emphasizes that the range resource will shrink throughout the world due to demands for other uses of the land. Furthermore, he concluded that "in developed countries, a major effort will be put in two areas -- increasing forage supply and intensifying animal management."

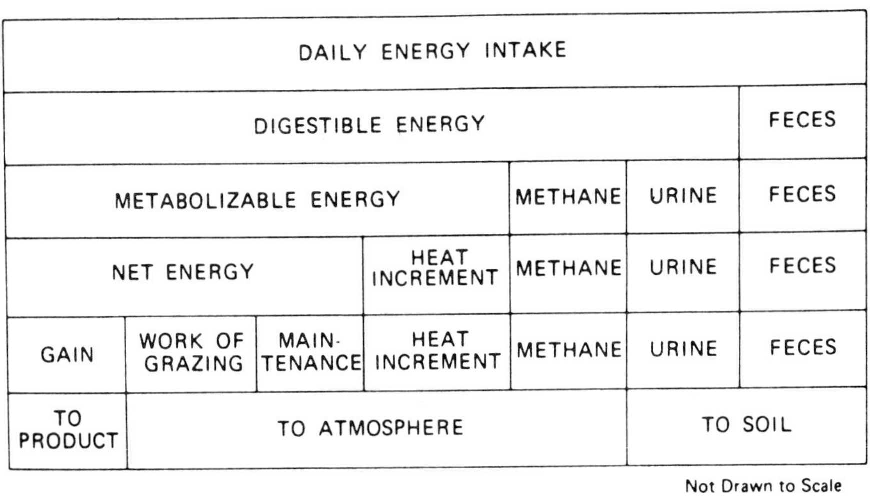

Fig. 2. Energy partitioning and disposition by ruminants (from Hyder, 1969).

Recently, new interest has been generated in the possibilities of game ranching. P.A. Garakis states:[12] "One approach to more effective use of material and energy resources in arid and semi-arid pastures is game ranching. Some wild species may provide considerable meat per unit area and per unit water. Perhaps because of their coevolution with the vegetation, less grazing damage might occur with such animals, and game ranching may be an ecologically sound alternative to some practices".

What are the energy implications of such a practice, and how can we divert more energy to those game animals which are subject to harvest?

To return to the energy-flow diagram, we should not neglect the direct transfer of energy to man by the eating of fruits, tubers, or other parts of range plants, and/or direct use of insects or micro-organisms for food. This was the accepted way of life for primitive cultures and is still important for some native tribes in Africa and Australia. The total biomass tied up in insect population at any given time may be quite variable since certain types of insects multiply rapidly under some trigger mechanisms and remain rather dormant under other situations. Certain aphids, for example, will reproduce every eight days if the conditions are right (even with no males in the population at their location). Also, some insects may consume three to four times their body weight daily, while cattle and bison consume only about three percent (dry forage) of their body weight. These factors must be considered in energy harvest by man. While we recognize that the primary source of food for humans on rangelands is at present the ruminant animal, with its ability to convert roughage to proteins and energy, the long-term needs of the world will force a reexamination of some of the old food-gathering customs. In addition, new research on the production of yeast and other micro-organisms from ground mesquite (research at Texas Tech) or tumbleweeds (research at New Mexico State University) opens up new opportunities to increase food yields on uncultivated lands. No possibility can be neglected as population pressure increases.

Dr. Thadis Box,[13] Past President of the Society for Range Management, states, "Efficiency of animals may dictate that we raise goats, camels, donkeys, antelope, rabbits or kangaroo rats on the rangelands if we are interested in greatest efficiency of protein production".

The "cultural energy" requirements for range livestock operations varies widely in different parts of the world. In many areas, most of this energy input comes from manpower, thus drawing on vegetation energy rather than fossil fuels. Comparative data for different range ecosystems in most desert range areas is very limited.

Recent research on range lands in the U.S. indicates that cultural energy requirements, *largely fossil fuels,* is much higher than is generally believed. Cook[14] reports that range meat production requires more subsidized energy than cereal crop production when compared on the basis of kilocalories of cultural energy expended for each kilocalorie of food produced. Choice grade lamb from U.S. western ranges can be produced at an expenditure of 2.25 kcal of cultural energy per kcal of dressed carcass compared to a weaner calf that requires about 2.09 kcal of cultural energy per kcal of dressed meat. Cook reports that the expenditure of cultural energy to produce a dressed carcass in feedlot operations was 5.33. Mechanized equipment on U.S. ranches consists primarily of motor vehicles, tractors and sometimes aircraft for brush or insect control. Cultural energy costs also include buildings, fences, fertilizers and water development.

The net energy analysis is not complete, however, until the livestock product reaches the consumer. This usually means fossil fuels for processing, transportation, storage, curing, cooking and delivery to the table. We are just beginning to assemble data on these "cultural energy" costs. Also, the range livestock industry, particularly in the U.S., is very dependent upon cultivated lands for supplemental feeds and certain other inputs. "In the developing countries of the world, production per animal unit -- whether it be for food or fiber -- is much lower than it is in the developed world, simply because an animal in a developing country uses a higher percentage of its feed for body maintenance...Thus, less of its feed intake is converted to products useful to man."(Hodgson)[15]

CULTIVATED LAND ECOSYSTEMS

A significant breakthrough in the efficiency of sunlight energy "diversion to man" was made when crop cultivation was developed. Man could now select and develop crop plants for his specific needs and concentrate cultural practices on maximum production. Under modern intensive cultivation systems, the efficiency of sunlight energy capture for some crops has exceeded 3 percent, while the theoretical efficiencies of conversion have been calculated at 5.3 percent for total energy and 12 percent for visible light radiant energy (Loomis and Williams).[16] This contrasts with capture capabilities for mechanical collectors which have tested efficiencies up to 50-70 percent. Mechanical collectors capture heat only, whereas vegetation captures the energy in complex chemical form. Laboratory studies at New Mexico State University foresee photosynthetic capture capabilities for selected strains of algae under greenhouse conditions and in a CO_2 enriched atmosphere at nearly 5 percent. This efficiency is possible in desert areas with high levels of solar radiation.

After capture by the vegetation, the food energy flow varies from country to country. Post-harvest losses take place during all stages of storage, processing and delivery to the consumer. "It is tragic that on a worldwide basis over one-third of the food produced is lost through insects and spoilage before it reaches the consumer." (Thomas, *et al.*)[17] The improvement of food storage, processing and distribution systems in all nations is an important as increasing production at the farm level. This fact is becoming more apparent as energy analyses, crop by crop, become more sophisticated.

An examination of energy flow patterns on cultivated lands reveal some interesting historical trends. In some of the developing countries of the world, about 30-40 percent of the energy input to run the system comes from manpower or oxenpower. Even in the United States, in the early 1900's, there were about 27 million horses and mules on the farms and ranches. Through the years, farmers in the United States have steadily mechanized and have substituted more than 5 million tractors and many other forms of power equipment for about 22 million of these horses and mules (see Fig. 3). As a measure of progress, we have released over 80 million acres of land that would have been required to feed the horses and mules and may now be used for direct food production for humans (Gavett).[18] In addition, we have increased efficiency and output per acre. Years ago, the man with a good team of horses would plow about two acres a day. Today, mechanized power makes it possible for him to plow over 100 times that much.

However, as a result of this increase in mechanization on croplands in the United States, energy flow patterns have been significantly changed. Horsepower, mulepower, oxenpower, and manpower operate on the solar energy collected by vegetation -- a continuing resource for all practical purposes. Tractors

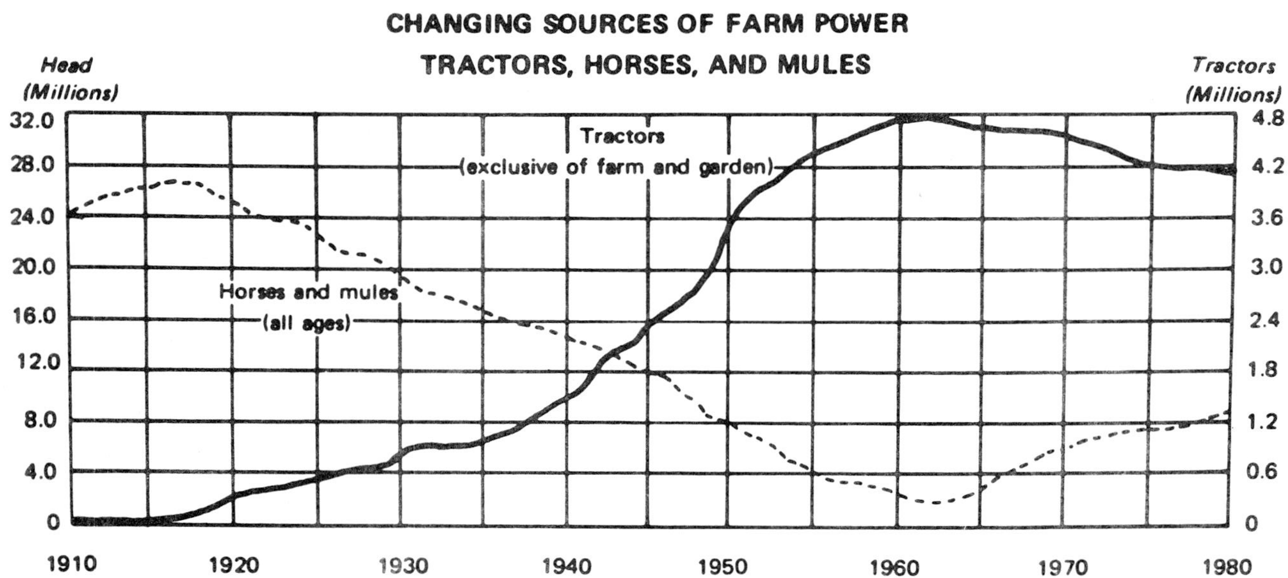

Fig. 3. Since 1918, American farms have moved rapidly from dependence upon horses and mules for power to tractors and other forms of mechanized equipment. The present number of horses and mules is back up to about eight million, nearly all dedicated to recreation purposes (USDA data).

and machinery utilize fossil fuel -- a finite and depletable resource.

The trend toward mechanization on farms is not confined to this country. In 1950, FAO estimated that there were about 6.1 million tractors in the world. By 1970, this number had exceeded 15.5 million (Brown).[19]

In view of the worldwide energy shortage, it appears logical to slow down the trend toward mechanization in agriculture -- particularly in the Less-Developed Countries. Also, most of these LDC's have an abundance of cheap labor. Therefore, a new term has appeared in international literature -- "appropriate technology." This term, appropriate technology, implies that the rest of the world need not adopt the high level of mechanization used in the U.S., but rather should develop "labor intensive" methods. This sounds good in theory, but, as a recent National Research Council Report states:[20] "no panaceas or quick remedies for the problems of development are to be found in the choice of particular types of 'intermediate' technologies."

This NRC report points to the need for additional research relating to energy saving approaches for the developing countries. Meanwhile, in the U.S., it will be difficult, if not impossible, to move back toward more labor intensive techniques. For example, one recent study pointed out that at present costs, in order to substitute labor for fossil fuels, the human labor must be valued at nearly zero.(Hill, Erickson)[21]

The energy crisis also rather shockingly focused the world's attention on fertilizer and the other petrochemicals needed for food and fiber production. Since the bulk of our nitrogen fertilizer comes from natural gas, both shortages and incresased costs have created problems worldwide. According to Pimentel,[22] nitrogen fertilizer is the largest single energy input in crop production. The genetic potential of the new crops developed as part of the the Green Revolution cannot be realized without optimum inputs of fertilizer and water. It has been estimated that the fertilizer shortage in 1973 in India alone reduced grain production by 10 million metric tons, despite good weather.(Pimentel)[22] World fertilizer use increased from 15.2 million metric tons in 1950 to over 70 million metric tons last year. (Brown)[19]

Figure 4 presents a schematic diagram of energy flow for cultivated ecosystems. Although solar energy capture takes place only at the farm level, fossil fuel energy subsidies are required throughout the system. Energy is consumed in the

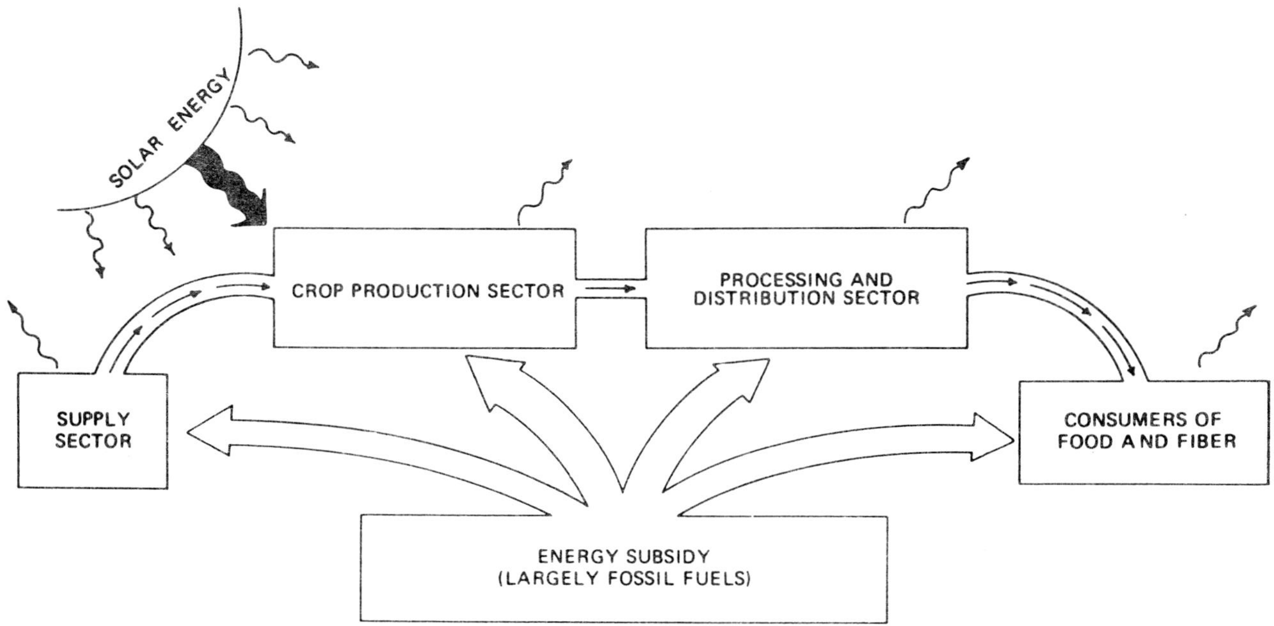

Fig. 4. An illustration of energy flow and utilization for cultivated land ecosystems. Fossil fuel subsidies are relatively high in order to capture and deliver solar energy to the consumer.

preparation of agricultural chemicals, fertilizers, farm machinery, seed and other supplies. Energy is also consumed in the process of planting, cultivation, and harvesting; and large amounts of energy are consumed in processing, storage, packaging, and transportation before the finished product reaches the consumer level.

In the United States, one estimate indicates that over 10,000 calories of fuel are required to help capture 3,000 calories of food and place it before the consumer. (Thomas)[23] Other estimates of fossil fuel subsidy range up to 10 or more calories of fossil fuel for each food calorie delivered to the table. (Odum)[24] Even in rural India, about twice as much fuel energy is used in cooking a unit weight of rice as there is food energy in the rice. (Revelle)[25]

One of the difficulties in arriving at accurate statistics on agriculture and energy relates to the definition of "agriculture." This accounts for part of the variability in estimates. However, for the "production" or "on-farm" sector, we do have some fairly good studies which show that in the relatively primitive rice cultures of the Philippines, about 16 calories of digestible energy (food) resulted from each calorie of "cultural" energy input. In this case, the "cultural" energy was hand labor. (Heichel)[26] As mechanization has developed, the ratio of "cultural" energy input to digestible energy output has increased and thus, the net caloric gain has decreased. In recent calculations, Heichel and Frink[27] calculated that the total cultural energy required to deliver our food only to the farm gate in 1972 was about 4944 kilocalories per person per day. They stated that 73 percent of this cultural energy requirement was used to produce the 1380 kilocalories per day of animal products as a part of our total diet which is now about 3400 kilocalories per day.

Energy inputs for corn production in the United States tripled from 1945 to 1970. (Kornberg)[28] An estimated 742 liters of fossil fuel equivalents currently are utilized to raise a hectare of corn. For crop "production" to feed a world population of 4 billion (employing modern intensive agricultural techniques), it would require an energy equivalent of 1700 billion liters if the off-farm sector is included. The Cornell group concludes that, "certainly, United States agricultural practices could not produce enough food to feed a world population of 7 billion with current energy limitations even if capital were unlimited. (Pimentel)[29]

Special mention should be made of the importance of "irrigated agriculture" and special dry farming techniques for water use and conservation. Where water is the most limiting factor in cultivated crop production, as is true of desert and semi-desert areas, more advanced techniques are being promoted to make more effective use of water. In general, most of these water conservation measures require larger inputs of "cultural energy" -- usually in the form of fossil fuels.

While land irrigated with pumped water represents only about 10 percent of the harvested acreage in the U.S., this land accounts for over 25 percent of the total farm sales. (Sloggett)[30] In the 50 states we now have about 40 million acres under irrigation. (Thomas)[17] With the advent of the rather sophisticated center-pivot sprinkler systems, irrigation has moved rapidly into the Mid-West, with Nebraska now being the third "most-irrigated" state. There seems to be no question that sprinkler systems increase the efficiency of water use. However, the "energy cost" of the center-pivot system alone is the equivalent of adding over 200 more feet to the depth of the well, which may double the fuel requirements.

BIOMASS CONVERSION AND "FUEL FARMING"

As the price of petroleum increases, the argument over whether to go the "food" route or the "fuel" route for the utilization of plant biomass and organic wastes is becoming hotter. Studies are now underway to examine tropical forests and brush lands for their fuel production potential. Also, certain cultivated crops under intensive management can be used for the production of an end product that may be competitive as a fuel or as a source of methane, ethanol or other alcohols, to power mechanized equipment.

Harvesting of forests and woodlands for fuel is certainly not a new concept. Historically, some of the earlier "energy crises," resulting sometimes in mass migrations of people, were assoicated with "over-utilization" of brush and trees for fuel. What is new, however, is the re-examination of the "sustained-yield" concept for wood energy production and new techniques for concentrating and packaging -- including possible refinement to alcohol.

The concept of "energy farms" has stimulated new enthusiasm over possible new crops, farm practices and bioconversion techniques. Agricultural scientists are particularly interested in alfalfa, sugar cane and some of the cereal grains.

Some people are projecting yields in excess of 200 tons per acre per year by maximizing photosynthesis. (Bassham)[31] New plants are also coming under examination, particularly the *Euphorbia* species. Nobel Laureate Melvin Calvin has estimated that such plants might produce the equivalent of 20-125 barrels of oil per hectare per year at a cost of $10 or less per barrel.[32] Our own research in cooperation with the Los Alamos Scientific Lab, points to good possibilities for algae production and bioconversion utilizing some recently discovered microorganisms which appear to be good digesters of cellulose. One of our scientists stated recently that if he could transpose his laboratory experiments to the field, he could produce all of the U.S. energy needs by growing algae in a lake in New Mexico which measured 40 by 200 miles (8000 square miles) and subjecting this plant material to aerobic and then anaerobic digestion to produce ethanol or methane.

Incidentally, the solar energy incident on U.S. cropland varies from a high of 260 watts/m^2/year in most of New Mexico, Arizona and parts of California, to a low of 150 watts/m^2/year in the dairy regions of upstate New York, Vermont and Oregon. (Heichel)[33] In a recent article in *Science*, James Bassham[31] of the University of California proposed a system of fuel farming for the U.S. Southwest using huge greenhouses with a carbon dioxide enriched atmosphere. Leaving aside "certain economic, engineering and physiological conditions," he estimates production of 200 metric tons of alfalfa per hectare with a 5 percent efficiency of solar conversion.

Before we get our hopes too high for the possibilities of fuel farming, I think it should be pointed out that most of the economic studies to date have shown that food farming is more economically sound than fuel farming. In other words, unless energy prices rise much more than at present, the best uses for our cultivated lands -- and most of our organic wastes -- is in the food system. And, as the population increases, world food shortages may be even more critical than energy supplies, shifting the emphasis on cropping systems back to food production. It is my hope that the studies of "energy farms" will help us find answers to the impending food production crisis.

THE ECOLOGICAL BALANCE

A word of caution must be inserted as a closing note. While man could undoubtedly tap more photosynthetic material for food and fiber products, or for fuel, there remains a serious concern over the environmental consequences of such major manipulations of natural ecosystems. Man lives in a delicate natural balance with other organisms and the physical factors of the environment. Desert environments are especially sensitive to man's influence. As more people are added to the population base, as the developing nations adopt modern technology, as man continues to deplete non-renewable resources, we move closer and closer to the ultimate limits of the environment. No one yet knows where those limits lie; but, it is our responsibility, as we examine energy alternatives, to be conscious of the fact that decisions made today may be critical to all future generations of mankind. We must find ways and means to sustain our total food and fiber complex on "renewable" rather than "depletable" resources. It is, indeed, later than we think, sooner than we thought!

REFERENCES

1. Lester R. Brown (1970). Human food production as a process in the biosphere. In, The Biosphere. Scientific American, Pub.
2. Gerald W. Thomas (1975). Environmental sensitivity and production potential of semi-arid range lands. In, Frontiers of the Semi-arid World. ICASALS Publ. No. 77-2. Lubbock, Texas.
4. George M. Van Dyne (1978). A grass-roots view of the grassland biome. Unpub. manuscript. Colorado State University, Ft. Collins, Colorado.
5. Rex D. Pieper, et al. (1974). Structure, function and utilization of North American desert grassland ecosystems. Manuscript. New Mexico State University, Las Cruces.
6. Time Magazine, June 19, 1978. War, famine and death: along with other problems, the desert locust is back again.
7. Donald N. Hyder (1970). The impact of domestic animals on the function and structure of grassland ecosystems. Range Science Series No. 2. Colorado State University, Ft. Collins, Colorado.
8. C. Wayne Cook (1970). Energy budget of the range and range livestock. Colorado State University Experiment Station Bulletin, TB 109.
9. Merton Love (1970). The rangelands of the western U.S. Scientific American 222(2).
10. B.P. Cardon (1974). Energy and the production of livestock products. Paper, AAAS annual meeting, February 1974, San Francisco.
11. Thadis W. Box (1974). Increasing red meat from rangeland through improved range management practices. J. Range Management 27:333-36.
12. P.A. Gerakis (1974). Environmental consequences of land use in arid and semi-arid areas. Journal of Soil and Water Conservation 29(4):160-64.

13. Thadis W. Box (1978). Food, fiber, fuel and fun from rangelands. Journal of Range Management 31(2).

14. C. Wayne Cook (1977). Use of rangelands for future meat production. Journal of Animal Science 45(6).

15. Harlow J. Hodgson (1978). Foreword, in Potential of the World's Forages for Ruminant Animal Production, Winrock Report, Winrock International, Morrilton, Arkansas.

16. R.S. Loomis and W.A. Williams (1963). Maximum crop productivity, an estimate. Crop Science 1.

17. Gerald W. Thomas, S.E. Curl and W.F. Bennett. Food and Fiber for a Changing World. Interstate Publishers, Danville, Illinois.

18. Earl E. Gavett. Agriculture and the energy crisis. Paper presented to Conference on Agriculture and the Energy Crisis, University of Nebraska, Lincoln.

19. Lester R. Brown (1973). Population and affluence: growing pressures on world food reserves. Population Reference Bureau 29:2.

20. National Research Council. Appropriate technologies for developing countries. National Academy of Sciences, Washington, D.C. 1977.

21. Lowell D. Hill and S. Erickson. Economic restraints on the reallocation of energy for agriculture. Proceedings of the 1975 Cornell Agricultural Waste Management Conference. Ann Arbor Science Publishers, Inc. Ann Arbor, Michigan.

22. David Pimentel, et al. (1974). Workshop on research methodologies for studies of energy, food, man and environment, Phase I. Cornell University, Ithaca, N.Y.

23. Gerald W. Thomas (1974). Agriculture faces the energy crisis. In, Exploration and Economics of the Petroleum Industry, Vol. 12. Southwestern Legal Foundation, Dallas, Texas.

24. Howard T. Odum (1971). Environment, Power and Society. Wiley-Interscience, New York.

25. R. Revelle (1976). The resources available to agriculture. Scientific American (September).

26. G.H. Heichel (1973). Comparative efficiency of energy use in crop production. Connecticutt Agricultural Experiment Station Bulletin 739.

27. G.H. Heichel and C.R. Frink (1975). Anticipating the energy needs of American agriculture. Journal of Soil and Water Conservation. 30(1).

28. Warren Kornberg (1975). All that unplowed land. In, Mosaic. National Science Foundation, Washington, D.C.

29. David Pimentel. World food, energy, man and environment. Proceedings of the 1975 Cornell Agricultural Waste Management Conference. Ann Arbor Science Publishers, Inc., Ann Arbor, Michigan.

30. Gordon Sloggett. Energy used for pumping irrigation water in the United States. Solar Irrigation Workshop, Sandia Laboratories, Albuquerque, New Mexico. July, 1977.

31. James A. Bassham (1977). Increasing crop production through more controlled photosynthesis. Science 197 (August).

32. Denis Hayes (1977). Biological sources of commercial energy. Bioscience 27(8). August.

33. G.H. Heichel (1976). Agricultural production and energy resources. American Scientist 64 (Jan.-Feb.).

PRODUCTIVE USE OF SOLAR AND WIND ENERGY IN DESERT AREAS

LLOYD P. SMITH
President, Desert Research Institute, Reno, Nevada

ABSTRACT

The combined use of solar and wind energy will be stressed in increasing the productivity of desert areas. The desirable products to be obtained from the desert are food and energy in the forms that can be most easily used locally or that can be economically transported to urban centers.

It is the object of this paper to suggest ways to use a combination of solar and wind energy to increase food production and produce liquid and gaseous fuel in addition to the production of electrical energy in desert areas. Solar energy converted to foods requires water in some form (liquid or vapor). Other conversion processes can be divided into two categories: (1) those requiring water and (2) those that do not. A list of these processes will be given.

Wind energy conversion to water or liquid pumping or to electrical energy will be based on an efficient vertical axis turbine invented and patented by the Desert Research Institute. A program will be suggested in which wind turbines can be used to pump sea water into the desert combined with pumping sewage from urban areas into the desert to provide nutrients and water for producing foods or to produce those types of biomass from which ethanol can be efficiently produced and the biomass residue can be converted to methane gas. The latter mentioned processes solve the energy storage and transportation problems.

Where some water can be made available in the liquid or vapor state a method for making efficient use of solar energy for producing food with the help of a sky radiator will be explained.

Integrated wind and solar systems for overall use in different geographic and hydrological desert conditions will be discussed.

INTRODUCTION

With the large area of Egypt (1,001,450 sq. km) of which only about 40,000 sq. km, i.e., only about 4% of the area, is cultivated and populated, it is no wonder that at this point in time Egypt would wish to make a serious effort to make the desert more productive by using present available technology or that which can be developed in the future specifically aimed at increasing the productivity of the desert region. The reasons for doing this are all the more appealing because of the magnitude of the solar and wind energy resources of the desert.

ENERGY RESOURCES OF THE WESTERN DESERT

It will be of interest to estimate the magnitude of the available solar and wind energy over the Egyptian desert. Since the solar flux is of the order of 1 kw per m^2, the total solar energy incident on the western desert region would be 960,000 sq km x 1 x 10^6 or 9.6 x 10^{11} kw. Assuming 8 hours of sunshine per day, then the total energy per year would be 9.6 x 10^9 x 365 x 8 = 2.8 x 10^{15} kwh per year which would supply the total energy needs of the United States, even if the conversion efficiency was only 1%.

It would also be instructive to make an estimate of the maximum amount of energy that could be extracted from the wind over the western desert based on the performance of the Desert Research Institute's patented vertical axis wind turbine. The maximum theoretical obtainable power from wind with a velocity V assuming that the turbine converted the wind energy to mechanical energy with an efficiency ε is*

$$P_W = .000364 \, V^3 \, d \, \varepsilon \text{ kw} \qquad (1)$$

where V is the wind velocity in meters/sec, d is the diameter of the turbine and ε is the length of the vertical vanes in meters. For estimating purposes, I shall assume that d = 10m and = 50m. Optimistically, the conversion efficiency ε might be taken to be 70% whereupon equation (1) becomes

$$P_W = .1274 \, V^3 \text{ kw} \qquad (2)$$

---

*The Energy Source Book, Center for Compliance Information, Aspen, Colorado, p. 405.

Computing the power developed for wind velocities of 5, 10 and 15 meters per second, i.e., 18 km per hour, 36 km per hour and 54 km per hour, we find that P_W (5 m/sec) = 15.9 kw, P_W (10 m/sec) = 127.4 kw, and P_W (15 m/sec) = 430 kw, respectively.

To estimate the magnitude of the wind power that might be obtainable from the desert, I shall consider an area of the western desert 900 km in the north-south direction and 480 km from Egypt's western boundary eastward. We shall consider a north-south row of turbines spaced 30 meters apart. There would be then 30,000 turbines per row. For each of the velocities assumed above, the total power that could be derived from such a row would be

$$
\begin{aligned}
P_{wr}(5\ m/sec) &= 47.7 \times 10^4\ kw \\
P_{wr}(10m/sec) &= 382.2 \times 10^4\ kw \qquad (3) \\
P_{wr}(15m/sec) &= 1290 \times 10^4\ k2
\end{aligned}
$$

If rows of turbines were spaced 2 km apart over the 480 km west to east so that the wind velocity could be restored from winds aloft, there would be 240 rows. The total power then developed for each of the wind velocities would be

$$
\begin{aligned}
P_{wr}(5\ m/sec) &= 1.14 \times 10^8\ kw \\
P_{wr}(10m/sec) &= 9.17 \times 10^8\ kw \qquad (4) \\
P_{wr}(15m/sec) &= 31.0 \times 10^8\ kw
\end{aligned}
$$

If wind over the desert blows on the average of 12 hours per day, then the array of turbines would convert the following amounts of energy:

$$
\begin{aligned}
E(5\ m/sec) &= 5.0 \times 10^{11}\ kw\ hrs\ per\ yr \\
E(10m/sec) &= 40.2 \times 10^{11}\ kw\ hrs\ per\ yr \qquad (5) \\
E(15m/sec) &= 135.8 \times 10^{11}\ kw\ hrs\ per\ yr
\end{aligned}
$$

Under the rough assumptions made, the total solar energy of 2.8×10^{13} kw hrs and the above total wind energy per year could be compared with the 10^{11} kw hrs per year expected from the Aswan Dam Power Station.*

---

*Area Handbook for Egypt, 3rd ed., p. 274.

POSSIBLE PRODUCTIVE USES OF DESERT ENERGY RESOURCES

The most productive use of wind energy would appear to be the pumping of water from underground aquifers or the transport of urban sewage to the desert to provide nutrients and water for growing plants to be converted to ethanol and methane. The latter use of wind energy combined with solar energy will be discussed later.

The water pumping capability of the vertical axis wind turbine can be readily computed. The energy required to raise v cm^3 or v gms of water to a height of h cm is vh gm cms, or the power required would be vh gm cms/sec which equals 980.7 x 10^{-10} kw. This translates into the fact that 1 kw = 102 kg m/sec = 102 liter m/sec = 27 gal m/sec = 1620 gal m/minute.

Using the turbine power from equation (3), the following pumping table can be calculated:

TABLE 1.

	V=5m/sec	V=10m/sec	V-15m/sec
gals/min	h meters	h meters	h meters
100	258	2064	6966
200	129	1032	3483
300	86	688	2320
400	64	516	1742
500	52	413	1393
600	43	344	1161
700	37	295	995

The results are shown graphically in Fig. 1.

Though I have no idea of how much water might be available in deep aquifers below the western desert, it is evident that with efficient vertical axis wind turbines a great deal of water could be pumped at a small capital cost and an almost negligible operating cost.

A COMBINATION OF SOLAR AND WIND ENERGY TO PRODUCE LIQUID AND GASEOUS FUELS

One of the best ways of converting solar energy into a more useful form is by means of biosystems. It takes a combination of water, sunshine and nutrients to grow plants. It has been shown that a number of crops such as sugarcane, sugar beets, wheat, corn and sorghum can be fermented to convert the plants' natural sugars to ethanol and the plant residue can be converted to methane by means of a bacterial digester. These are valuable fuels in themselves and the process stores the solar energy in chemical form. Such conversion of solar energy into ethanol by means of a biosystem assumes even greater importance because of a new process developed by the Mobil Oil Company which converts ethanol into gasoline directly.

Dr. John J. Wise, vice-president of planning at Mobil Research and Development, explains the conversion process in this way:* Think of ethanol as having a hydrocarbon-like part and a water-like part. The Mobil process wrings out the water and rearranges the remaining hydrogen atoms and the carbon atoms to make gasoline molecules. The key to the process is a unique zeolite catalyst. Zeolites are porous crystalline substances made up of silicon, oxygen and aluminum. An ethanol molecule can easily penetrate the pore system and molecules with 3 to 10 carbon atoms (i.e., in the gasoline molecular range), have no trouble getting out but molecules with more carbon atoms cannot be accomodated by the pore system so they are never formed. The final product is about 85% high octane gasoline (about 95 unleaded), 13.6% LPG and 1.4% light fuel gases. This new process makes the biomass conversion of solar energy very attractive.

The Desert Research Institute is making a feasibility study concerned with the piping of municipal wastewater to reservoirs in the Nevada desert where a biomass conversion installation for producing gasoline and methane would be investigated as well as the growing of animal feed. Irrigation techniques which minimize water consumption and reduce evaporative losses will be utilized. Wind energy may eventually be utilized in this project to reduce pumping costs. This activity would stimulate botanists and plant physiologists and geneticists to develop more rapidly growing plants with a higher sugar content with possible reduction of water.

---

*Popular Science, June 1978, p. 90.

Such a project might be productive in Egypt. As a start, municipal wastewater (sewage, etc.) could be pumped, using wind energy, from urban centers to the near desert where biomass conversion installations could be built to convert the wastewater, the nutrients and the solar energy to gasoline and methane. Should food be more beneficial to the economy of Egypt, the wastewater could be used to convert solar energy into animal foods without any deleterious effects on humans.

At distances too far west of the Nile river to pump urban wastewater economically, wind energy could be used to pump water from deep aquifers to either grow plants to provide sugar and convert this into gasoline by means of the Mobil process, or plants to provide food. In this case chemical fertilizers would in all likelihood have to be provided. It is important to note here that plants are being developed at the University of California at Davis that will flourish in a saline environment and other plants wich will thrive in a dry environment because the leaves inhibit transpiration. The most important plant is the tomato, especially cultivars of *Lycopersicon esculentum*. Because of the large volume produced (8,755,950 short tons produced at a value of $914 million, the largest vegetable crop in the United States) and the large consumption, it ranks at the top in total vitamins and minerals in the U.S. diet. The great diversity among the various species of tomato make it possible to develop species with special characteristics. Of interest here is the fact that *L. Cheesmanii* has a high tolerance to salt. It thrives on the shores of the Galapagos Islands scarcely two meters above high tide.* Tests at the University of California have demonstrated that these forms will survive in sea water. Also, another species flourishes in exceedingly dry habitats in western Peru where the sources of water are fog and mist.

The excellent progress being made as a result of research being carried out by the Desert Institute in the Arab Republic of Egypt** on the saline irrigation and growth of cereal crops in an arid environment is commendable and should by all means be continued.

Dr. Fritz Went, an eminent botanist long associated with the Desert Research Institute has recommended that the growing

*"The Tomato," Scientific American, August 1978, p. 78.

**Cereal Crops on Calcareous Soils. Soil Science Division, ACSAD/SS/P4, Damascus, 1977.

of the gas plant or burning bush (*dictamnus albus*) in the desert be investigated. The plant is a member of the orange family or Rutaceae. It is native to Eastern Europe and is hardy in Nevada. It produces such quantities of volatile oil that on a warm day it may become enveloped in a flame when a lighted match is held some distance away, without harming the plant. In this plant a considerable amount of solar energy is converted into hydrocarbons by means of photosynthesis. Covering the field of such plants would not only permit controlling the amount of solar energy absorbed, the production of terpenes, the reduction of water loss, but would provide the means of collecting the volatile oils. The whole process would not only have the advantage that the plant itself need not be destroyed during its growth cycle while the volatile oils are produced and collected but could continue to produce terpenes throughout the lifetime of the plant.

FURTHER RESEARCH ON DESALINATION METHODS

Methods of desalinating large amounts of sea water are still economically unattractive from an energy standpoint. However, in those arid countries having a long sea coast, research and development of more economical methods of desalination should be persistently pursued, especially those involving solar and wind energy sources. Because of advances made in the synthesis of membranes, reverse osmosis should be kept in mind where the water pressure required would be furnished by pumps driven with wind turbines, thus keeping the cost of the required energy low.

Because of the fact that the water molecule has a considerable electric dipole moment, most salts are dissolved in water as ions. For example, sodium chloride exists in water as sodium positive ions and chlorine negative ions and thus will not precipitate out. From an energy point of view it is cheaper to evaporate the water than it is to provide the forces to remove the sodium and chloride ions. Research should be conducted to pass salt water through an unpolarized liquid such as carbon tetrachloride in which the sodium and chloride ions would recombine and precipitate out.

Evaporated water in the desert could be collected, leaving the salt behind. Such a process could be used to produce potable water pumped by wind energy from the sea coast to areas where there is an abundance of solar energy.

THE PRODUCTION OF HYDROGEN USING SOLAR ENERGY

Research should be conducted for the purpose of storing solar energy by using it to convert water (sea water or otherwise) to hydrogen and oxygen through the following reaction:*

$$H_2O \xrightarrow{\text{sunlight}} H_2 + 1/2\ O_2 \qquad (6)$$

This conversion would be especially appropriate because of the fact that water at least on the sea coast is abundant and cheap. When energy is stored it should be done chemically if possible because it usually represents more energy stored per unit weight. As Wrighton says, "A promising mechanism for accomplishing the above reaction involves the use of so-called photoelectrochemical cells, in which electrode systems play the role of the photoassistance agents."

As a practical matter one uses a photoelectrochemical cell using suitable semiconductors as electrodes. Light shines upon one semiconductor which instigates the electrolysis. Hydrogen is liberated at the dark electrode and oxygen at the illuminated electrode. Remarkably enough electric current is produced in this process and not consumed as in the normal electrolysis process. The details of the process according to Wrighton are as follows. When current flows through such a cell, chemistry must occur at each of the electrodes. The resulting reaction in water is that

$$2H_2O \longrightarrow 2H_2 + O_2 \qquad (7)$$

The key feature of such a cell is the photoelectrode. Research conducted in 1976 demonstrated that a $SrTiO_3$ electrode had a light-to-chemical-energy conversion efficiency rivaling that of any man-contrived photochemical conversion device and estimates show that such a photoelectrolytic cell can produce fuel with an efficiency close to that of schemes exploiting plant photosynthesis.

CONCLUSION

As can be seen from the foregoing, there are a great number of scientific and technical developments which could,

*Mark S. Wrighton, Technology Review, May 1977, p. 31.

with proper application, be used to increase the productivity of arid lands. I look forward to a time when desert lands can be used to produce food and especially fuels to substitute for the dwindling world supply of fossil fuels.

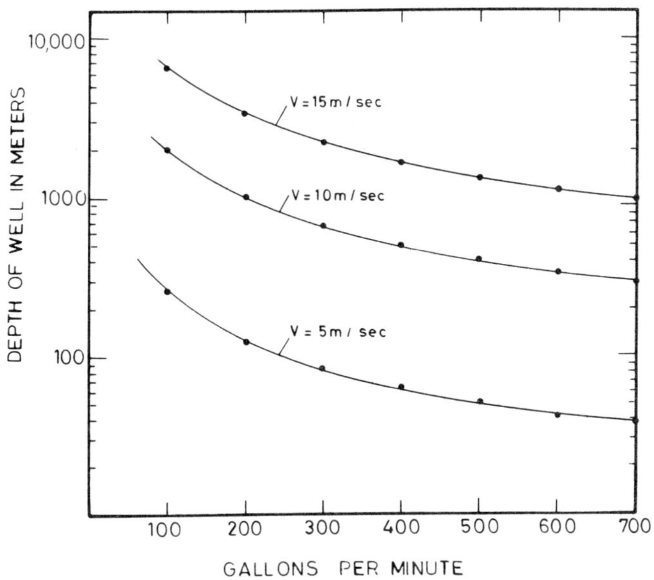

Fig. 1.

THE POTENTIAL OF HYBRID SOLAR IRRIGATION SYSTEMS FOR DESERT APPLICATIONS

EDWARD LUMSDAINE
Acting Director, New Mexico Solar Energy Institute, Las Cruces, New Mexico, U.S.A.

ABSTRACT

The biggest obstacle to desert development is the availability of water. In areas where underground water is available, pumping of this water requires large quantities of energy. Desert areas borderlining the sea could obtain water by desalination and pumping of the water inland. The potential of using a hybrid system for irrigation pumping as well as desalination in desert regions is discussed in this paper, and the design and thermal analysis of a desalination plant along with the wind energy conversion system are presented.

INTRODUCTION

The most important element in desert development is water for both irrigation and drinking. The agricultural economics of most countries are heavily dependent on irrigation. The energy consumed by irrigation worldwide is staggering; for example, in the United States of America alone, water pumping for irrigation during 1974 consumed the equivalent of 8¼ billion kwh.

Another example is Egypt. Egypt covers an area of 1,020,000 square kilometers supporting a population of 40 million. Yet the Nile River Valley which has a total area of only 30,000 square kilometers contains 99 percent of the population, and with the exception of some oases, the rest of the country is barren desert. In order to help solve the economic problems of Egypt and reduce the population density along the Nile river, the government has felt that there is an urgent need to develop the Eastern and Western deserts. To accomplish this task, an adequate supply of potable water must be available.

Establishing windpowered solar distillation units along the Eastern shore (Gulf of Suez and the Red Sea) and the

Northwestern shore (Alexandria to Mersa Matruh) would be a starting point towards the development of the desert. Solar water pumps can also be used in the inland oases.

The purpose of this paper is to discuss some hybrid solar systems that are prime candidates for this type of application.

A rather detailed summary of the history of solar irrigation systems is given in Reference 1. Past attempts at solar-powered pumping achieved efficiencies of 1 Bhp/272 ft^2 (1.0 kw/33.86 m^2) using focusing collectors and low pressure steam engines. A block diagram of the plants is shown in Figure 1.

Since the time of the successful Schuman-Boyce irrigation pump in Egypt,[2] little was done in the past sixty years because of the low cost of fossil fuels. However, solar-powered irrigation is again becoming attractive with the increasing prices of fossil fuels and in remote regions. Solar-powered irrigation systems have the potential of being an alternate to fossil energy in aiding food production in the future, especially in arid regions which generally experience abundant insolation.

Many concepts and ideas have developed during recent years on solar irrigation systems, generally following past developments but applying contemporary technology and new materials. Examples of these systems are the thermal irrigation plant near Willard, New Mexico, and the photovoltaic plant in Mead, Nebraska (U.S.A.). The Willard plant began operation in June 1977 (see Figure 2). The Rankine cycle operates at a peak temperature of 436° K for the secondary fluid (Refrigerant 113) and a temperature of 489° K for the primary fluid (Caloria HT-43 is also used as the heat storage fluid in a 22.7 m^3 tank). The system develops 25 Hp with a Rankine cycle efficiency of 15 percent. The water output of 2.6 m^3 per minute from a 34 m deep well can provide irrigation for up to 10^5 m^2 of land. A schematic diagram of the system is shown in Figure 3.

A large-scale application of photovoltaic cells to generate electricity for pumping irrigation water was completed near Mead, Nebraska in August of 1977 (see Figure 4). The electric system consists of 100,000 individual silicon cells mounted on 28 flat panels arranged in two rows. Each row develops a power of about 25 kw. Lead acid batteries with a total capacity of 85 kwh provide about 6 hours of energy storage. Inverters are used to convert direct current into alternating current to power the irrigation pump motor. A

10 Hp motor runs 12 hours a day pumping 3.8 m^3 of water per minute out of a reservoir. However, these systems are expensive and complicated.

Recently Rao[3] developed a constant-volume vaporization technique to pump water by the saturated pressure of a low boiling point fluid (pentane) which is immiscible with water. The diagram of the system is shown in Figure 5. The pentane liquid will be circulated through the solar collectors and vaporized in the tank. For a period of three hours of sunshine (depending on the size of the tank), the whole tank of pentane will be vaporized and then flashed into the water tank in the well to lift the water out from the water tank. The disadvantages of the system are that the collectors have to be pressurized and the fluid will not be able to circulate back to the flash tank without a booster pump in the line.

An open Brayton cycle with a closed type vapor generation topping cycle is presented here. This will give better thermal efficiency without pressurized collectors. This basic thermal cycle is linked to either a photovoltaic system or wind system to form hybrid systems.

PRINCIPLES OF OPERATION

A sketch of an open Brayton cycle system is shown in Figure 6. It consists of a heating chamber for increasing the temperature (the enthalpy of the air), an air compressor (preferably a reciprocating type in order to give higher pressure with low cost), an air turbine, water pump, vapor generator and condenser. The heating fluid in a closed cycle circulates through the solar collectors, heating chambers, and vapor generator. The fluid can be water, air or some other fluid with good thermal properties. The first working fluid in the vapor generator can be pentane or a petroleum fraction having a low boiling range of 35-40°C. The saturated vapor will be introduced into the reciprocating engine to compress the second working fluid, air, to designed pressure. The exhaust vapor will be condensed into liquid in a condenser and moved back to the vapor generator by a feed pump. After passing through the heating chamber, air with high temperature and high pressure will expand through the turbine. A centrifugal water pump directly shafted with the turbine will pump the water from the well to the field.

SYSTEM ANALYSIS

Thermodynamics

Closed Cycle of the First Working Fluid (Pentane or Petroleum Fraction): The working cycle of the first working fluid, such as pentane (pure substance) is shown in Figure 7. State 1 corresponds to the condensate temperature, and state 2 is the compressed pentane liquid. With continued heating, liquid pentane reaches state 3, the saturation temperature corresponding to the pressure at which the liquid was pumped. The thermal efficiency of the cycle is defined as

$$\eta_1 = \frac{\bar{W}net}{\dot{Q}h} \tag{1}$$

where

η_1 = thermal efficiency of the pentane cycle

Wnet = net work output, Kw/kg

$\dot{Q}h$ = heat energy input, Kw/kg

with the net work output given by

$$Wnet = \bar{W}_e - \bar{W}_p = (h_3 - h_4) - (h_2 - h_1)$$
$$= (h_3 - h_4) - \nu_1 (P_2 - P_1)$$

where

$\bar{W}_e$ = reciprocating engine output, KJ/kg

$\bar{W}_p$ = feed pump input, KJ/kg

h_i = enthalpy at state i, KJ/kg

ν_i = specific volume of pentane, m^3/kg

P_i = pressure at state i, kpa

Therefore,

$$\eta_1 = \frac{(h_3 - h_4) - \nu_1 (P_2 - P_1)}{(h_3 - h_2)} \tag{2}$$

Open Brayton Cycle (Air): The working cycle of the second working fluid, air, is shown in Figure 8. State 1 corresponds to the ambient temperature, and state 2 is the compressed air at high pressure. As the air passes through the heating chamber, it will reach design temperature at state 3, then expand through the turbine to state 4. The thermal efficiency of this cycle is defined as

$$\eta_2 = \frac{\bar{W}net}{Qsp} \tag{3}$$

where

$\bar{W}net$ = turbine output, compressor input

$= \bar{W}tb - \bar{W}cp.$ KJ/kg

$\bar{W}cp$ = compressor work input, $\bar{W}net$, KJ/kg

$Qsp = h_3 - h_2$, KJ/kg

or

$$\eta_2 = \frac{\bar{W}tb - \bar{W}cp}{Qsp} \qquad \frac{(h_3 - h_4) - (h_2 - h_1)}{(h_4 - h_2)} \tag{4}$$

Centrifugal Pump Efficiency: The mechanical efficiency of a centrifugal pump is defined as

$$\eta_3 = \frac{\text{pump power output}}{\text{air turbine power input}} = \frac{\dot{\bar{W}}_{pp}}{\dot{\bar{W}}_{tb}} \tag{5}$$

where

$\dot{\bar{W}}$ = pump power output = $\dot{\bar{V}}$ WH, kw

$\dot{\bar{V}}$ = volumetric flow rate, m^3/hr

W = water density, kg/m^3

H = total hydraulic head, m

$\dot{\bar{W}}_{tb}$ = air turbine power input = $\dot{m}_a \bar{W}_{tb}$, kw

$\dot{m}$ = mass flow rate of air, Kg/hr

Heat Transfer

Heat Transfer Through the Solar Collectors: The useful heat gain through the solar collector is defined as (from Ref. 4)

$$Qu = A_c F_r \ [(\zeta\alpha) - U_L \ (T_{f,i} - T_a)] \qquad (6)$$

where

Qu = total useful heat gain, kw

A_c = collector area, m^2

F_r = removal factor

$(\zeta\alpha)$ = transmittance and absorptance product

U_L = loss coefficient, $kw/m^2 - {}^oC$

$T_{f,i}$ = fluid entering temperature, ^oC

T_a = ambient temperature, ^oC

Equation (6) can be rewritten as

$$Qu = \dot{m}_f \cdot C_p \ (T_{f,e} - T_{f,i}) \qquad (7)$$

where

$\dot{m}_f$ = mass flow rate of fluid in collectors, kg/hr

c_p = specific heat of fluid, $Kcal/kg \ {}^oC$

$T_{f,e}$ = exiting fluid temperature, ^oC

$= Ta + I/U_l + [T_{f,i} - T_a - I/U_L]e - [U_L \ FWL/\dot{m}_f c_p]$

(from Ref. 4)

Heat Transfer in the Vapor Generator: The heat transfer in the vapor generator is a saturated boiling process. McAdams and Levy[5,6] developed an experimental equation which gives the maximum available heat flux through the heating coils as

$$\dot{Q}_f = 2,253\ A_t (T_w - T_{sat})^{3.86} \quad \text{for } 0.2 < p < 0.7\ MN/m^2$$

or (8)

$$= 0.061\ A_t P^{4/3} (T_w - T_{sat})^{3.0} \quad \text{for } 0.7 < p < 14\ MN/m^2$$

where

$\dot{Q}_f$ = heat flux, kw

A_t = heat transfer surface area of the heat coils, m^2

T_w = wall temperature of the heating coil, ^oC

T_{sat} = saturation temperature of the fluid, ^oC

P = saturation pressure, N/m^2

Heat Transfer in the Heating Chamber: The heating chamger is acting as a heat exchanger. With the NTU-method,[7] the heat gain by the cold air can be expressed as

$$\dot{Q}sp = \dot{m}_a\ C_{pa} \times E_c \times \Delta T_h \quad (9)$$

where

$\dot{Q}sp$ = heat supply in heating chamber, kw

$\dot{m}_a$ = air flow rate, kg/hr

C_{pa} = specific heat of air, $KJ/kg\text{-}^oC$

E_c = effectiveness of heat exchanger

ΔT_h = high temperature fluid temperature drop, ^oC

Correlation Between Pumps and Collectors

The correlation between the total hydraulic heat or the volumetric flow rate of the water pump with the solar collector fluid temperature is established as follows. The total hydraulic heat or the volumetric flow rate can be expressed as a function of the solar collector fluid temperature under the following assumptions:

(a) The wall temperature of the heating coil is equal to the fluid temperature inside the coil, i.e. $T_{f,e} = T_w$.

(b) Air can be considered as an ideal gas. From equation (5),

$$\eta_3 = \frac{\dot{\bar{W}}_{pp}}{\dot{\bar{W}}_{tb}} \tag{5}$$

or

$$\dot{\bar{W}}_{pp} = \eta_3 \dot{\bar{W}}_{tb} \qquad \dot{\bar{V}}WH$$

or

$$H = \frac{\eta_3}{\dot{\bar{V}}W} \dot{\bar{W}}_{tb} = \frac{\dot{M}_a \eta_3 \bar{W}_{tb}}{W\dot{\bar{V}}} \tag{10}$$

and from equation (4),

$$\eta_2 = \frac{\dot{\bar{W}}_{tb} - \bar{W}_{cp}}{Q_{sp}} \tag{4}$$

or

$$\bar{W}_{tb} = \bar{W}_{cp} + \eta_2 Q_{sp} \tag{11}$$

Equations (10) and (11) together yield

$$H = \frac{\dot{M}_a \eta_3}{W\dot{\bar{V}}} (\bar{W}_{cp} + \eta_2 Q_{sp}) \tag{12}$$

From equation (1) the compressor work input $\bar{W}_{cp}$ is equal to the net work output of the closed pentane cycle, i.e. $\bar{W}_{cp} = \bar{W}net$, or

$$\eta_1 = \frac{\bar{W}net}{Q_h} \tag{1}$$

$$\bar{W}net = \bar{W}_{cp} = \eta_1 Q_h \tag{13}$$

A substitution of equation (13) into equation (12) gives

$$H = \frac{\dot{M}_a \eta_3}{W\dot{\bar{V}}} [\eta_1 Q_h + \eta_2 Q_{sp}] \qquad (14)$$

From equations (8) and (9), equation (14) can be rewritten as

$$H = \frac{\dot{M}_a \eta_3}{W\dot{\bar{V}}} [2.253\, \eta_1 A_t (T_{f,e2} - T_{sat})^{3.0} + \dot{M}_a C_{pa} E_c \eta_2 (T_{f,el} - T_{f,e2})]$$

From equations (8) and (9), equation (14) can be rewritten as

$$H = \frac{\dot{M}_a \eta_3}{W\dot{\bar{V}}} [2.253\, \eta_1 A_t (T_{fe2} - T_{sat})^{3.86} + M_a C_{pa} E_c \eta_2 (T_{fe,1} - T_{fe,2})]$$

for $0.2 < P < 0.7\ MN/m^2$

or

$$H = M_a \eta_3 [0.061\eta_1 A_t (T_{fe,2} - T_{sat})^{3.0} + M_a C_{pa} E_c \eta_2 (T_{fe,1} - T_{fe,2})]$$

for $0.7 < P < 14\ MN/m^2$ (15)

where

T_{fel} = collector fluid temperature entering the heating chamber, ^oC

T_{fe2} = collector fluid temperature leaving the heating chamber, ^oC

Then the relation between the volumetric flow rate with collector fluid temperature can be written as

$$\dot{V} = \frac{\dot{M}_a \eta_3}{WH} [2.253\eta_1 A_t (T_{fe,2} - T_{sat})^{3.86} + M_a C_{pa} E_c \eta_2 (T_{fe,1} - T_{fe,2})]$$

for $0.2 < P < 0.7\ MN/m^2$

or

$$\dot{V} = \frac{M_a \eta_3}{WH} \; 0.061 \eta_1 A_t (T_{fe,2} - T_{sat})^{3.0} + M_a C_{pa} E_c \eta_2 (T_{fe,1} - T_{fe,2})$$

$$\text{for } 0.7 < P < 14 \text{ MN/m}^2 \qquad (16)$$

System Efficiency

Overall system efficiency is defined as

$$\eta_{sys} = \frac{\text{Water pump power output}}{\text{Total insolation on collectors}}$$

$$\eta_{sys} = \frac{\dot{W}_{pp}}{A_c I} \quad \frac{\dot{V}\,WH}{A_c I} \qquad (17)$$

or

$$\eta_{sys} = \frac{\dot{M}_a \eta_3}{A_c I} [2.253 \eta_1 A_t (T_{fe,2} - T_{sat})^{3.86} + \dot{M}_a C_{pa} E_c \eta_2 (T_{fe,1} - T_{fe,2})]$$

$$\text{for } 0.2 < P < 0.7 \text{ MN/m}^2$$

and

$$\eta_{sys} = \frac{\dot{M}_a \eta_3}{A_c I} [0.061 \eta_1 A_t (T_{fe,2} - T_{sat})^{3.0} + \dot{M}_a C_{pa} E_c \eta_2 (T_{fe,1} - T_{fe,2})]$$

$$\text{for } 0.7 < P < 14 \text{ MN/m}^2 \qquad (18)$$

ANALYTIC CALCULATIONS

Figure 9 shows the solar collector fluid exiting temperature versus collector length in the flow direction of a typical flat plate water-cooled collector. When this temperature is given, the irrigation system can be designed to have maximum system efficiency through correct choice of the associated equipment such as air turbine, pump, reciprocating compressor as well as the working fluid pentane or petroleum fractions. Figure 10 shows that the analytic overall system efficiency versus the overall mechanical efficiency of the associated equipment, the overall mechanical efficiency η_{me}, is defined as

$$\eta_{me} = \eta_1 \times \eta_2 \times \eta_3$$

where

η_{me} = overall mechanical efficiency

η_1 = thermal efficiency of the pentane cycle

η_2 = thermal efficiency of the Brayton cycle

η_3 = mechanical efficiency of the centrifugal pump

After the system has been set-up, the total hydraulic heat or the total volumetric flow rate of the irrigation pump can be calculated as a function of the fluid temperature difference according to equations (15) and (16).

Figure 11 shows the total hydraulic head in meter (feet) versus the temperature difference between the exiting fluid from solar collector and the saturation temperature of the pentane for different overall system thermal efficiencies, and Figure 12 gives the total volumetric flow rates in liters/min (gallons/min.) versus collector area. Therefore, by knowing the design volumetric flow rate and the total hydraulic head, one can easily determine the fluid temperature difference and the collector area required.

ADVANTAGES OF THE SYSTEM

(a) The fluid in the solar collectors circulates in a closed loop; no pressurization of the collectors will be required. This will reduce the system cost dramatically.
(b) The working fluid pentane or petroleum fraction operates in a low temperature range. This will reduce flammability and toxicity.
(c) Air can be heated to high temperatures and high pressure by proper design of the heating chamber; this will give better system efficiency.

(d) The system is suitable for both high temperature operation and low temperature operation.

(e) The compressed air can be used directly in an air lift pumping system. This will reduce moving mechanical parts to a minimum as well as lower the capital cost of the system.

HYBRID

Photovoltaic - Thermal Hybrid System

A schematic of a typical thermal - photovoltaic irrigation system is shown in Figure 13. Since the efficiency of the photovoltaic cells decreases as the plate temperature increases, increasing the thermal output by increasing the fluid flow rate decreases the cell temperature, thus resulting in higher system efficiency. Thus a hybrid system combining the two will be most attractive. Figure 14 shows the efficiencies of a photovoltaic system, a thermal system and the combined system. In the 200-250°C temperature interval, which is a convenient range for solar energy photothermal systems, the output is nearly doubled in the hybrid system compared to the efficiency of either system alone.

Wind-Thermal Hybrid System

The available wind power can be attached to the thermal system for water irrigation by two means. One way is to generate the electricity by wind power then go through a DC (or AC) motor to drive the water pump. The other is to connect the wind power directly to the thermal irrigation system. Since one of the characteristics of the wind power system is operation at considerably slower rotating speed, larger drive shafting and bearings would be required to accomodate the greater torque involved. This may make the hybrid system uneconomical. One alternative method is to have the wind power drive a reciprocating air compressor, then direct the compressed air to the Brayton Cycle which has been discussed previously. This will have the following advantages:

(a) The system is inherently simple compared with other hybrid systems.

(b) With the horizontal axis rotor, the system would be self orienting and self starting. The air compressor cylinders can easily be unloaded for no load rotor starting.

(c) Location of the windpower unit is not critical. Compressed air can be conveyed considerable distances through pipes without serious pressure losses.
(d) Power from two or more units can be readily coupled for applications requiring increased power levels. Power systems could be developed on a modular plan.
(e) In some locations, underground formations exist in which compressed air might be economically stored.

The amount of wind energy that can be added to a solar thermal irrigation system varies from location to location and season to season. A typical monthly distribution of available wind power and solar radiation for an era at 33° N latitude is shown in Figure 15. This shows that wind power can be used as a back-up energy source in addition to increasing the power to the irrigation system.

EXAMPLE OF HYBRID DESALINATION PLANT USING WIND/SOLAR POWER

As an example of a hybrid plant an analysis of a modular wind/solar distillation plant is presented (see Figure 16). The design is based on a constant year-round demand of 50,000 gallons per day.

A computer modeling program has been developed to calculate the daily output of the solar basin still, taking into account all the parametric variations.

A comprehensive cost analysis is underway to assess the feasibility of windpowered solar distillation plants for actual locations.

Analytical Flow Diagram

Figure 17 shows the analytical flow diagram incorporating the following steps:

(a) Requirement Analysis and Scale Need: This part identifies the needs for distilled water and gives the variation of gallons per day (gpd) versus month. Information input here will have a direct impact on plant size and the most economical way of choosing the type of distiller. In this analysis, however, the demand will be kept constant year-round and fixed at 50,000 gpd.

(b) Weather Data: A complete survey of solar insolation for the proposed location must be available along with the variation in ambient temperature. In addition, meteorological data of wind speed and direction is needed.

(c) Site Evaluation: For still installation, the site will have to be near saline water and economically feasible.

(d) System Synthesis: System synthesis includes: (1) Still type selection; (2) a preliminary estimate of still size, (3) design steps (theory), (4) monthly supply and demand, (5) power requirement, and (6) storage requirement.

(e) Power Supply: Power is supplied through a wind energy conversion system (WECS). The WECS study deals with:
   (1) wind machine type selection (horizontal or vertical axis machines), number of blades (for horizontal axis machines), Darrieus or Savonius type (for vertical axis machines).
   (2) machine size
   (3) number of machines
   (4) location
   (5) transmission

(f) Cost Estimates: Cost analysis for the wind-powered solar distillation plant in U.S. dollars is compared with the cost of a conventionally powered still.

(g) Comparative Evaluation: A comparative feasibility study will have to include still productivity, capital cost of the installation, its reliability, cost of power supplied, and amortization and interest rates.

Theoretical Analysis of the Still

Still Type Selection: A recent study on solar distillation[8] has found that solar stills with a capacity of about 50,000 gallons can produce water more economically than other desalination processes. Since the most common type of solar still is the basin-type (also known as roof-type) it was chosen for this study.

Basin or Still Theory: A complete theoretical analysis of the basin type solar still is presented in Reference 9. In that analysis, heat and mass transfer relationships which govern the operation of the solar still in the unsteady state have been solved graphically by means of a characteristics chart.

Because of inaccuracies resulting from using the charts, a computer modeling program has been developed to solve the system of equations more efficiently and accurately, and also to make a complete parametric study possible.

In this section, the governing equations are presented and solved using an implicit continuous modeling computer program to determine the output of the solar still under the daily changes of solar insolation and ambient conditions.

The following equations can easily be written for the thermal behavior of the solar still shown in Figure 18.

$$\alpha_g H_s + \alpha_w \tau H_s \quad q_{ga} + q_b + c_{wg} \frac{dTw}{dt} \tag{18}$$

$$q_{ga} = q_r + q_c + q_e + \alpha_g H_s \tag{19}$$

$$q_b = k_b (T_w - T_a) \tag{20}$$

$$q_r = 0.9\, \sigma\, [(T_w + 460)^4 - (T_a + 460)^4] \tag{21}$$

$$q_c = 0.128\, [T_w - T_g + (\frac{p_w - p_{wg}}{39 - p_w})(T_w + 460)]^{1/3} (T_w - T_g) \tag{22}$$

$$q_e = 0.0254\, [T_w - T_g + (\frac{p_w - p_{wg}}{39 - p_w})(T_w + 460)]^{1/3} I p_w - p_{wg}) h_w \tag{23}$$

$$q_{ga} = 0.9\alpha\, [(T_g + 460)^4 - (T_a + 460)^4] + h_{ga} (T_g - T_a) \tag{24}$$

Let:

$TG = T_g + 460$

$TA = T_a + 460$

$TW = T_w + 460$

$TGA = T_g - T_a$

$TWA = T_w - T_a$

$TWG = T_w - T_g$

$$A = (p_w - p_{wg}) \; / \; (39 - p_w)$$

$$B = (p_w - p_{wg}) \; h_w$$

Now equations (1) through (7) can be combined to give:

$$\frac{dTw}{dt} = \frac{1}{c_{wg}} \left[\alpha_g H_s + \alpha_w \tau H_s - 0.9 \, \sigma \, (TG)^4 - (TA - 20)^4 - h_{ga} \, (TGA) - k_b \, (TWA) \right] \qquad (25)$$

and

$$0.9 \, \sigma \, [(TG)^4 - (TA - 20)^4] + h_{ga} \, (TGA) = 0.9 \, \sigma \, [(TW)^4 - (TG)^4]$$
$$+ 0.128 \; [(TWG) + A \, (TW)]^{1/3} \, (TWG) + 0.0254 \, [TWG - A(TW)]^{1/3} B \qquad (26)$$

where

c_{wg} = thermal capacity of water, still and ground, Btu/^oF ft^2

h_{ga} = convective heat transfer coefficient of glass cover to air, Btu/hr ft^2 ^oF

H_s = solar radiation on horizontal surface, Btu/hr ft^2

h_w = latent heat of vaporization of water, Btu/lb

k_b = thermal loss coefficient of base of still to surroundings, Btu/hr ft^2 ^oF

p_w, p_{wg} = partial pressure of water vapor at T_w, T_{wg}

q_b = heat loss from base of still, Btu/hr ft^2

q_e = heat transfer of brine to glass by evaporation, Btu/hr ft^2

q_{ga} = heat transfer of glass to cover to surroundings, Btu/hr ft^2

q_r, q_c = heat transfer of brine to glass by radiation and convection, Btu/hr ft^2

t = time, hr

T_a = ambient temperature, ^oF

T_g, T_w = temperature of glass cover and saline water, ^oF

α_g, τ = solar absorptance and transmittance of glass cover

α_w = solar absorptance of brine and trough system

σ = Stefan - Boltzman constant, 17.2 x 10^{-10} Btu/hr ft^2

Equations (25) and (26) are solved simultaneously for T_w and T_g. Figures 19 and 20 give the variation of T_w and T_g with time during a winter day (December) and a summer day (June). Variations in ambient temperature, solar insolation, and partial pressure of water have been accounted for. In the analysis it is assumed that

c_{wg} = 16 Btu/^oF ft^2

h_{ga} = 2.6 Btu/hr ft^2 which corresponds to a wind velocity of 5 mph

h_w = 1020 Btu/lb

k_b = 1.0 Btu/hr ft^2 ^oF

α_g = 0.1

t = 0.8

α_g = 0.9

The output of the still is calculated using the relationship

$$m = 0.89 \frac{\eta_o}{h_w} \int_o^{24} q_e \, dt \quad \text{lb/ft}$$

η_o drainage efficiency of still = 0.98

For a typical summer day Figure 20 gives the daily distribution of q_e from which $m_s = 0.30$ lb/ft^2 approximately, and for a typical winter day Figure 19 gives the daily values of q_e and the output of still as $m_w = 0.05$ lb/ft^2.

Design Data Capacity: Surface area of the still ≅ 1.4×10^6 ft^2

Power Requirements:

(a) Saline Water pump ≃ 1 Hp

(b) Product Water pump ≃ 7 Hp

(c) Total = 8 Hp

Thus a maximum power of approximately 6 kw is needed to operate the plant.

Wind Machine Selection

Two units are selected to deliver the required power. Specifications are listed below:

Trade Name: Jacobs

Type and Number of Blades: Horizontal Axis Three-Bladed Rotor

Output kwh/Year = 6,000

Blade Diameter - 14 ft.

Rated kw = 3.0

Operating Speeds: Rated = 23 mph

Cut-in = 7 mph

Cut-out = 60 mph

CONCLUSIONS

(1) An open Brayton cycle is more feasible than a conventional Rankine cycle for solar thermal irrigation applications.

(2) The overall system efficiency of the solar thermal is mainly dependent on the efficiencies of the associated equipment, such as turbine, pump, heat exchangers, and solar collectors. Highly efficient equipment should be underdesigned for solar thermal system application.

(3) Photovoltaic-thermal hybrid systems will reach a much higher efficiency than each system alone; thus they appear to be the most attractive system for irrigation.

(4) Wind power can be used as a back-up energy source in additional to boosting the power to the irrigation pump in a wind-thermal hybrid system, thus resulting in a solar thermal irrigation system having a fairly steady output.

REFERENCES

1. J.T. Ptylinski. Solar energy installations for pumping irrigation water (to appear in Solar Energy).

2. F. Shuman (1913-14). Sun Power Plant. Proceedings, Manchester Association of Engineers. Manchester, England.

3. D.P. Rao and K.S. Rao (1976). Solar water pump for lift irrigation. Solar Energy 18:405-411.

4. J.A. Duffie and W.A. Beckman. Solar Energy Thermal Processes. John Wiley & Sons, New York.

5. W.H. McAdams, et al. (1949). Heat transfer at high rates to water with surface boiling. Ind. Eng. Chem. 41:1945-1955.

6. S. Levy (1965). Generalized correlation of boiling heat transfer. J. Heat Transfer 810:37-42.

7. J.P. Holman (1976). Heat Transfer. 8th ed. McGraw-Hill, New York.

8. United Nations (1970). Solar distillation as a means of meeting small-scale water demands. New York.

9. R.N. Morse and R.W. Read (1968). A rational basis for the engineering development of a solar still. Solar Energy 12:5-17.

10. Solar water distillation: the roof type still and a multiple diffusion still. International Development in Heat Transfer, Colorado, 1961.

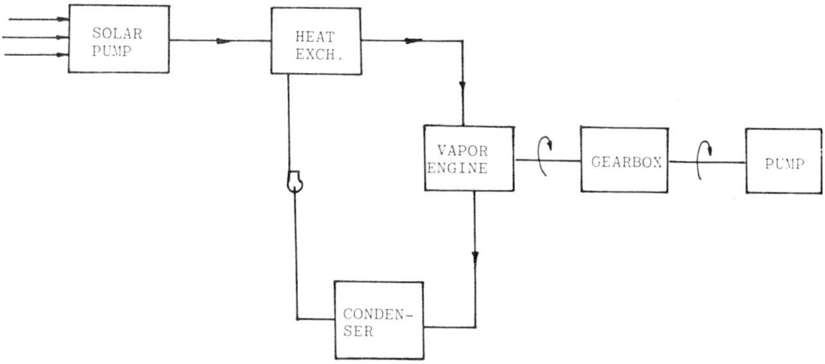

Fig. 1. Schematic of solar irrigation system.

Fig. 2. Solar irrigation plant at Willard, New Mexico.

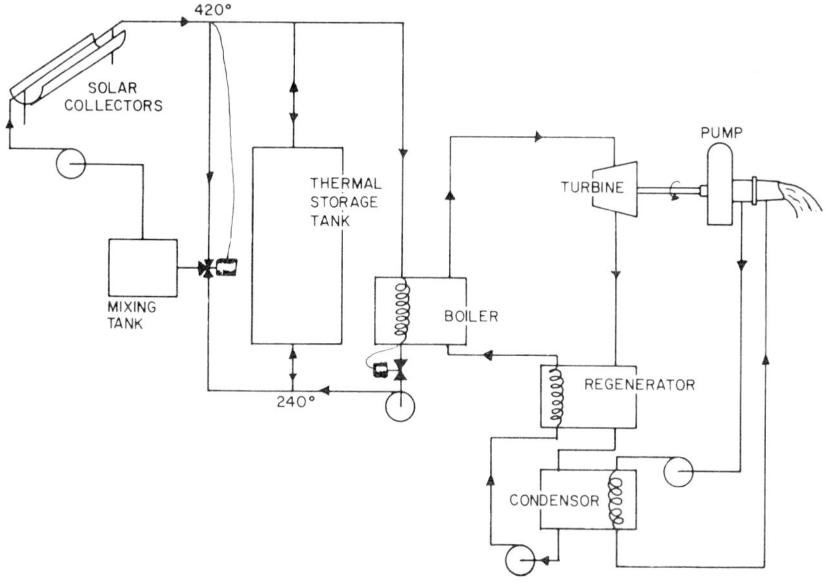

Fig. 3. Schematic of Rankine cycle, Willard Plant.

Fig. 4. Mead, Nebraska photo cell irrigation system.

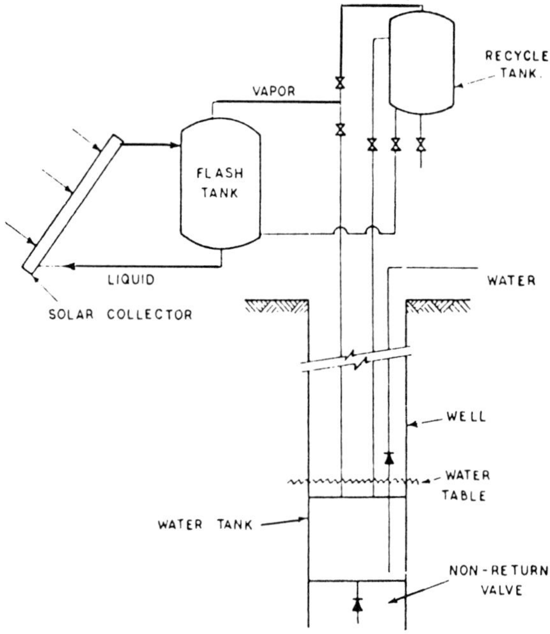

Fig. 5. Sketch of constant volume vaporization system.

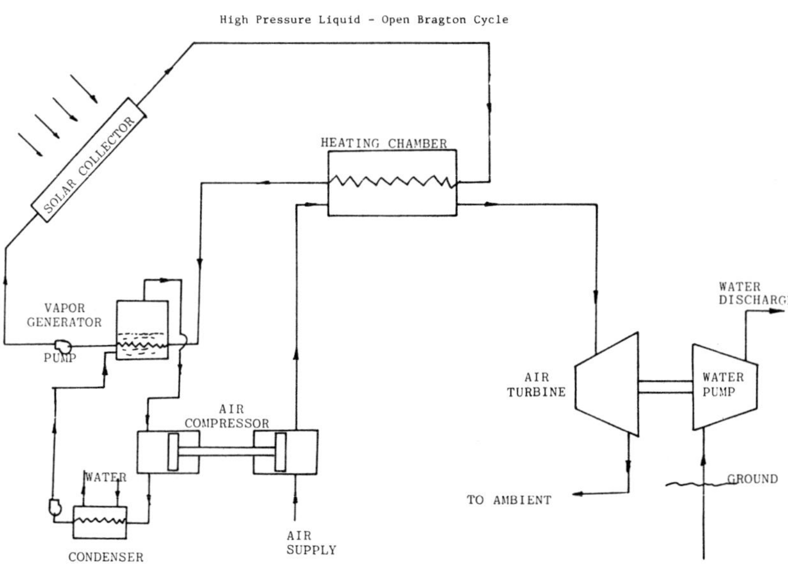

Fig. 6. Sketch of an open Bragton cycle system.

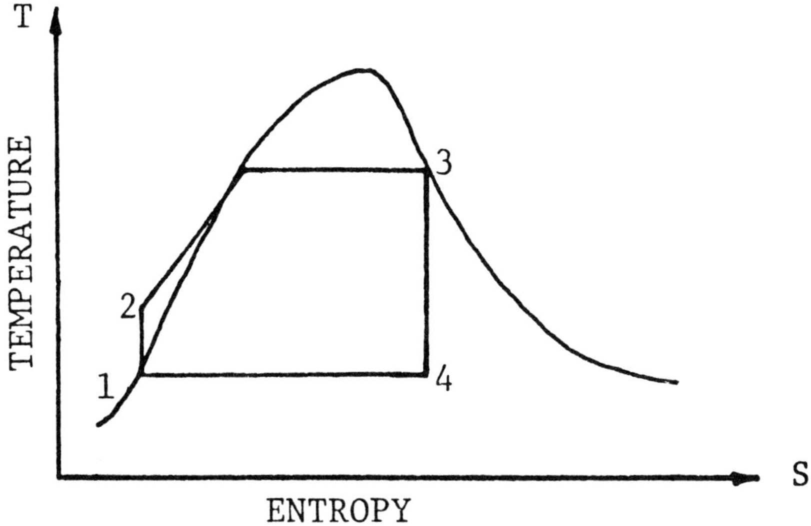

Fig. 7. T-S diagram for closed cycle with pentane fluid.

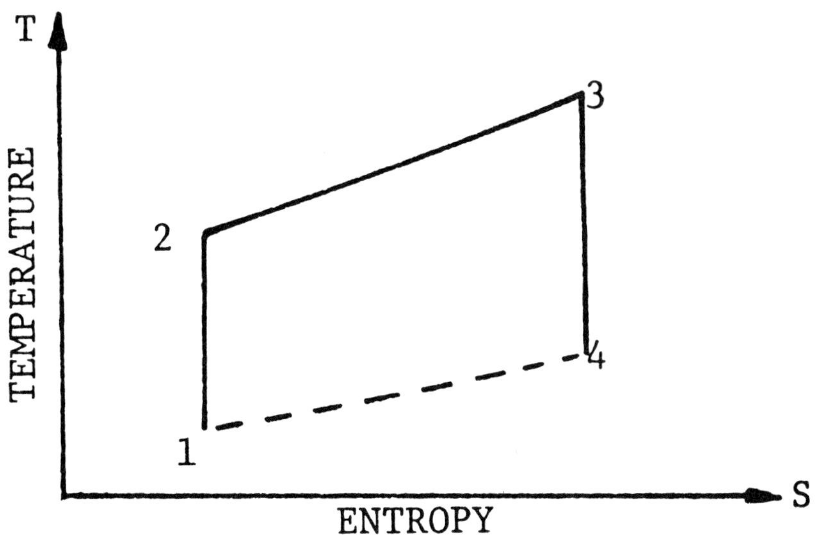

Fig. 8. T-S diagram of open cycle (air).

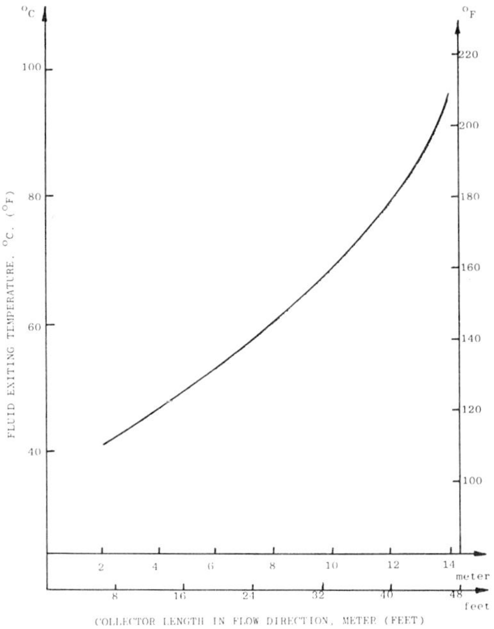

Fig. 9. Fluid exiting temperature in collectors.

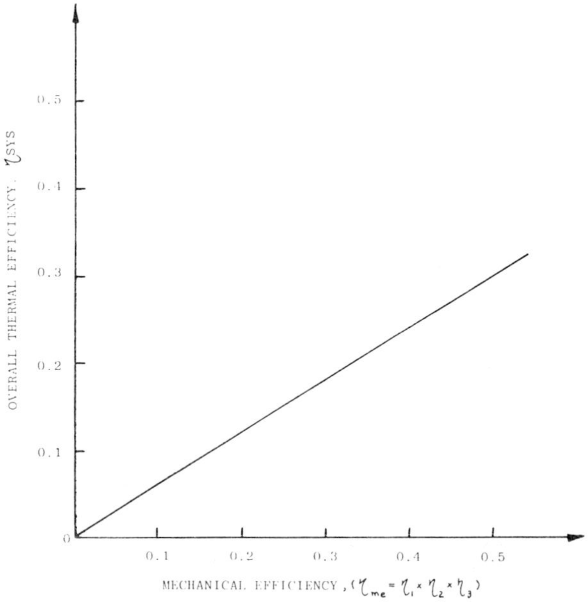

Fig. 10. Analytic prediction of system overall thermal efficiency.

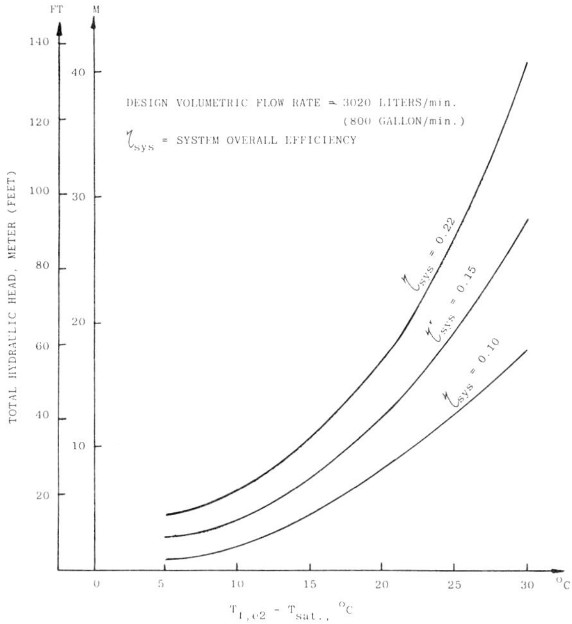

Fig. 11. Total hydraulic head of an irrigation pump.

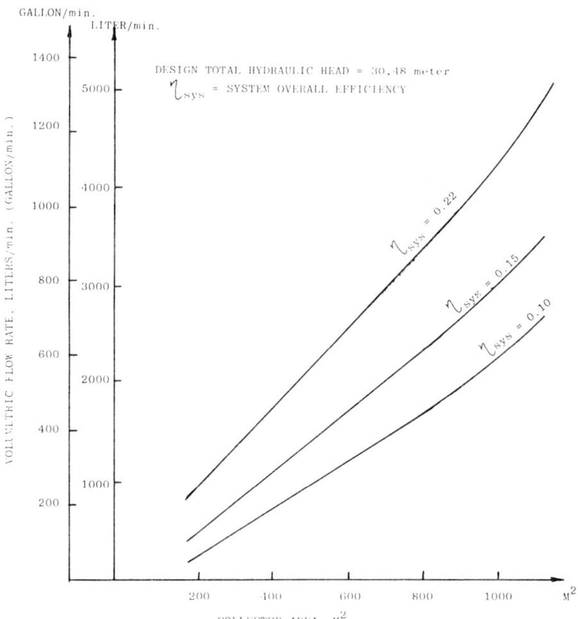

Fig. 12. Total volumetric flow rate of an irrigation pump.

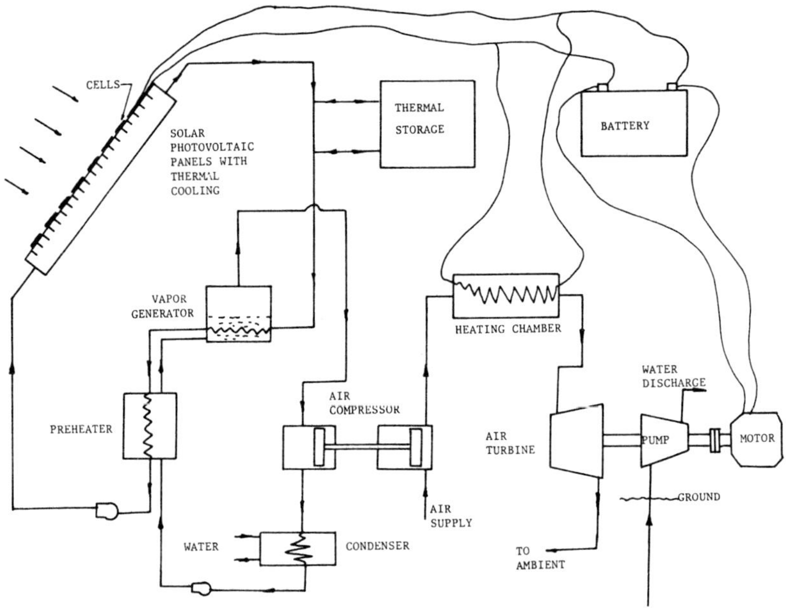

Fig. 13. A sketch of photovoltaic-thermal hybrid system.

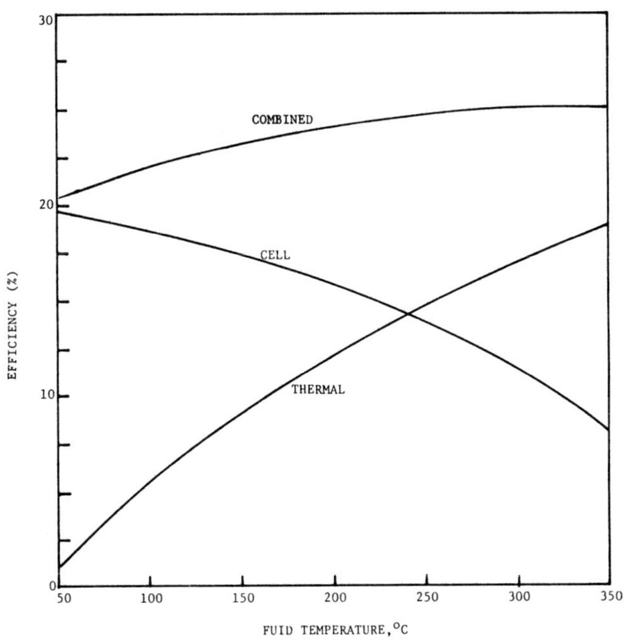

Fig. 14. Photovoltaic, photothermal, and combined efficiencies versus working fluid temperature.

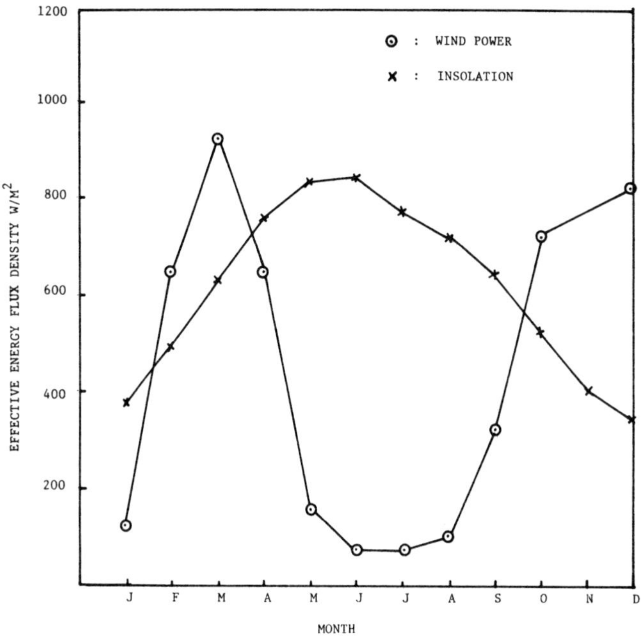

Fig. 15. Wind power and insolation distribution (33° north latitude).

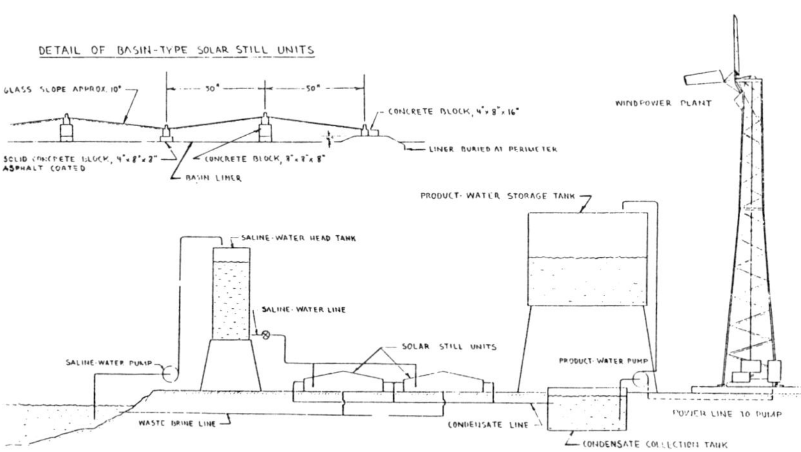

Fig. 16. Windpowered solar distillation plant.

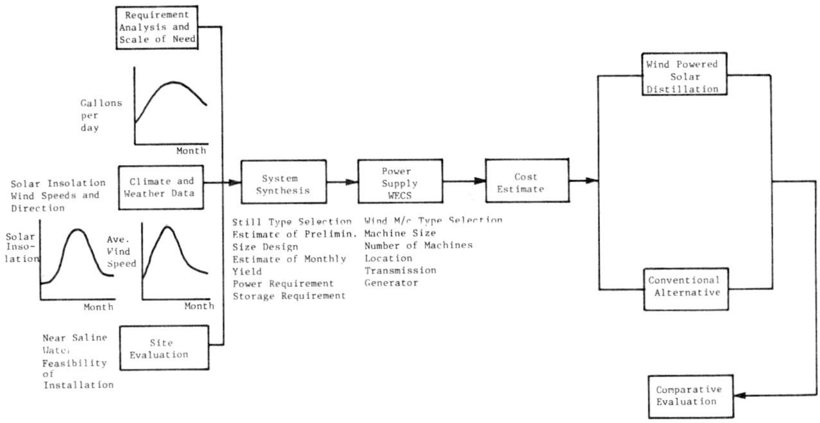

Fig. 17. Windpowered solar distillation plant analysis.

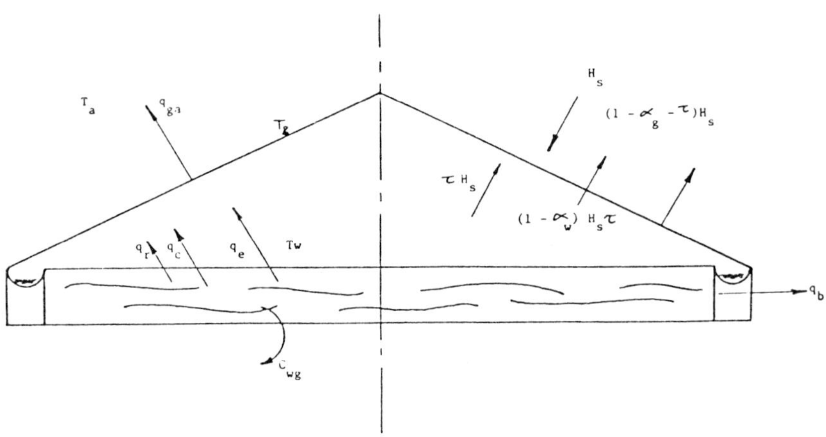

Fig. 18. Heat fluxes for a solar still.

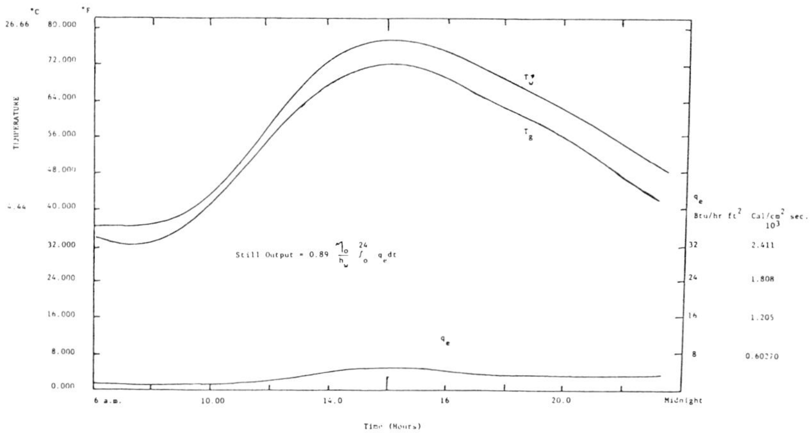

Fig. 19. Solar still thermal performance (winter).

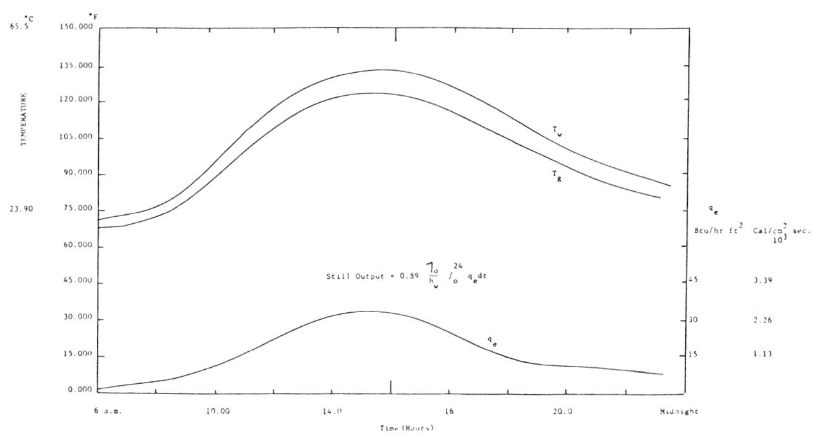

Fig. 20. Solar still thermal performance (summer).

POTENTIAL APPLICATION OF RENEWABLE ENERGY TECHNOLOGIES FOR DESERT DEVELOPMENT (WITH SPECIAL EMPHASIS ON EGYPTIAN DESERT)

A. EL-SAWY, JACK J. FRITZ, AND MARTIN SCHOLL
The MITRE Corporation, Mclean, Virginia, U.S.A.

ABSTRACT

A mix of municipal effluents, photovoltaics and wind machines can be used to meet the total energy requirements of a small community in a tropical or arid region of the world. The potential proportion of energy available from each of the three sources is a function of wind velocity, solar radiation, wastewater flow and characteristics as well as the manner in which the capital costs are divided among the three systems. Controlled photosynthesis in organic wastes offers a viable system for supplying both energy and food in the desert. Biomass (e.g. unicellular algae and similar living organic matter), a renewable non-fossil carbon material, provides indirect use of stored solar energy and is an attractive and inexpensive source of nutritional protein. The idea is to use animal, human, and agricultural wastes as substrates for mass cultivation of algae in the desert. In the warm and sunny desert climate, the algae trap radiant energy of the sun and convert it into cell material with high protein content. By suitable processing the algae can also be converted into gaseous fuels, liquid fuels, or a host of basic chemicals including fertilizers and other valuable products for desert development.

INTRODUCTION

Most of the developing countries in Asia, Africa and South America are threatened by a population explosion, food shortages, a dwindling and costly supply of fossil and conventional fuels and an acute problem of disposal of residuals. In these countries the development of desert areas at the outskirts of big cities offers a potential solution to modernization that should not be overlooked.

Increased emphasis on desert development will create a significant increase in energy requirements, threatening to overburden the energy supply systems in most developing countries. Experience has shown[1] that the age old techniques of using exclusively manual and animal labor, animal wastes and firewood as the main sources of energy are highly regressive in human as well as environmental terms. The prohibitive costs of large central generators and massive transmission and distribution systems discourage hopes that desert development energy needs can be met with national electric grids.

This paper discusses one potential alternative, namely application of renewable energy sources (e.g., solar radiation, wind energy and biomass*) in desert development. Developing countries are usually well endowed with one or more of these resources and, for the most part, they can be utilized with the locally available intermediate technology and human resources. Moreover, the use of renewable energy sources has few waste disposal problems and is less amenable to fuel or technological monopolies.

POTENTIAL OF RENEWABLE ENERGY TECHNOLOGY IN DESERT DEVELOPMENT

Various technologies that exploit renewable energy sources could be useful for a variety of tasks in desert development. Table 1 lists technologies[2] which are expected to be potentially useful in their estimated order of likely success. Of these technologies, at least three are either currently available or can be developed and adapted to local conditions within a few (1-5) years for use in isolated desert areas. These promising technologies include (1) windmill generators, (2) photovoltaic power generators, and (3) biogas generation and controlled photosynthesis in organic wastewaters. Potential applications of these technologies in desert development discussed below.

Windmill Generators

The potential of wind energy is presently receiving serious consideration in those parts of the world where appropriate meteorological conditions prevail. Egypt, for example, has two windy regions,[3] the Mediterranean and the Red Sea, both vast coastal areas. For the average wind speed (20 kph) and duration, usable annual electric energy output in these

---

* Biomass = Non fossil organic matter.

TABLE 1*

POTENTIAL USEFULNESS OF VARIOUS RENEWABLE ENERGY TECHNOLOGY

Table 1. Solar energy applicability matrix. Animals are included as a solar technology (Based on photosynthesis). For many villages the use of animals would represent a modernizing step. It includes the use of dung for burning or fertilizing. Source: adapted from a presentation by J. R. Williams at the Tanzanian workshop. Symbols: + +, applicable; +, somewhat applicable; -, not applicable.

Solar technology	Energy use													
								Heating						
	Water pumping	Lighting	Cooling	Communications	Water desalting	Spinning	Sawing	Cooking	Space	Domestic water	Grinding	Drying	Transport	Fertilizer
Solar cells (flat plate)	++	+	+	++	-	+	-	-	-	-	++	-	-	-
Flat-plate collectors	++	-	++	-	-	-	-	-	++	++	-	++	-	-
Concentrating collectors	++	+	++	++	+	+	+	++	+	-	++	-	-	-
Solar, stirling	++	+	+	-	-	+	+	-	-	-	+	-	-	-
Solar, rankine	++	+	+	-	-	+	+	-	-	-	+	-	-	-
Wind (mechanical)	++	-	-	-	-	+	+	-	-	-	++	-	-	-
Wind generator	++	++	+	++	-	+	+	-	-	-	++	-	-	-
Water (mechanical)	++	-	-	-	-	++	++		-	-	++	-	-	-
Hydroelectric	++	++	++	++	-	++	++	-	-	-	++	-	-	-
Bioconversion wood/pyrolysis	-	+	-	-	+	-	-	++	++	++	-	++	-	+
Biogas	++	++	+	-	-	-	-	++	-	+	-	+	+	++
Solar cooling equipment	-	-	++	-	-	-	-	-	-	-	-	-	-	-
Draft animals	++	-	-	-	-	-	-	+	-	+	++	-	++	++

*Source: Norman L. Brown and James W. Howe, "Solar Energy for Village Development", Science, Volume 199, 10 February 1978.

areas is estimated to be about 650 kwhr per square meter of wind swept area. In cooperation with Oklahoma State University, Egypt is exploring the development of a special system (field modulated generator, FMGS) for generating electricity efficiently and economically from wind. These electricity generating windmills can be used in the desert lands near the Mediterranean and the Red Sea coastal areas to pump water from underground aquifers and to generate electricity for lighting and other household purposes. The estimated minimum energy requirement[4] per capita per day is 1 kwhr distributed among various uses as shown in Table 2. With this in mind, the energy requirement for a desert community of 100 families (each with 5 persons) living near the windy regions of Egypt could be supplied by a set of eighteen, 4.5 meter rotor diameter windmills spaced at a distance of 45 meters apart. In addition, a lead-acid battery storage system of 250 kwhr capacity could provide reliable household power during low wind periods when windmill generators would be idle. In Table 3, the estimated total capital investment for both the windmill set and the back-up lead-acid battery storage system is $127,000 (i.e., $705/kw).

Assuming 7.5 percent interest rate and 0.14 loading factor*, the cost of supplying electricity by the wind-electric system is estimated to be 85 mills/kwhr. This apparently high cost of energy generation compares favorably with the expected cost of electricity transmitted to villages within 20 km of the present high dam transmission lines.[5]

Even more instructive is the comparison of the busbar cost of energy generated by small-scale diesel generator to that of the wind-electric system. In this situation the windmill generation cost (i.e., 107 mills/kwhr) would be competitive with conventional diesel generation plants using oil priced at 11.6 dollars per barrel or higher. This oil is not unreasonable for remote desert areas in Egypt where the transportation cost can become very high.

Photovoltaic Power Generators

In photovoltaic conversion solar radiation as light rather than heat is converted by semiconductors (e.g., silicon

---

* Loading factor = K = Annual Average energy production factor

$$= \frac{\text{Annual kWh Energy output}}{8760 \times (\text{rated kW})}$$

cells) to electric energy. The most important limitation here is the fact that about half the solar radiation occurs at long wavelengths in the infrared where the energy is insufficient to bring about a photovoltaic action. Silicon solar cells, with an experimental efficiency of 14 percent, are the most promising semiconductor currently available. The limited theoretical efficiency of photovoltaics and the low-energy density of sunlight require large areas of photovoltaic arrays in order to generate appreciable amounts (e.g., 1 MW) of power. In addition, a photovoltaic energy system requires an energy storage system or ensure energy supply during the night hours and during cloudy or stormy weather when sunlight is not available.

However, photovoltaic power generators are of interest for low power output in remote areas requiring small capacity pumps for drinking water and irrigation, and where no inexpensive alternatives are available. Under these circumstances a combination of relatively small solar cell arrays and lead-acid battery sets can be used to provide energy to operate educational television sets, radio beacons or other telecommunication systems. In Niger, this concept motivated the construction of an experimental educational TV station[6] as far back as 1968. More recently, (1973), the approach to the airport of Medina in Saudi Arabia was facilitated by illuminating the seven mountain peaks leading to it with light beacons powered by photovoltaic generators.[7] Today, the Egyptian authorities are negotiating the installation of a photovoltaic power generation system whose energy output will be used to operate a pumping station in Wadi-Sudr in Sinai as well as to charge batteries to be used for telecommunications and for radio and television reception.[8] The application of photovoltaic power generation in remote desert areas is thus already established in several developing countries.

In the wind regions of the Egyptian desert the average annual solar insolation[9] is 800 kj/cm^2 and the total average annual hours of sunshine is 3800 hours. These figures indicate that a silicon solar cell array with a peak power capacity of 160 kwp would be necessary to supply 100 families (five persons each) with their daily energy requirement of 500 kwhr. Assuming an installed cost of solar cell array of $30 per peak watt

TABLE 2

MINIMUM PER CAPITA ENERGY REQUIREMENT IN KWH PER DAY

ENERGY USAGE	KWH PER CAPITA/DAY
(1) Domestic water pumping and water supply ($0.12m^3$ per person per day; total pumping head 35m; overall efficiency 33 percent)	.05
(2) Water sanitation	.04
(3) Potable water supply and purification	.006
(4) Educational and Communication devices	.012
(5) Street Lighting (15 lamps of 100 watt each for 4 hour per day; distribution efficiency 70 percent)	.034
(6) Household electricity use:	
(a) Lighting (120 watt for 4 hours a day per family of 5; distribution efficiency 70 percent)	.138
(b) Cooking (1.0 kw for 2.5 hours a day per family of 5; distribution efficiency 70 percent)	.72
TOTAL	1.000

TABLE 3

ESTIMATED TOTAL CAPITAL INVESTMENT FOR WINDMILLS AND LEAD-ACID BATTERY STORAGE SYSTEM

SYSTEM COMPONENT	INSTALLED COST
Windmill generators 18 units each 10 kw at $400/kw	72,000
Wiring and controls	18,000
Installation	18,000
Battery storage $60/kwhr & 85% efficiency	19,000
TOTAL COST	127,000

the installed cost of 160 kw_p solar cell array alone would amount to 4.8 million dollars, or 66 times more expensive than the set of windmill generators of the same energy output. At an interest rate of 7.5 percent and a loading factor of 0.14 the estimated cost of energy by photovoltaics is estimated to be 4.0 dollars per kwhr. This busbar cost of energy by photovoltaics is 47 times more than the equivalent cost of wind generated energy. This indicates that the use of current photovoltaic technology in the Egyptian desert is likely to be limited to very low power output of the order of 1 kw* suitable for powering TV sets, light beacons and other telecommunication systems.

Controlled Photosynthesis Using Organic Wastewater

The generation of methane gas by anaerobic digestion of human, animal and agricultural wastes is a process that is finding more and more use in developing countries[10]. Recently, the technology of gas production from or ganic wastes has been promoted in India and other Asian countries[11]. It is reported[12] that 50,000 such plants were in operation in India in 1978. Farm scale methane generation plants have also been built in significant numbers in the Peoples Republic of China, Korea, and Taiwan[13]. The technology of anaerobic digestion has been available for many years for sewage sludge treatment. Only recently has emphasis been placed on anaerobic digestion technology for energy production.

Anaerobic digestion is a complex biochemical process in which organic matter is fermented in the absence of oxygen to produce methane, carbon dioxide, lower fatty acids, stabilized residue and trace quantities of other compounds. Despite the complexity of the biochemical reactions, anaerobic digestion is a natural process that can gasify organic material without the addition or use of fuel. Moreover, the solid residue produced is an odor-free stable material that does not pose a health hazard, has substantial fertilizer value and, could conceivably, with additives, have sufficient food value as an animal feed. In addition, the digested residue slurry could be used to produce algae by photosynthesis in special ponds (i.e., oxidation, facultative or maturation). Organics in waste-

* See Table 1.

water received by the ponds are oxidized by existing bacteria which derive oxygen through algal synthesis. Algae in turn use the decomposition products of bacterial oxidation (basically carbon dioxide) to grow. Approximately 2 grams of carbon dioxide are required per gram of algae synthesized[14]. Subsequent anaerobic digestion of algae can be used to produce more methane, fertilizer and animal feed than would otherwise be obtained by conventional (i.e., anaerobic fermentation) sewage sludge treatment. A simplified diagram of the integrated waste treatment algae bioconversion nutrient recycling system is shown in Figure 1. Controlled photosynthesis in organic wastewater thus offers a feasible system for supplying energy, fertilizer and animal feed.

However, there are many difficulties in the mass production of algae[15]. Briefly, they are: (1) At low light levels the rate of photosynthesis is controlled by the intensity. In the bright daylight of the desert however, growth is controlled by the rate at which carbon dioxide is produced or fed into the pond. The required high concentration of carbon dioxide is no bottleneck to increased photosynthesis if it can be increased by nearby combustion processes (i.e., combustion of biomass derived fuels); (2) The outdoor algae suspensions may become contaminated with other organisms or toxics which consume or destroy the algae and do not themselves bring about photosynthesis. Nevertheless, Tamiya in Japan[16] and others have been successfully growing algae in large ponds of over an acre; (3) The harvesting is a problem because algae suspensions are not easily filtered and the separation by centrifuging is expensive. The best approach appears to be the selective cultivation of those large microalgae that can be cheaply harvested (½¢/lb of algae). Present species control emphasis is on the filamentous blue-green algae which are easily separated through rotary filters (microstrainers). The growth of such blue-green algae is best promoted under the high water temperature in the desert. A pond heated to 23-25°C produced 25% more biomass than a non-heated pond. An integrated waste treatment algae bioconversion nutrient recycling system would thus be appropriate for desert development.

In experimental work the maximum energy conversion of sunlight into algae was 3 percent[17], and conversion of algae into methane was about 66 percent, corresponding to an overall efficiency of 2 percent for the conversion of sunlight to easily combustible and storeable fuel. Furthermore, one acre of pond could supply 10 to 30 acres of agricultural land with organic fertilizer. Thus, on the average only 5% of any given agricultural area would be required to supply all the necessary fertilizer for that area. The water used in algae grown can be subsequently utilized for irrigation.

It is estimated that one acre of algae pond could give 15 kw of electricity. In regions of bright sunlight and ample available wastewater, with reasonable technical improvements a cost of 11¢/kwhr appears to be possible for large-scale plants. This cost is much higher than the estimated current cost of electricity from windmill generators in the windy regions of the Egyptian desert. A critical assumption is the productivity of microalgae and the necessary solar energy input required to achieve that productivity.

APPLICATION IN DESERT DEVELOPMENT

System Configuration

To put the foregoing technologies in perspective, windmills, photovoltaics and biomass energy farm are combined in a conceptual integrated energy center for desert development. In high wind level regions, windmill generators may be used to supply the energy needed for water pumping and household electricity. Where wind energy is not available, low level electric loads may be supplied by photovoltaic cells (e.g., TV receivers, telecommunication and light beacons). The biomass energy farm may provide water for irrigation, cattle feed, fertilizers and intermediate BTU gas for cookings, and for firing in industrial kilns (e.g., to produce bricks). Biomass synthesis will also aid in the treatment of nearby city wastewater effluent. The geographical location of the chosen desert site and the cost of delivered energy will eventually determine the application priorities and extent of utilization of these renewable energy technologies. A schematic block diagram of the integrated energy system is shown in Figure 2.

Implementation

It is envisaged that the integrated energy center could be developed in three phases. The first phase would be intended to provide the energy needed to improve the basic living environment. This involves provision for adequate drinking water supply, educational facilities and communication systems as well as household electricity. In the second phase energy would be supplied to improve agricultural productivity. Thus the energy needs for irrigation, agricultural processing machinery, fertilizer production and cold storage of perishables would be provided. Energy requirement for long range industrial and community level usage would be provided in the third phase. These requirements will include cooling of public buildings such as schools, hospitals, and other community buildings, as well as the necessary energy to operate small-scale rural industries.

Basic Assumptions

The basic assumptions underlying the conceptual design of the integrated renewable energy system are

1. A desert community that has 100 families consisting of 5 members each.
2. Only one day of water storage is provided.
3. Biomass gaseous fuel storage will be provided to take care of 3 days of low wind.
4. Each family is supported by farming 10 acres of desert land.
5. One acre of algae is necessary to provide the fertilizer needs for 10 acres of farm land.
6. The efficiency of an internal-combustion engine operating on biomass derived gas is 30 percent.
7. The average annual productivity of one acre of aquatic fresh water algae is 20 DTE/acre/year.
8. The annual average wind speed is 23 kph.
9. The average duration of low wind per day is 10 hours.
10. The annual average solar insolation is 800 Kj/cm^2 for 3800 hours per year.
11. Potable and domestic water need is $0.12m^3$ per capita per day.
12. Energy content of biogas (roughly 60% methane) is 5000 BTU per pound of dry algae.

13. Algae is 60% digestible into biogas consisting of CO_2, CH_4 and H_2.
14. Wastewater flow is 10,000 m^3/day.
15. Wastewater chemical oxygen demand, COD, is 1000 mg/l.
16. Cost of algae pond construction and algae processing equipment $10,000/ha[19].

Resource Requirements for the Integrated Energy Center

(a) Land Requirement: Land is required for housing, agriculture, algae farms, energy center, access roads, and community services. In the present context, the agriculture and algae farm land need is the decisive factor in the estimation of total land requirement. These farming areas will cover 1200 acres. Other facilities and access roads will require an additional 200 acres at most. The total land requirement is thus estimated as 1400 acres.

(b) Water Requirement: The water supply must be adequate to provide the water needs for 1000 acres of conventional desert crops, 100 acres of fresh water algae ponds, potable, cooking and other household water requirements for 100 families as well as community service facility water requirements. The controlling factor in the estimation of the total water requirement is the make up water for evaporation from algae ponds. Because of the desert, high ambient temperature (annual average water temperature 28°C), high wind velocity (23 kph) and the vast open area (100 acres) covered by the algae pond, it is estimated that the average evaporation rate from the pond will be .008 m/day (i.e., 3.0 meters per year)[20]. For 100 acres of algae ponds the daily water requirement to make up for evaporation is thus 3238m^3. This amount of water plus the necessary flow providing waste organics for algae synthesis can be supplied by the daily wastewater flow from a nearby city of approximately 70,000 inhabitants. A flow of 10,000m^3 per day is selected (assuming 0.15m^3 of sewage per capita). What remains after evaporation, roughly 70% of the influent flow rate, is more than sufficient to meet the irrigation needs of 1000 acres of conventional desert crops as well as provide for some groundwater recharge. Artesian wells operated by wind-driven mechanical water pumps can be used to supply both domestic and potable water. A resident of our hypothetical desert community will get his water supply from community taps located at several points in the village.

(c) Algae Growth Requirement: For each pound of algae synthesized, some 1.62 pounds of oxygen must be consumed in bacterial oxidation to provide enough carbon dioxide to support algae growth. In other words, to grow 2000 tons per year of algae, 3240 tons of COD must be available in the annual influent. With the assumed influent COD of 1000 mg/l, some 4000 tons of COD are available. The resulting loading is 217 pounds COD per acre per day with a hydraulic residence time of 40 days, values which are typical for tropical or arid sites.

(d) Energy Requirement: For the first phase of desert development the estimated minimum daily per capita energy required to improve the basic living environment is itemized in Table 1, showing a total of 1 kwhr/capita/day. This daily requirement however, does not include the necessary energy for growing, harvesting and processing of algae. The algae farming and processing is a net producer of an amount of energy equal to 2.0×10^{10} BTU/year.

The mass and energy balance diagram, Scheme 1, of the integrated energy system is shown in Figure 3. From the balance we can see that all 2000 tons per year of algae biomass is to be 60% digested[21] yielding a gas of roughly 30-32% carbon dioxide, 61-64% methane and 1 to 2% hydrogen by volume. The balance, some 800 tons per year of slurry, is spread on the agricultural land as fertilizer. Following digestion, 1200 tons per year of gas is cleaned to separate carbon dioxide for return to the algae pond as a carbon source for photosynthesis. The remaining portion, the combustible gas (mainly methane), some 500 tons per year, is available for conversion to electricity or as a fuel for process heat.

If electrical energy is unavailable from the windmills or photovoltaics, approximately 50 tons per year of methane fuel gas are fed to the community energy center for conversion to electrical power at roughly 30% efficiency with the internal combustion engine.

Since 90% of the methane is available for commercial combustion processes, some 1376 tons per year of carbon dioxide can be created. When this is added to the already existent 700 tons per year from the gas separation step, we have some 2076 tons of carbon dioxide available for stimulation of algae growth in the pond. To produce 2000 tons of algae per year, approximately 4000 tons of carbon dioxide must be available. Most of this requirement is met through bacterial oxidation; however in the case of low wastewater flows or low organics concentration, reintroduction of combustion products will increase productivity. Scheme 2 (Figure 4) shows the energy possibilities when only some 20% of the algae biomass is digested with the remaining 1600 tons per year spread on the land in the agricultural development as fertilizer. Yields per unit mass and conversion efficiencies are the same as with Scheme 1; however the energy available for commercial combustion processes is less as is carbon dioxide generation. This particular scheme is more advantageous if agricultural productivity is stressed and fewer outlets for surplus energy are available. The energy requirement for the second and third phase of desert development will depend on the nature and extent of industrialization envisaged and cannot be estimated here.

(e) Capital Requirement: Based on current unit equipment costs, the total plant cost is estimated as 2.8 million dollars itemized as shown in Table 4.

For phase 1 only, the cost of supporting a desert community of 100 families thus amounts to 2000 dollars (i.e., L.E. 1400) per acre of desert developed (with no industrialization involved). This cost of desert development seems reasonable if compared to current experience of land improvement in Egypt (e.g., in Mudiriet El Tahrir).

An even more realistic comparison is the value produced by algae bioconversion systems versus conventional agricultural production. In Egypt the nationwide gross annual production amounts to less than $100 per acre. With bioconversion systems one could probably expect about 200 million BTU's of methane per acre worth about 400 dollars. In addition one acre of algae biomass would produce two tons of nitrogenous vertilizer worth 300 dollars at current market price. Thus, the gross revenue for algae ponds is 7 times that of normal farm land. As shown in Table 5 the total net revenue of both the algae pond (i.e., 100 acre) and the cultivated desert land (i.e., 1000 acre) amounts to 75,500 dollars per year which corresponds roughly to a 37 year payout period. This apparently long payout period is quite reasonable on the socioeconomic basis as discussed below.

Socio-Economic Impact of Desert Development

As envisaged, the application of the integrated renewable energy system of desert development with its ability to provide basic human needs and to create a stepwise agro-industrial environment in remote desert areas will make available a large number of living and work opportunities in these areas. These opportunities should provide for a healthy and dynamic rural growth that will mitigate the problems of mass unemployment and mass migration into the cities of many developing countries. The integrated energy center with its algae pond will also significantly reduce the wastewater disposal problem of overpopulated cities in developing countries.

CONCLUSIONS AND RECOMMENDATIONS

Conclusions

The basic conclusions which can be drawn from this brief analysis is that a mix of municipal effluents, photovoltaics and wind machines can be used to meet the total energy requirements of a small community in a tropical or arid region of the world. The potential proportion of energy available from each of the three sources is a function of wind velocity, solar radiation, wastewater flow and characteristics as well as the manner in which the capital costs are divided among the three systems. The added benefits of such a scheme are the possibility of wastewater disposal from a large adjacent urban area, aquifer recharge, the development of an agricultural area due to the availability of organic algae based fertilizer and irrigation water, as well as net storable energy generation in the form of methane.

A process of phased development is vital to such a scheme since wind machines are initially considered the most cost effective in providing for immediate community energy needs. It is during this first phase that the wastewater conduit from an adjacent urban area, the photovoltaic array, the pond and associated digestion system, can be installed at the site. Once this second phase of energy system construction is complete, the third phase of light industrial development can follow. It is during this last step that all the available energy resources can be totally utilized resulting in a multiplier effect bringing permanent economic development.

TABLE 4

CAPITAL COST REQUIREMENT

ITEM	COST (x10^3 $)
Windmill Generators 18 units 10 kw each at 400 $/kw	72
Pond Stirring Equipment	2.4
Storage Battery System	40.0
Photovoltaic Array 1 kw_p at $30 per w_p	30.0
Algae Pond Construction and Processing Equipment	400.0
Housing and Infrastructure 480 m^2/family at $30/$m^2$	1,680.0
Kiln (0.2 m^3)	1.0
Internal Combustion Engine (gas)	14.0
H_2O and gas storage	15.0
Contingency 25%	563.6
	2,818.0

It is anticipated that a number of such communities can be established on the periphery of large cities in arid or tropical countries. As fossil energy resources become dearer and ecosystem stress increases around urban complexes, such peripheral development becomes increasingly attractive. Increased agricultural productivity through use of algae fertilizer will make these communities self reliant with possibilities for exporting surpluses to urban markets.

The opportunity for developing countries to use every resource available in their path to modernization must be exploited, if their dependence on foreign sources of materials and energy is to be minimized. Therefore soft energy paths using local labor and materials must be developed. The concept outlined in this paper represents one such approach.

TABLE 5

NET REVENUE AND PAYOUT TIME OF THE INTEGRATED WASTE TREATMENT-ALGAE BIOCONVERSION SYSTEM

Biogas Gross Revenue (2×10^{10} BTUs x \$s/MBTU)	\$ 40,000
Nitrogen Fertilizer (2t/acre x 100 acre x \$150)	30,000
Conventional Desert Crop Revenue (\$50/acre x 1000 acre)	50,000
GROSS REVENUE	\$120,000
MINUS	
Operating Cost at (\$15/acre year of pond)	1,500
Cost of Conventional Agriculture (\$15/acre)	15,000
Cost of O&M for energy center (5 percent of equipment cost)	28,000
TOTAL OPERATING COST	\$ 44,500
NET REVENUE	\$ 75,500
Payment - time = $\frac{2.8 \times 10^6}{75,500} \cong 37$ years	

Recommendations

Two principal recommendations are proposed. Initially a detailed technical and economic feasibility study of the described concept or similar concept should follow. The schemes outlined are based on preliminary information using conservative estimates. It is hoped that through a detailed study, both economic and technical feasibility will be determined.

The second recommendation is the securing of financial backing in order to develop a pilot project in a desert area. The project should be of a suitable size and character to clearly determine the feasibility of the concept.

In summary, it is felt that more emphasis should be given to the development of renewable energy sources particularly in areas of the world undergoing modernization. The economic and social constraints are vastly different than in the developed world. Therefore the path of energy development and utilization must be consistent with those constraints.

REFERENCES

1. R. Ramakumar, Technical and Socio-Economic Aspects of Solar Energy and Rural Development in Developing Countries. Solar Energy, 9: 643-649.
2. Norman L. Brown, and James W. Howe, Solar Energy for Village Development, Science, Volume 199, 10 February 1978.
3. W.L. Hughes, Egyptian Energy Resources Study Project - Final Report, (Oklahoma State University, July 25, 1975), NSF Grant Number GF-38938, June 1973.
4. R. Ramakumar, op.cit., 1977.
5. Norman L. Brown and James W. Howe, op.cit., 1978.
6. Ibid., 1978.
7. La Radiotechnique Completec (a brochure), 1977.
8. M.M. Abaza, M.A.B. El Koshairy and I.A. Sakr, Solar Energy In Egypt, A paper presented to the International Conference on Solar Energy organized by Ministere de l'industrie et de lu Recherche, France, 1977.
9. Ibid., 1977.
10. Norman L. Brown and James W. Howe, op.cit., February 1978.
11. R.T. Skrinde, Biomass to Energy Technology - International Solar Energy Congress, U.S. Department of Energy, Contract Number ET-78-X-01-2404-0, 1978.
12. Ibid., 1978.
13. Ibid., 1978.
14. D.W. Hendricks, W.D. Pate, Thermodynamic Analysis of a Primary Oxidation Pond, Journ. Wat. Poll. Control Fed., Volume 46,

15. John R. Benemann, Joseph C. Weissman, Ben L. Koopman and William J. Oswald, Energy Production by Microbial Photosynthesis, Nature Volume 268, July 1977.

16. H. Tamiya, Growing Chlorella for Food and Feed, Proceedings World Symposium on Applied Solar Energy, Phoenix, Arizona, pp. 231-42, 1955.

17. C.G. Golueke and W.J. Oswald, Power from Solar Energy Via Algae Produced Methane, Solar Energy, 7: 86-92, 1963.

18. E.G. Foree, P.L. McCarty, The Decomposition of Algae in Anaerobic Waters, Tech. Dept. No. 95, Stanford University, Palo Alto, California, (1968).

19. J.R. Benemann, et al., 1977.

20. Sabbay Associates, David A. Crane and Partners, Marcel Brener and Associates, Parsons Brinckerhoff International, Inc., Peat, Marwick, Mitchell and Company and Warner Burns Toan Lunde, The Planning of Sadat City, Status Report 1, Preliminary Investigation, Chapter 6, p. 29, October 1976.

21. Foree and McCarty, op. cit., 1968.

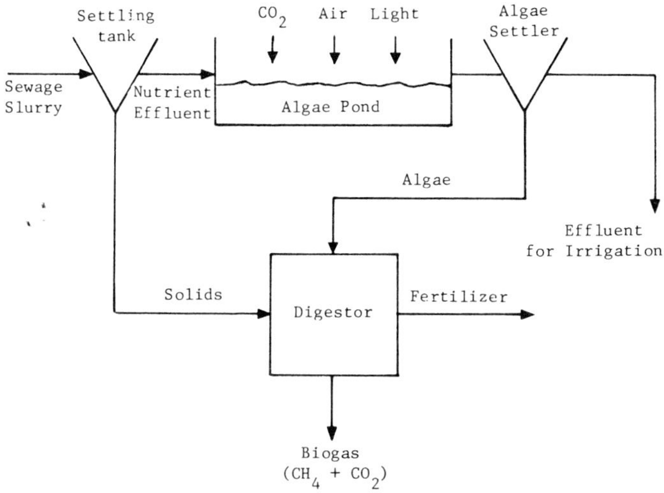

Figure 1. Integrated waste treatment - algae bioconversion system.

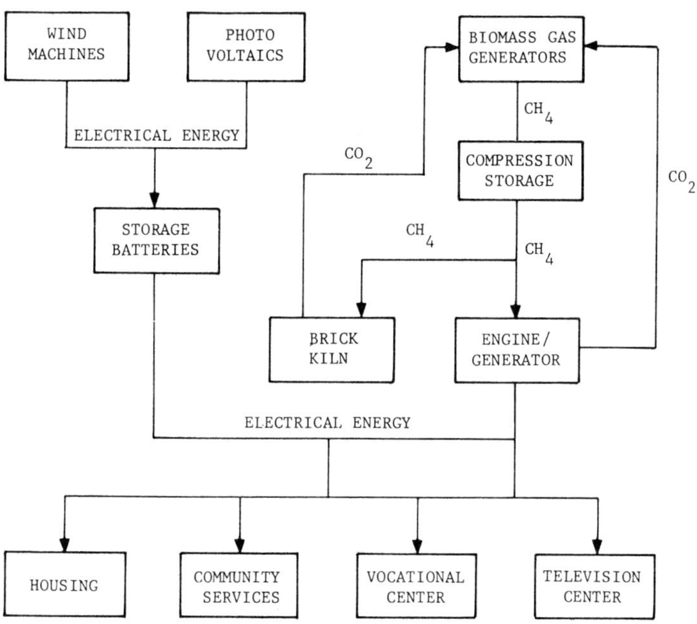

Figure 2. Integrated energy system.

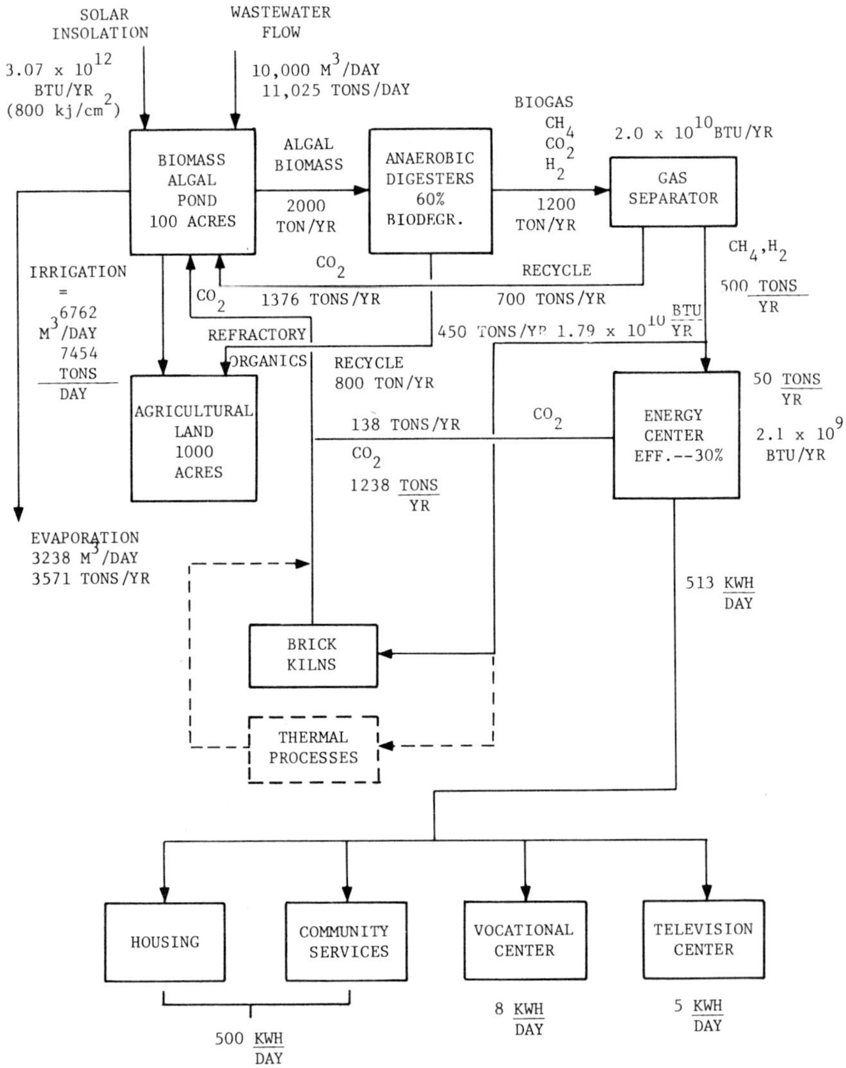

Figure 3. Energy balance diagram. Scheme 1: 100% algae digestion.

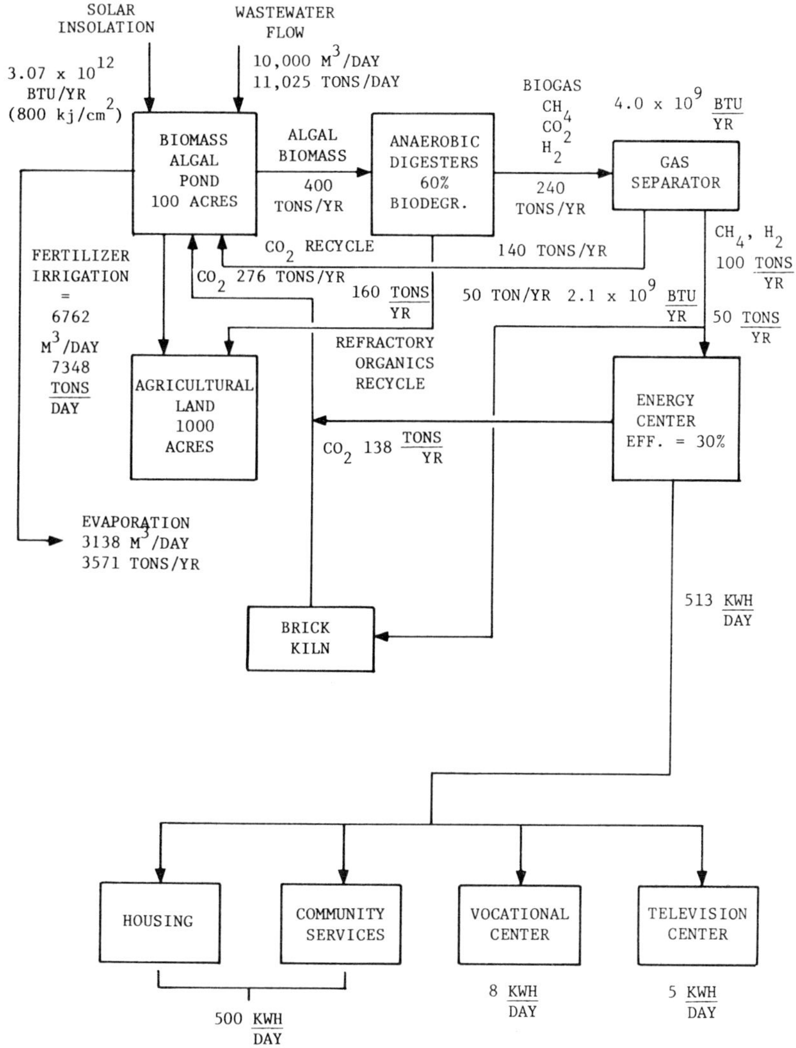

Figure 4. Energy balance diagram. Scheme 2: 20% algae digestion.

ENERGY FORECAST FOR THE WESTERN DESERT OF EGYPT UNTIL THE YEAR 2025

A.I. IBRAHIM ABDEL FATTAH
Prof. Chemical Engineering, Alexandria University

MUSTAFA M. KAMEL
Assoc. Prof. Engineering, Cairo University

ABSTRACT

Settling the Western Desert of Egypt is a rational goal. It is projected that 84 new towns, with a population reaching 23 millions, are to be there by the year 2025 in the four geographical zones: northern Mediterranean coast, the Qattara Depression area, the New Valley and the western banks of Lake Nasser. In this paper the energy needs of such new communities, whether electric or thermal, are forecast. Different forms of energy resources are considered and estimates of the overall costs of new energy projects are presented.

The Egyptian dilemma lies in the fact that more than 95% of its population inhabits less than 4% of its land of one million km^2. To illustrate this imbalance, we find that the population density of the Western Desert, comprising three quarters of the total area of Egypt, is only 0.26 persons per km^2. In the fertile Nile Valley, on the other hand, this figure jumps to 1,300 persons per km^2. This, coupled with a high rate of population increase of 2.5% per year, puts a great burden on Egypt's agicultural land in the Nile Valley due to its utilization for building purposes.

To alleviate this burden, new population centers must be created in areas outside the Nile Valley, particularly in the Western Desert, along the Red Sea coast, in the Suez Canal zone, and in Sinai, to exploit their potentials for agriculture, chemical and mining industries, as well as tourism. These new communities will, of course, create new demands for energy. Long-term forecasting of these demands is therefore necessary.

This paper reports the results of a study conducted in 1972-1975 to estimate the demand for energy during the 50-year period till the year 2025 that will be required to support a potential population of 23 million inhabitants in the Western Desert. For the purpose of this study, the Western Desert was divided into four geographical zones (Fig. 1), namely, (1) the northern Mediterranean Coast; (2) the Qattara Depression area; (3) the New Valley, and (4) the Western banks of the Aswan High Dam lake. Each zone was studied separately to estimate its potential number of inhabitants and hence, its energy requirements. In addition several postulates were made, based on other studies, for use in estimating energy demands. These include: (1) An annual consumption of electric energy of 1,000 KWH per person; (2) An installed electric power equalling 250 W/person; (3) A number of inhabitants in new towns ranging from a minimum of 100,000 to a maximum of 750,000 persons, and (4) A potential for land reclamation in the Western Desert of 2.3 million feddans, including 500,000 feddans in Zone I, 1,300,000 feddans in Zone III, and 500,000 feddans in Zone IV.

ESTIMATED DEMAND FOR ENERGY

According to a study conducted by the Egyptian Ministry of Energy and Electricity, a total of 46,200 km^2 can be utilized for housing in the Western Desert. This area has a potential for support between 14 and 23 million persons, with a population density of less than 500 persons/km^2. The estimated potential population distribution in the four geographical zones of the Western Desert is detailed in Table 1. To house this projected population, a total of 84 new towns are needed, with a number of inhabitants ranging from 100,000 to 750,000. The distribution among the four zones is indicated in Table 2.

Establishing and developing these new towns requires both electric and thermal energies for lighting, industry, agriculture and locomotion, as well as other household usage. Reports have been issued by the Egyptian Ministry of Energy and Electricity regarding expected loads and demands of electric energy for the new communities of Table 2 which were used in the present study. These are summarized in Table 3. Table 4 gives the total installed power required for the four geographical zones, which amounts to 5,765 MW in the year 2025 for a projected population of 23 millions.

Thermal energy requirements, on the other hand, were statistically estimated by the author to an average of 14 KWH per day per person, or 5,100 KWH/year per person. Thus, again

TABLE 1. Estimates of Potential Population Distribution in the Western Desert

Zone	I	II	III	IV	Total
Possible housing area km^2	4200	4200	33600	4200	46200
Population, millions	3.35	0.7	8(17)	2	14(23)
Population density/km^2	800	166	240 (500)	475	300 (500)

TABLE 2. Projected New Towns to be Constructed in the Western Desert

Zone	Population	100,000	200,000	300,000	500,000	750,000
I	3.35 mill.	7	4	2	1	1
II	0.7 mill.	7	-	-	-	-
III	8*	28	5	4	3	2
IV	2	20	-	-	-	-

*Can be increased in the second phase to house 9 millions more

TABLE 3. Electric Power Demand for Each Town

Town Population	100,000	200,000	300,000	500,000	750,000
No. of towns	62	9	6	4	3
Installed power per town in MW	25	50	75	125	200
Total required MW	1540	450	450	500	600

based on a projected population of 23 millions, the total demand for energy in the Western Desert is summarized in Table 5.

POSSIBLE ELECTRIC ENERGY RESOURCES

To meet the projected demand for energy, electric energy should be generated through all available means, namely hydro-thermal, nuclear, solar and wind power plants. An optimal amalgamation of all these methods should be conscientiously chosen.

Hydro-power Resources

If we do not take into account the existing hydro-power plant of the Aswan High Dam, the most promising project being studied now is the Qattara Depression hydro-solar plant with a projected peak load of 2,400 MW. It utilizes the difference in water level between the Mediterranean Sea and the depression's artifical lake.

Another hydro-power plant of promise is that at the entrance to the proposed Kharga Canal. Having a base load station of 70 MW, it is to utilize the drop in water head from 175 m upstream of the High Dam to 140 m at the canal's entrance.

Nuclear Power Stations

One nuclear power station is to be constructed at Sidi Kreir, west of Alexandria, with a single unit of 600 MW capacity of electric power, in addition to 20,000 m^3/day desalinated sea water. Future plans are also made to double the electrical output to 1,200 MW and the desalinated water to 40,000 m^3/day, which is equivalent to 15 million m^3/year.

It is also proposed to add another nuclear power station west of Mersa Matruh, also on the Mediterranean coast, to bring the total installed nuclear power to 2,400 MW base load with an average cost of $1,250 per KW of installed power.

Solar Energy

The average annual radiation intensity or solar energy in the Egyptian Western Desert is about 1,800 KWH/m^2 in the northern zones and 2,500 KWH/m^2 in the southern zones. Assuming an average radiation of 2000 KWH/m^2 annually, the amount of solar energy falling on the Western Desert surface area of 0.75 million km^2 will be about 1.5×10^{15} KWH/year. If

TABLE 4. Total Installed Electric Power for the Four Zones

Zone	I	II	III	IV	Total
Population in millions	3.35	0.7	8(17)	2	14 (23)
Installed power in MW	840	175	2000 (4250)	500	3500 (5765)

TABLE 5. Energy Demands for 23 Million New Inhabitants

Type of Energy Needed	Demand/annum	Units
Electric	20 million	KWH
Thermal	11.7×10^{10}	KWH

only 1% of this energy is to be utilized at an efficiency of 10%, the available net energy would be about 1,500 M KWH/year.

Wind Energy

It is estimated that the average wind power blowing annually over the North-Western coast of Egypt is about 100 watts per m^2 of wind mill area. Since this coast stretches for 500 km with an average inhabitable depth of 10 km, it is estimated that an area of 5,000 km^2 can utilize wind energy along the coast.

The efficiency of wind turbines is rather high relative to solar systems, reaching 35% with a theoretical maximum of 60%. By comparison, the efficiency of solar electric systems reaches 5%, and when power is produced through driving a shaft the efficiency is reduced to 1.5%. When installing wind mills one has to keep them at a certain height to reduce ground effect. In addition, if wind mills are to be cascaded, one has to make sure that the wake of one mill does not cause a drop in the efficiency of the following one. Since installing such a large number of metallic wind mill blades can cause massive interference with television reception, the blades should be constructed from fiber glass which, in addition, may reduce the cost and weight of the blades.

With an average wind power of 100 watts/m^2, one may be able to recover 100 KW from a mill of about 36 meters diameter which would be sufficient to fill the power needs of a small community of 400 persons. Assuming one such mill is to be placed every 10 km^2 in the area bounded by Alexandria to the east and Sallum to the west, the total wind power to be generated can reach 50 MW, supporting a population of 200,000 persons.

THE INTIAL COST FOR SOLAR AND WIND UNITS

Solar water heaters are now available on the market at prices comparable to that of propane-butane heaters. For example, a 120-liter solar heater costs about LE 100. Solar stills may now produce 1 m^3 distilled water at a cost of 60 piasters, which is economical for use in small communities in isolated areas where the cost of transporting water is prohibitive. Solar cookers and pumps, however, are still uneconomical for use on a large scale.

Solar electric power generation is also rather uneconomical and initial costs vary widely. For example, a 100 MW solar power plant was suggested to cost an estimated \$175,000,000, or \$1,750 per KW. Another, 1 MW, unit was suggested costing \$12,000,000 or \$12,000 per KW. This large variation in estimated initial cost indicates that the required technology is still in a state of flux.

Three large experimental wind mill power generators were constructed in the United States. They are designed to generate 100, 1,500 and 2,500 KW and have wind mill diameters of 38, 60, and 90 meters, respectively. The initial cost per KW installed for the first and second mills were \$5,500 and 2,000. No estimate is available for the third. It is also reported that a 175 MW wind mill generator is now operating in the United States that was designed for an average wind speed of 27 km/hr and a maximum speed of 45 km/hr. It has three fixed pitch steel blades of 25 m diameter, installed on a 25 meter tower. Smaller units producing 2-3 KW also exist that have an initial cost of \$1,100 to \$1,800 per KW. Other units of 15, 25 and 30 KW are manufactured which have an estimated initial cost of about \$1,300 per KW.

Although the above estimates per KW are high, it is thought that mass producing these units would reduce the cost to \$750 per KW and \$500 per KW for 8 and 40 KW units, respectively.

FOSSIL FUEL

Several companies are searching for oil in the Western Desert with some encouraging results. For a project population of 23 millions, oil, of course, will be the main type of energy required for transportation. Assuming a need for 250 liters of oil per person per year, 4.6 million tons will be required annually. This is equivalent to 5.33×10^{10} KWH per year, which is only 45% of the required thermal energy.

Another 10 million tons of oil, costing over one billion dollars, are required annually for household usage. Thus, extensive use of solar cookers, water heaters, air conditioning and solar refrigeration should save a large percentage of the needed oil.

TOTAL ELECTRIC ENERGY RESOURCES

Taking into account all the above types of energy, the total resources capable of producing electric energy in the the Western Desert can be summarized as follows:

The Qattara hydro-solar power station	2,400 MW (peak)
Nuclear power station at Sidi Kreir with extension	1,200 MW (base)
Nuclear power plant west of Mersa Matruh	1,200 MW (base)
10 solar power stations, each 100 MW	1,000 MW
500 wind mill power stations, each 100 W	50 MW

Thus, a projected total of 5,850 MW electrical power should be generated to cover the required 5,765 MW as indicated in Table 4. The above energy projection does not include any additional fossil fuel thermal power stations, which can also be used to generate further amounts of electrical energy.

TOTAL THERMAL HEAT RESOURCES

As indicated in Table 5, the projected population in the Western Desert will require 11.7×10^{10} KWH annually. This amount can be covered by tapping just 7.8% of the 1.5×10^{12} KWH solar energy available annually. For transportation, however, fossil fuel resources should be used.

TOTAL ENERGY COST

The electrical energy cost of the whole project of settling 23 million inhabitants in the four zones of the Western Desert can be summarized as follows:

	MW	$/KW	Total, $
Hydro-solar	2,400	1,430	$3,268 \times 10^6$
Nuclear	2,400	1,250	$3,000 \times 10^6$
Solar	1,000	1,750	$1,750 \times 10^6$
Wind	50	1,000	50×10^6
TOTAL	5,850	5,430	$8,068 \times 10^6$

The thermal energy required for transportation, 4.6 million tons of oil at $120 per ton, will total $\$552 \times 10^6$ annually. Solar water heaters and cookers for 5 million families, comprising the projected population, will add another one billion dollars.

Thus, the total energy cost would add up to \$9.068 x 10^9, which is equivalent to LE 6.347 x 10^9 at the current exchange rate of 0.70 LE/\$.

MINERAL RESOURCES AND DESERT DEVELOPMENT: A RATIONAL APPROACH

RUSHDI SAID
Consulting Geologist, Cairo

ABSTRACT

One of today's pressing problems is, how much can the sea bed and the deserts of the world give in the way of resources which are becoming depleted and polluted by the unrestrained and exponential growth of production and population. This paper attempts to evaluate the resources potential of the Egyptian deserts, outline the physical and social difficulties of their development, and propose actions to preserve and utilize these resources rationally.

Foremost among the mineral resources of the deserts of Egypt is the reservoir of ground water in the Western Desert, estimated to hold about 230,000 cu. km. of fresh water. The daily discharge is estimated to be 3,375,000 m^3*. The concensus is that the recharge is minimal; it has been held in the pore spaces of the Nubia Sandstone reservoir since the pluvials of the Pleistocene and Holocene epochs. Other mineral resources include, in addition to large quantities of phosphate reserves and industrial and building materials, small reserves of iron ores and base metals. Traditional energy resources are lacking in most desert areas. Except for limited quantities of coal known in Sinai, fossil fuels in the form of oil and gas are known around the Gulf of Suez, the Delta and the northern reaches of the Western Desert. Until an energy source becomes available on an economic basis, the future of the desert will necessarily be limited. Pumping of water, industrialization and use of mineral resources all depend on the availability and cost of energy. Until such non-traditional sources of energy are developed, the reclamation of the desert will be limited to centers around areas of some mineral resource, primarily water.*

The development of these centers has its challenges. Among the physical difficulties are the ever drifting sands which accumulate around new structures, the limited infrastructure and the harshness of the environment. Foremost among the social difficulties are the attitudes of new settlers to the new and necessarily higher quality of life which will entail better health conditions and higher expectancy of human life. If older attitudes continue to prevail, the proliferation of human life will damage all benefits that could ensue from development.

By the year 2025 the population of Egypt is estimated to reach about 85 million, provided that the annual rate of increase is 2% for the coming twenty years and 1% for the following twenty-five years. In other words, forty-five million new individuals will be added to the present day population. The provision for these of a space and a standard of living equal to that which present-day Egyptians have presents a formidable task which can be accomplished only with serious planning by serious authorities well versed in scientific and rational thinking and armed with knowledge of the environment and the limits of its use. To provide even this minimum space requirement of 700 m^2/individual, below which conditions would be barely human, needs the full utilization of almost all the potential of Egypt without any allowance for waste, mismanagement or faulty planning. Six million acres of land have to be reçlaimed during the coming forty-five years. In addition, nine million new housing units, twenty million new jobs and an estimated total energy source of 70 to 80 billion kilowatt/hours per year have to be provided.

In the following paragraphs a summary survey of the most important and critical resources of Egypt is given and the limits of their use is outlined.

WATER

Water represents the most critical resource in Egypt. Rainfall is insignificant; only the northern stretches of the land and for a short distance inland receive about 200 mm/year and this is used for sporadic cultivation along the coasts.

As will be seen from the review of the ground water potential of Egypt, the Nile represents the most important source. It does supply Egypt today and will continue for a long time to supply Egypt with almost all its water requirements. The Nile is a humble river. Its average discharge is 86 billion m^3/year, an amount which makes this longest river in the world go to the bottom of the list of the major rivers with regard to discharge (Table 1).

Table 1: The Nile Compared to Other Major Rivers

River	Length Km	Drainage Area Km^2	Annual Discharge Billion m^3
Nile	6,825	3,100,000	86
Niger	4,100	1,890,000	180
Congo	4,700	3,700,000	1400
Zambesi	2,700	1,300,000	500
Mississippi	3,970	3,220,000	600
Amazon	6,700	7,050,000	3000
Danube	2,900	1,165,000	200
Rhine	1,320	162,000	80
Mekong	4,200	795,000	400

The discharge of the Nile fluctuates from year to year and varies from 95.5 to 80 billion m^3/year. Egypt's share of this water is 55.5 billion m^3/year (or 1760 m^3/second) which is consumed around the year at the rate of 926 m^3/second during the winter months and 2660 m^3/second during the summer months. Recent studies by several experts indicate that about 15% of the water reaching the lands of Egypt can be saved without involving dramatic changes in the habits of the Egyptian farmer and his methods of flood irrigation. This surplus water in addition to the share that will accrue to Egypt after diverting the Upper Nile in Southern Sudan to avoid the Sudd region along the Jongli Canal, represents all the surplus that Egypt can obtain from the Nile.

The main groundwater aquifer system lies in the Western Desert and consists of a thick sequence of sandstones with clay lenses overlying basement rocks. The system is confined by a thick clay sequence in the oases and their vicinity. This produces artesian conditions and results in the free-flowing wells in the depression areas. The regional hydraulic gradient is from the southwest.

Recently a mathematical model was constructed to simulate the aquifer system in an area 86,000 km^2 surrounding the oases. The polygon network was devised to fit the known external and internal boundary conditions and extraction pattern. The system was modeled with two layers representing the upper production zone tapped by shallow wells and the lower production zone tapped by recently drilled deep wells. The model was calibrated first against the pre-1956, quasi-steady state piezometry, and then using the history of discharge and head charges from 1956 to 1975. Thus calibrated, the model was used to provide forecasts of the response of the system to alternative extraction patterns. The model showed that it would be possible to increase extractions to about 150 million m^3/year in Kharga and 630 million m^3/year in Dakhla through 10-20 year implementation programs, without the water level falling to more than 100 m below ground level in the irrigated areas and without having to resort to major lateral transfer of water at the surface.

For about the last twenty years, groundwater extraction in the Kharga and Dakhla depressions has exceeded natural recharge with resultant head decline and falling free flow discharges. The groundwater reservoir is immense, but the availability of water is essentially limited by the pumping head which is viable, that is, by the energy costs. The continuation of the present water allocation to the lands now cultivated (275 million m^3/year) would leave about 30,000 million m^3 which can be extracted during the coming fifty years with pumping lifts about fifty meters but not exceeding 100 meters. This order of magnitude of the "economically available" water corresponds to an annual extraction rate ranging from 575 to 800 million m^3/year depending on the rate of utilization. The development of new lands could be extended to over 5000 feddans in Kharga and 28,000 feddans in Dakhla.

Water resources available to Egypt are, however, mainly from the Nile whose proper use would permit the addition of *3 million* feddans of agricultural land.

ENERGY

Energy is another critical resource in Egypt. At present the peak load is about 12 billion kilowatt hours and the per capita share of electricity is 340 kilowatt hours. Of the total available energy, close to 75% is consumed in industry, 12% in agriculture, and 10% for domestic purposes. The High Dam contributes today 10% of the total electricity available in Egypt and the remainder comes from thermal stations. Future energy requirements are estimated to rise to 1200 kilowatt hours per capita by the year 2000, if the majority of the Egyptians still live in urban industrialized areas. The total energy requirement would then be 70 to 80 billion kilowatt hours/year or seven times the amount generated today. The securing of this enormous energy presents a great challenge that will need imaginative thinking and the use of modern technology to supply Egypt with its needs in the future.

MINERAL RESOURCES

Egypt's mineral resources have been sought after since time immemorial. Both ancient and modern Egyptians are known to have made enormous efforts exploring for minerals and using them in a most rational way. During the past fifteen years alone, an extensive program of mineral exploration using sophisticated techniques has been carried out. It has involved the employment of scientific personnel (approximately 50,000 man months of whom about 8000 were foreign experts from six nations), 100,000 man months of drilling, 500,000 geochemical analyses, airborne and ground geophysics, tunnelling and detailed mapping. The mineral wealth of Egypt is reasonably well known, and one can safely say that Egypt has a few occurrences of base metals, a limited quantity of iron ores, but is rich in phosphates, evaporites, salines, limestones, building materials, glass sand, kaolin and other earthy materials. These could represent the base upon which industry in Egypt should be established. Further, many industries, when properly managed, can afford to import their raw materials, and these possibilities must be studied.

FUTURE OUTLOOK

A survey of the three most critical factors in the future development of Egypt, namely water, energy, and mineral resources, shows that agricultural expansion in Egypt is limited. Since only about 16 billion m^3 of water can be saved or gained from future Nile control projects and groundwater reservoirs, it is difficult to conceive that future Egyptians will be able to reclaim more than three million feddans for agriculture. Considering present day standards of living and the density of population, only 9 to 10 million people can make a living out of these. The remaining thirty-five million expected to live in Egypt fifty years from today will have to settle in urban areas and earn a livelihood from industry and services. If the present-day space requirements are used as a measure, it is estimated that the thirty-five million urbanized Egyptians of the year 2025 will need a space of three million feddans, about three billion additional m^3 of water/year (100 m^3 per capita/year) and about 3600 million kilowatts hour a year.

Table 2 shows the expected space, water and energy requirements of the additional Egyptians who will live in Egypt in the year 2025 as compared with present day statistics.

Table 2: Space Occupied, Water and Electricity Consumed by Egyptians in the Year 1975 as Compared to 2025

		Space occupied		Water consumed		Electricity consumed	
A: 1975	No. of people living millions	Total (million m^2)	Per individual (m^2)	Total (billions (m^3)	Per acre (m^3)	Total billion Kw/h	Per individual Kw/h
Rural (Mainly Agriculture)	18	6	1400	54	9000	3	166
Urban (Mainly Industrial)	18	½	116	1.5	80*	9	500
	36	6½		55.5		12	
B: 2025							
Rural (Mainly Agriculture)	27	9	1400	54	6000	6	220
Urban (Mainly Industrial)	58	2	144	6	100	58	1000
	85	11		60		64	

* Per individual

From the above table it can be shown that if the present fertility rate continues for the coming fifty years, a very difficult situation would result, since the expansion of agriculture is limited by the amount of available water and the development of industry is hampered by the lack of energy resources. However, the table shows clearly that Egypt has to turn to industry and to greater urban expansion. There is no doubt that the planners of Egypt face an immense task which requires maximum efficiency, imagination and vision and it is perhaps high time to initiate appropriate programs for reforming the social structure in order to induce zero population growth.

TECTONIC CONTROLS UPON THE PRESENT-DAY EGYPTIAN DESERTS

DAVID R. LAWRENCE AND WILLIAM H. KANES
Professors, Department of Geology, University of South Carolina, U.S.A.

EL SHAZLY M. EL SHAZLY
President, Nuclear Materials Corporation, Cairo, Egypt

ABSTRACT

An appreciation of present-day desert environments must include knowledge of their past history and the geologic record responsible for their formation. This geologic history can be quite ancient. Cretaceous-Holocene (last 100-125 million years) tectonic events have most strikingly affected the Egyptian deserts, yet many of these events represent reactivation of Pan-African (500-700 m.y.) and/or Proterozoic (over 1000 m.y.) zones of weakness in the earth's crust.

A Cretaceous thermal event included uplift and arching south of Egypt's present deserts. This uplift resulted in deposition of the widespread, non-marine Nubian facies sediments, which now include the water-bearing sandstones beneath the Western Desert. Regional subsidence and sea level rise followed the Nubian interval; a Cretaceous marine transgression, from the north, is manifest in the overlying phosphate deposits.

This thermal arching progressed northward through the earth Cenozoic, following the NNE-SSW axis of the Proterozoic rocks. Extensional (tensional) tectonics accompanied and followed this arching; basinal, marine, early Cenozoic deposits were formed in downthrown crustal blocks (e.g. Nile Valley, Duwi zone.

Cenozoic rifting, associated with the opening of the Red Sea, also progressed northward through time. The relative uplift of the Red Sea Hills, during rifting, now confines the Egyptian deserts on the east. The Red Sea phosphate deposits are preserved and mineable because the Duwi zone was relatively depressed during the uplift, which stripped the sedimentary cover from the main of the basement rocks.

Continuing modeling of tectonism and ancient sedimentary environments should materially aid the development of Egypt's iron ore and phosphate industries. Continuing and expanded regional geologic syntheses should help to bring these regions into full perspective view.

INTRODUCTION

The Red Sea region continues to draw geologic attention for a variety of reasons. Many of these reasons involve the youthful nature of this oceanic area and the belief that, by coming to understand tectonism in this recently and still active zone, we may better comprehend the early and formative periods of the earth's major ocean basins. Since 1975 an international group of earth scientists has been examining the geologic record of eastern Egypt, with these thoughts in mind. Primary aims of this work have included: (1) analyzing tectonism and resultant rock structures in the more than billion-year geologic record of this region, (2) examining the influence of more ancient events upon the striking Cretaceous-Holocene (last 100-125 million years) tectonism, (3) evaluating sedimentary responses to tectonism, in this area, during the Cretaceous-Holocene interval, and (4) providing appropriate results to Egyptian authorities for use in planned resource development. Results to date suggest that (1) the major Cretaceous-Holocene tectonic events represent reactivation of older zones of weakness in the crust, (2) tectonic controls over sedimentation have been profound, in the Cretaceous-Holocene sequences of strata, and (3) continued work and modeling of tectonism and sedimentary basins should materially aid the development of Egypt's phosphate, iron ore, water, and petroleum resources.

Recent tectonic models (including Dewey and Bird;[1] Falvey[2]) suggest that ocean basin formation and sea floor spreading may result from a sequence of events spanning 50 million years (my) or more in time. Important events include: (1) regional thermal uplift and arching which produces erosion, crustal thinning near the axis of arching, and a widespread blanket of nonmarine and shallow marine sediments on the flanks of the uplifted zone, (2) block collapse of the up-arched area and the formation of a rift valley basin whose deposits may be both sedimentary and volcanic, and (3) break-up, with the generation of new crust along the rift axis giving rise to sea-floor spreading (Falvey).[2] Numerous lines of evidence indicate that this sequence of events has progressed northward through time in the Red Sea area (Hutchinson and Engels;[3] Lowell and Genik;[4] Soliman et al;[5] Swartz and Arden.[6]) Evidence for new oceanic crust and sea-floor spreading exists in the southern REd Sea (Girdler and Styles);[7] data from the northern Red Sea are meager and less conclusive but point towards continental crust underlying the offshore areas. The mechanics of the opening of the northern Red Sea is one major unresolved problem in Middle East geology.

The Egyptian region does include evidence of both the thermal arching and rift valley phases, with evidence manifest in both rock sequences and rock structures of the Egyptian Red Sea area.

ROCK SEQUENCES IN EGYPTIAN RED SEA AREA

Three major sequences of rocks crop out in the central and eastern parts of Egypt-- basement, Cretaceous-Eocene, and Neogene units (Figure 1). Basement rocks, largely igneous and metamorphic, cover much of the Eastern Desert; the ages of these rocks are mainly Proterozoic (over 1000 my) and Pan-African (500-700 my) (El Shazly).[8]

A second and Cretaceous-Eocene sequence (Figure 2) involves desert and Nile Valley exposures but also crops out close to the Red Sea in a narrow strip near Quseir (Duwi trough or zone of Lawrence and Meguid;[9] see Figure 1). This sequence documents thermal uplift and the earliest phases of tensional tectonics. The sequence begins with the widespread Nubia Formation. In the Eastern Desert, the Nubia consists of two major subdivisions (Ward and Mansour)[10]: (1) Basal crossbedded sandstones and conglomerates are overlain by a sequence of shales and burrowed, fossiliferous sandstones; the oolitic iron ores near Aswan (Attia)[11] occur near the top of this sequence. These units record an initial pulse of uplift (block-faulting) followed by fluvial deposition and a marine incursion. (2) Overlying crossbedded sandstones and conglomerates are succeeded by siltstones and shales. This subdivision documents regional uplift to the south and the northward flow of meandering streams which spread alluvial deposits over wide areas. As erosion proceeded and arching waned or moved northward, the influx of coarser sediment decreased; subsidence and sea level rise brought deltaic and coastal zone sediments progressively southward over the alluvial deposits (Ward and Mansour).[10] Continued modeling of Nubia deposition will provide additional information about the origin of this important aquifer. Continuing detailed studies of the Aswan iron ores will provide a sound basis for interpreting the depositional environment in which they formed; these studies will, in turn, be a major source of information for the interpretation of other oolitic iron deposits.

The Nubia and its uppermost variegated shales are overlain by a sequence (Duwi Formation of Red Sea area) of sandstones, shales, oyster beds, and phosphorites; these phosphate deposits are being mined or developed in the Duwi zone, Nile Valley, and Western Desert regions. Coastal zone conditions continued during this deposition; exposures near Quseir were important in the development of a shallow marine, perhaps estuarine, model of phosphorite deposition (Yousef).[12] Although Duwi facies are coastal zone ones, reconnaissance suggests that the phosphorite may have been formed elsewhere and transported into the depositional sites (R. E. Garrison, pers. comm., 1977). Additional field studies and modeling of phosphorite deposition should significantly aid the development of this important Egyptian industry.

The phosphorites are overlain by a Late Cretaceous-early Eocene series of shales with minot chalk. These rocks show many attributes of moderate to deep water marine deposition; these same environments are reflected in the succeeding Thebes Formation of early Eocene age (Figure 2).

Recent work in the Nile Valley (Snavely, Garrison, and Meguid)[13] suggests that Thebes carbonates were there deposited in a deeply submerged and tectonically stable environment. By contrast, the Thebes in the Duwi zone suggests sedimentation in a submerged, tectonically unstable setting. Intraformational conglomerates and shell-hash intervals occur within sequences of deep-water cherty limestones and chalks; the presence of shallow-water fauna in the redeposited carbonates indicates proximity to a platform setting (Snavely, Garrison, and Meguid).[13] These rocks grade upward to shallow-water, perhaps subaerial, facies and the Thebes, in the Duwi zone, is overlain by nonmarine conglomerates and marls normally assigned to the Nakheil Formation.

The Late Cretaceous-early Eocene sequence records responses to the tensional forces which accompanied and followed thermal doming. Depressions formed in zones of prior weakness and clastics were deposited there while adjacent areas were raised; as arching moved northward, platform settings surrounded the structural lows and carbonate deposition ensued.

Instability was greater in the Duwi trough because of the area's proximity to the axis of arching. Northern Egypt became a high and uplift and emergence were conspicuous there during the following Oligocene time (El Shazly).[8]

The third major sequence, a Neogene one, is limited to a narrow strip bordering the Red Sea and Gulf of Suez. Clastics, carbonates, and evaporites are all prominent rock types (Figure 3). During most of this time, the Red Sea margin probably looked much as it does today: alluvial fan (wadi) deposits in structural lows, alluvium grading seaward to beach deposits, and a complex reef system offshore from basement structural highs (Johnson).[14] One major exception in the middle Miocene Gulf of Suez-Red Sea evaporite basin is as yet incompletely understood. Subsurface data are avilable and excellent surface exposures are present along the basin margin, making this area virtually unique in the possibility of studying both marginal and basinal depositional areas of an evaporite setting simultaneously. Our continuing evaporite basin analysis will provide interpretations of interest to Egypt's petroleum industry and will yield a process-response model for this and other similar evaporite basins of the world.

This Neogene sequence records tensional failure of the uparched area and represents a rift valley phase of tectonics. Sedimentation has been affected not only by NW or NNW basement trends (faults) but also by transverse (NE-SW or ENE-WSW) growth faults (Johnson)[14] which have been active throughout the Neogene. These transverse faults control the positions of major present-day wadi systems and Pb-Zn mineralization, in middle Miocene clastics (El Akkad and Dardir),[15] occurs in transverse fault zones along the Red Sea coast. Other structural events are best seen in historic perspective.

STRUCTURES IN THE EGYPTIAN RED SEA AREA

El Shazly[8] has suggested that the major structural elements of Egypt are old rigid crustal plates bounded by NNW-SSE faults, with the Red Sea segment acting as a low and the Eastern Desert segment as a high. The availability of LANDSAT imagery has increased our knowledge of structures within the individual plates.

Oldest basement structures show E-W to ESE-WNW trends but have been largely obliterated by younger events. Subsequent Precambrian structures show WNW-ESE to NW-SE and NNE-SSW (sometimes NE-SW) trends. The details of Paleozoic structures are not well known in Egypt but following Mesozoic tectonics is dominated by NE-SW to ENE-WSW ("Syrian arc") trends (El Shazly).[8] Additional, less clear, NW-SE to NNW-SSE Mesozoic structural trends may have resulted from the block faulting which preceded thermal uplift.

All of these older structural trends reoccur in Cenozoic tectonism and the NNW-SSE and NW-SE trends are especially evident along the Egyptian Red Sea margin (Figure 4).

The Duwi zone displays the older trends quite strikingly. Because of the phosphates in the Cretaceous-Eocene sequence, detailed mapping and structural analyses have occurred in the Duwi zone (for example, Issawi et al.;[16] unpublished reports on file with Red Sea Phosphate Co.). Uplift and extension of the Duwi trough, beginning in early Eocene-middle Miocene time, have resulted in the preservation of the Cretaceous-Eocene sequence in over 14 discrete blocks or units. Block boundaries display the older structural trends. The blocks were tilted and rotated and often include marginal drag folds. Block tilting and rotation do largely obscure horizontal motions in this zone.

Structural style within the Duwi blocks is relatively simple and pervasive (Figure 5) and includes step faulting, local grabens and horsts, and collapse structures. Post-Thebes uplift of the zone was greater in the north than in the south; the northern, succeeding intermontane basin deposits (Nakheil Formation) give way southward to coastal zone sands overlain by basic (pillowed?) volcanic rocks. Despite uplift, the Duwi zone has remained a relative depression since Eocene times.

Recently Lawrence and Meguid[9] proposed a structural and historic kinship between the Duwi zone and the Gulf of Suez and suggested that the Duwi zone may be our best outcrop window into the Gulf of Suez floor. Egypt's phosphate and petroleum industries will both benefit from additional studies in the Duwi zone.

SUMMARY AND DISCUSSION

Although major events in forming the present-day Red Sea may span but 40 my (Girdler and Styles),[7] rock structural controls over the Red Sea and eastern Egypt are much older. We interpret the Cretaceous-Holocene history of eastern Egypt via two stages in Red Sea formation: (1) early, northward-progressing thermal arching and (2) following tensional tectonics which culminated in rift valley formation. Additional complexities are present because northern Egypt has also been affected by the Cenozoic transformation of Tethys into the Mediterranean Sea.

Basement outcrop patterns suggest that thermal arching became more confined as it proceeded northward and followed a NNW-SSE trend in the northernmost Red Sea area. Important aquifers, iron ores, and phosphate deposits formed in response to arching. Early extension is preserved in depressions whose axes are oblique to present-day Red Sea trends-- depressions including the Duwi trough near Quseir. Rift valley settings have characterized the Egyptian Red Sea and Gulf of Suez margins since middle Miocene times. These tensional settings contain both phosphate and petroleum reserves. The means of separation in the northern Red Sea remain obscure. Agreement on the amounts, dates, and types of displacement along the Dead Sea trend will be necessary before holistic models of Red Sea tectonics can be developed.

ACKNOWLEDGMENTS

We thank H. A. El Sayed for administrative support and S. El Akkad, H. A. Hassan, A. A. Meguid, the Remote Sensing Center in Cairo, and the Egyptian Atomic Energy Establishment for scientific and/or logistic support. Discussions with R. E. Garrison, P. D. Snavely III, and W. C. Ward have been especially fruitful. Funded by NSF grant OIP75-07943.

REFERENCES

1. J.F. Dewey and J.M. Bird (1970). Mountain belts and new global tectonics. J. Geophys. Research 74:2625-2647.
2. D.A. Falvey (1974). The development of continental margins in plate tectonic theory. Austr. Petr. Expl. Assoc., p. 95-106.
3. R.W. Hutchinson and G.G. Engels (1972). Tectonic evolution of the southern Red Sea and its possible significance to older rifted continental margins. Bull. Geol. Soc. Am. 83:2989-3002.
4. J.S. Lowell and G.J. Genik (1972). Sea-floor spreading and structural evolution of southern Red Sea. Bull. Am. Assoc. Petr. Geol. 56:247-259.
5. M.S. Soliman, et al. (1965). Geologic setting of the Gulf of Suez province during the Eocene period. Fifth Arab Petr. Cong., Cairo, vol. 2.
6. D.H. Swartz and D.D. Arden, Jr. (1960). Geologic history of the Red Sea area. Bull. Am. Assoc. Petr. Geol. 44:1621-1637.
7. R.W. Girdler and P. Styles (1974). Two stage Red Sea floor spreading. Nature 247:7-10.
8. E.M. El Shazly (1977). The geology of the Egyptian region. In, A.E.M. Nairn, W.H. Kanes and F.G. Stehli, Eds., The Ocean Basins and Margins, Vol. 4A, The Eastern Mediterranean. Plenus Press, New York. p. 379-444.
9. D.R. Lawrence and A.A. Meguid (1978). Cretaceous-Holocene sedimentation in Egyptian Red Sea and Gulf of Suez area - tale of four troughs. Bull. Am. Assoc. Petr. Geol. 62:536.
10. W.C. Ward and S.E.I. Mansour (1977). Nubia formation of the Eastern Desert, Egypt: depositional framework. Abst. with Prog., Geol. Soc. Am. 9(7):1217.
11. M.I. Attia (1955). Topography, geology and iron-ore deposits of the district east of Aswan. Geol. Survey of Egypt, Ser. No. 76.
12. M.I. Youssef (1965). Genesis of bedded phosphate. Econ. Geol. 60:590-600.

13. P.D. Snavely III, R.E. Garrison and A.A. Meguid (1978). Regional variation of depositional environments in Thebes Formation (Eocene) of eastern Egypt - response to prerift crustal doming? Bull. Am. Assoc. Petr. Geol. 62:563.

14. M.A. Johnson (1977). Stratigraphy and paleogeographic-paleotectonic history of the middle Miocene Gebel El Rusas Formation, Red Sea coast, Egypt. Unpub. M.S. thesis, Univ. South Carolina, Columbia, S.C.

15. S. El Akkad and A.A. Dardir (1966). Geology of the Red Sea coast between Ras Shagara and Mersa Alam. Geol. Survey Egypt, Paper 35.

16. B. Issawi, M. Francis, M. El Hinnawy and M. Mehanna (1969). Contribution to the structure and phosphate deposits of Quseir area. Geol. Survey Egypt, Paper 50.

See also:

M.F. El Ramly (1972). A new geological map for the basement rocks in the Eastern and Southwestern Desert of Egypt. Ann. Geol. Survey Egypt. 2:1-18.

E.S. El Tarabili (1966). General outlines of epeirogensis and sedimentation between Safaga, Quseir and southern Wadi Qena area, Eastern Desert, Egypt. Bull. Am. Assoc. Petr. Geol. 50:1890-1898.

H.M.E. Schurmann (1966). The Pre-Cambrian along the Gulf of Suez and the Northern Part of the Red Sea. E.J. Brill, Leiden.

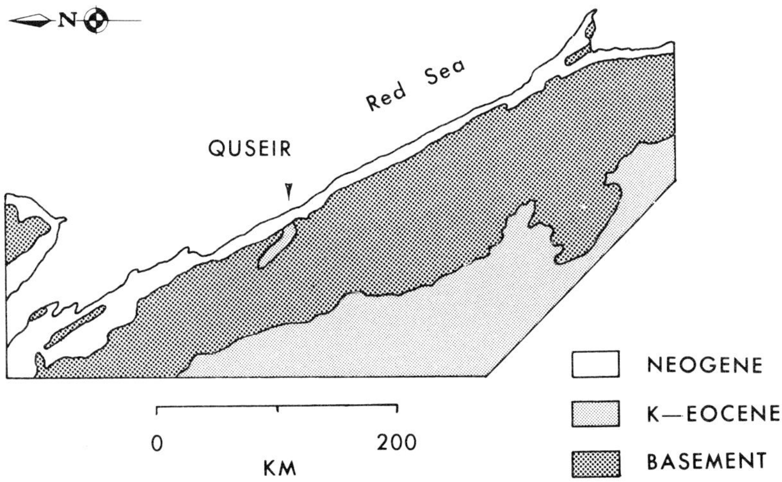

Figure 1. Geologic setting and major provinces in the Egyptian Eastern Desert. Simplified from El Ramly, 1972.

Age	Unit
Eocene	Thebes Formation - cherty limestones and marls
Eocene / Paleocene	Esna Shales
Paleocene	Tarawan (Chalk) Formation
Paleocene / Cretaceous	Dakhla Formation - shales
Cretaceous	Duwi (Phosphate) Formation
Cretaceous	Quseir Variegated Shale
Cretaceous	Nubia Sandstone

Figure 2. Cretaceous to early Eocene stratigraphy in the central Nile Valley and Eastern Desert of Egypt. After El Tarabili, 1966.

Age	Unit
Pleistocene	raised terraces and organic reefs
Pliocene	Shagra Fm.-sandstones and grits; reefal limestones and marls
Pliocene	Gabir Fm.-sandstones, some reefal marls, conglomerates
Miocene	Samh Fm.-marls, sandstones, shales, conglomerates
Miocene	Evaporite Fm.
Miocene	Gebel el Rusas Fm.-conglomerates, sandstones, limegrits

Figure 3. Miocene and younger stratigraphy along the Red Sea coast of Egypt. Modified from El Akkad and Dardir, 1966.

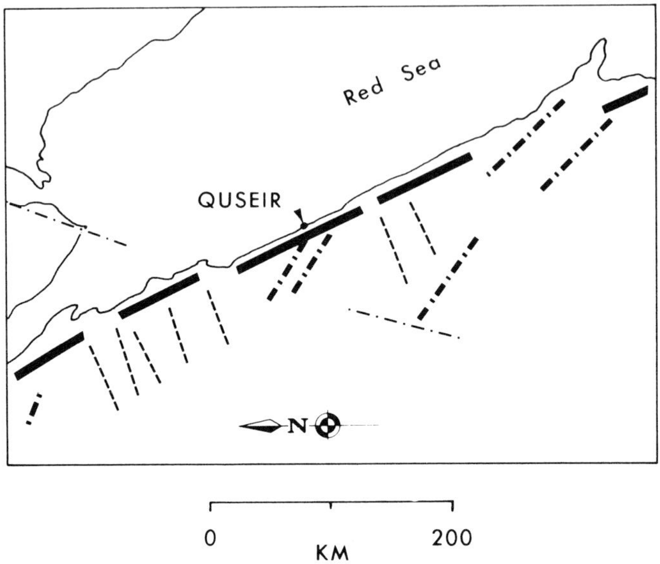

Figure 4. Major structural trends in the eastern Desert of Egypt. Greatly simplified from Schurmann, 1966.

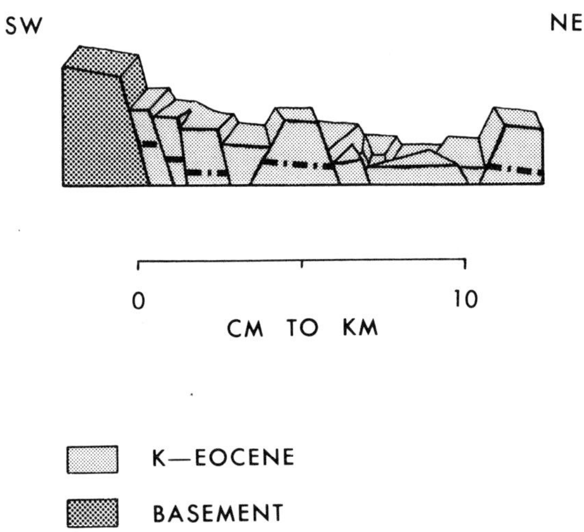

Figure 5. Pervasive structural styles in the Duwi zone near the Red Sea coast of Egypt. From an unpublished report on file with the Red Sea Phosphate Company.

MONITORING THE DESERT ENVIRONMENT FROM SPACE

FAROUK EL-BAZ
Research Director, Center for Earth and Planetary Studies, Smithsonian Institution, Washington, D.C., U.S.A.

ABSTRACT

The immensity, remoteness and inaccessibility of deserts make difficult their detailed study by conventional means. Space platforms provide a unique opportunity to study and monitor the desert environment, especially because deserts must be studied on a regional basis and because climatic conditions favor their observation and photography from Earth orbit. The Space Shuttle program will allow the acquisition of mapping quality photographs of world deserts in the early 1980's. This will be done using the Large Format Camera, which will obtain color, stereo and high resolution photographs.

In addition, space age technology can remedy the scarcity of meteorological data on the desert environment. Automated stations can be placed in remote areas to collect the necessary data and beam them to orbiting satellites. The latter can in turn re-transmit the data from orbit to receiving stations on the ground for analysis and synthesis. Monitoring the deserts from space in this way will help us better understand and utilize more of the land area of the Earth for the benefit of mankind.

INTRODUCTION

The term desert comes from the Latin verb *desero*, to abandon, thus *desertum* and *desertus* meaning abandoned, relinquished or forsaken. It is an exceedingly descriptive term since most desert regions once hosted greater numbers and varieties of flora and fauna than they do today. When their weather conditions changed, these tracts were forsaken by the biota for other regions where life-sustaining water was more plentiful.

A desert is commonly defined as a land area that receives less than 25 cm of precipitation per year. It cannot, therefore, hold much vegetation and remains dry most of the time. This definition includes the polar ice of the frozen tundra, which is characterized by perpetual snow cover and intense cold. Regions of "cold deserts" include one-sixth of the landmass of the Earth, or over 23,000,000 km^2 (Figure 1).

However, the term desert is here applied to hot and dry regions of the Earth which are concentrated between 10 and 40 degrees north latitude and 20 and 30 degrees south latitude. These are belts of land fairly close to the Tropic of Cancer and the Tropic of Capricorn, where rainfall is scarce and where the wind is very dry. The "hot deserts" constitute over one-fifth of the landmass of the Earth, or nearly 31,000,000 km^2 (Figure 1).

In general, hot deserts are formed in areas where subtropical high pressure atmospheric belts prevail. These belts produce cold dry winds in the winter, and hot dry winds in the summer. The aridity of these regions varies depending on the type of winds and the topographic characteristics of the terrain.

The sizes of individual deserts are staggering. For example, the North American Desert is approximately 1,300,000 km^2; the Arabian Desert is 2,600,000 km^2; the Great Australian Desert is nearly 3,800,000 km^2; and the Great Sahara Desert is over 9,000,000 km^2. The immense size of these deserts has precluded their detailed study by conventional means, particularly in view of the many difficulties encountered in desert travel. Consequently, deserts remain as the least understood of all the landforms of the Earth.

Two additional factors have limited geological knowledge of deserts. First, the Earth sciences had their origins in Europe, which is the only continent without a desert. Therefore, unlike mountain ranges, river basins, and even glaciers, deserts and arid lands received no attention in the classical literature. Second, by the nature of their training, geologists looked for solid rock layers to decipher the history of the Earth. Flat desert terrain that is covered by sand and soil attracted only a handful of investigators.

In 1965, when astronauts first observed and photographed the Earth from orbit, they did not fully realize the enormous usefulness of this new vantage point. Visual observations from orbital altitudes were made of the Earth and the Moon. Given the extensive dynamic range of the human eye, astronauts

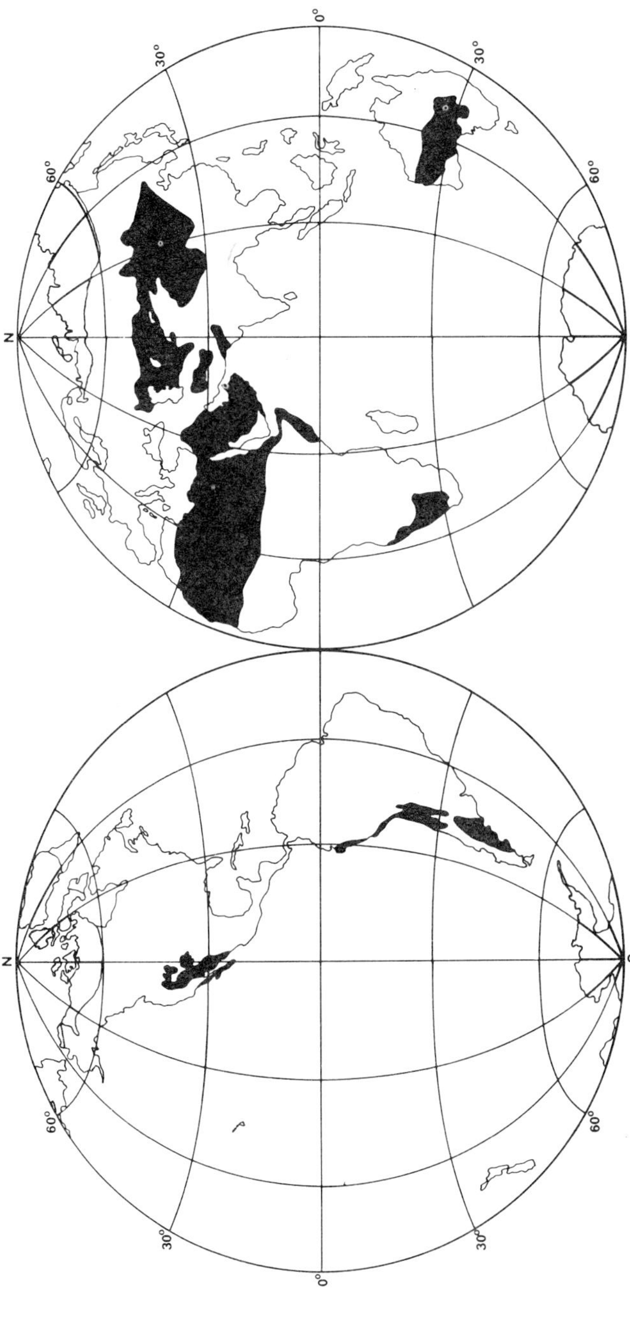

Figure 1: World map on an equal area projection showing the distribution of "cold deserts," north and south of 60° latitude and "hot deserts," particularly near 30° north and south. The largest expanse of desert is that occupied by the geographic region of the Arab World in North Africa and the Arabian Peninsula.

could see more color and textural variations than could be recorded on the most sensitive instrument or film. The astronauts were also able to document their observations by photographs and to test instruments for later use by manned and/or unmanned spacecraft.

During the Apollo lunar program the utility of orbital photographs in making accurate topographic maps was also established. On Apollo lunar missions 15, 16 and 17 two camera systems were used to photograph the Moon's surface from 100 km altitude. These systems included a 76 mm metric camera and a 609 mm panoramic camera. From these systems orthophotomaps of approximately 15% of the lunar surface have been made at 1:250,000, 1:50,000 and 1:10,000 scales (Masurky *et al.*).[1] These maps of the Moon are better than those available for many areas of the Earth.

New techniques were also learned in photographing the planet Mars by Mariner and Viking spacecraft. Concerning the exploration of Mars, Gilbert Grosvenor wrote: "To look upon this rust-red alien world with clear eyes of modern space techniques is an experience that ancient mythology reserved only for gods. Then, to probe and assess the Martian soil for signs of life, as far away as 235 million miles, so enlarges human capabilities that we must wonder why we cannot apply some of that power to problems that plague so many earthly cultures."[2]

The truth of the matter is that we can apply much of the power of space technology to solving certain earthly problems. This paper deals with two such applications to the study of world deserts and their environment via orbital observations and meteorological monitoring.

ORBITAL OBSERVATIONS

Photographs of the Earth, used chiefly for land surveys and geological mapping, are usually obtained from airplanes. However, during the past decade, Earth scientists have recognized the value of observations from space. For the purpose of photographing the Earth, orbiting spacecraft have three advantages over airplanes: 1) a spacecraft travels at higher altitudes (several hundred kilometers) and can, therefore, view a larger area and depict broader features; 2) since spacecraft travel above the atmosphere, they avoid gusts of wind and air pockets that make aircraft relatively unsteady; and 3) a spacecraft follows an orbit around the Earth with a precision that increases the cartographic value of the photographs.

The Gemini, Apollo, Skylab, and Apollo-Soyuz missions established the utility of color film in studying the Earth (El-Baz).[3] Color photographs from these missions have been used to map major fault systems and subsidiary fractures (Figure 2), to classify soil types in desert regions, and to determine sediment patterns at the mouths of major rivers. However, the stereo-coverage and the resolution of these photographs are below what is required for detailed study and mapping. These drawbacks are also applicable to Landsat data, which provide very limited stereo-coverage and an effective ground resolution of approximately 200 m.

Knowledge gained from previous space missions will help us to better photograph the Earth from orbit during the Space Shuttle program. The Space Shuttle is the major new program of the National Aeronautics and Space Administration (NASA). It is scheduled to start Operational Flight Test (OFT) missions in mid 1979. Following the OFT program, the Space Shuttle will be considered operational and will be termed the Space Transportation System (STS). It is anticipated that eventually there will be a total of four Shuttle vehicles with 30 to 50 missions flown each year.

Launching of the Shuttle will initially take place from the Kennedy Space Center. Range safety conditions there restrict the maximum orbital inclination to 57°, which will permit the Orbiter vehicle to cover a large part of the landmasses of the Earth, especially the desert belts. By 1983, launch operations may be available at the Vandenberg Air Force Base, California. From that locality polar orbits can be attained. Circular orbits from 200 to 1,200 km high can be achieved depending upon payload weight and orbit inclination. Mission duration will be from three to thirty days (Doyle).[4]

It is planned to operate the Shuttle in two different modes. In the sortie mode, experiments will be mounted in the Orbiter cargo bay, operated for the mission duration, and then returned to Earth. The cargo bay is 18.3 m long and 4.6 m in diameter and can carry a maximum of 30,000 kg payload (Doyle).[4] The payload will consist of a combination of pressurized modules in which crew members can work in a shirt-sleeve atmosphere and a number of external pallet modules on which experiments can be mounted.

In the second mode of operation the Shuttle will be employed to carry individual spacecraft into space, to place

Fig. 2: South-looking, Apollo-Soyuz photograph of the eastern Mediterranean region showing the fractures of the Levantine Rift zone.

them in appropriate orbit, and to service them on demand. A Remote Manipulator System (RMS) will extract the payload from the cargo bay and release it into its own orbit. Subsequently, the Shuttle Orbiter can rendezvous with the free-flying satellite, the RMS can retrieve it to be serviced in the cargo bay, or else returned to Earth for major refurbishment (Doyle).[4]

In addition to astronaut observations, the main tool that will be utilized in obtaining photographs of the Earth in the Shuttle era is the "Large Format Camera" or LFC (Figure 3, top). This camera will allow the acquisition of mapping quality, stereo, color, and high-resolution photographs from Earth orbital altitudes. Photographs from this camera can be used for mapping using conventional techniques and instruments without costly electronic and digital enhancement or image correction, which are necessary when dealing with Landsat data.

This new camera is designed to satisfy the operational requirements of cartographers, geologists, land-use planners, agricultural experts, environmentalists, and other Earth scientists. It features high photographic resolution on a choice of black-and-white or color film. It also features stereoscopic observation, orthographic (vertical) perspective, metric fidelity, and large areal coverage of each frame. Geometric/topographic maps may be made from its orbital photographs at scales of 1:100,000 and 1:50,000. These maps can be used for photogeologic interpretations and the compilation of data acquired by other means.

The Large Format Camera derives its name from the size of its individual frames, which are 45.7 cm in length and 22.9 cm in width, with a post-mission calibrated format option of 22.9 x 22.9 cm. It has a 305 mm, f/6 lens with a $40^{\circ} \times 74^{\circ}$ field-of-view. Both the rotary shutter and camera controls are derivatives of the Skylab S-109A camera. The camera back includes supply and take-up spools with a 1200 m film capacity. The film will be driven by a forward motion compensation unit as it is exposed on a vacuum platen, which will keep the film perfectly flat. Moving rollers will insure the synchronization of film advance into the vacuum platen and in and out of the take-up and supply spools (El-Baz and Ondrejka).[5]

The film calibration reference will be provided by 12 edge fiducial marks and 45 reseau marks. The radial distortion across each frame will be ± 10 micrometers. The operating frame rate will be 80% (Figure 3, bottom), which provides the required base/height ratio for topographic mapping 20 m contours. The framing rate will vary from 5 to 45 seconds to

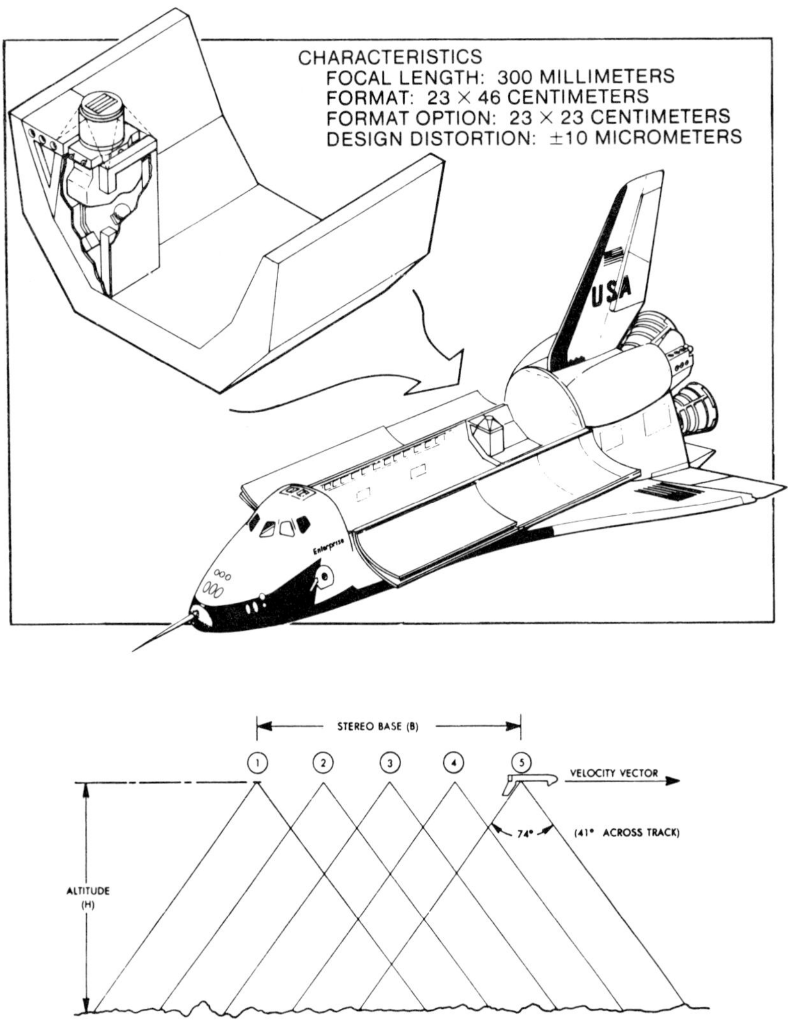

Fig. 3: Drawings showing the utilization of the large Format Camera on the Space Shuttle (top) and the mode of its stereo overlap photography (bottom).

allow the use of the camera various spacecraft altitudes. The illumination uniformity within each frame will be 10% to minimize vignetting (Doyle).[6]

The spectral range of the Large Format Camera will be 400 to 900 nanometers and the system resolution will be 100 lines/mm (1000:1 contrast) to 88 lines/mm (2:1 contrast). This means a photo-optical resolution of 10 to 20 m from an altitude of 260 km. The camera will have the ability to utilize a number of films, particularly Kodak high resolution black-and-white (3414), color (SO-356), and color infrared (SO-131) in magazines with a capacity of 1,200 m of film. An electronic filter changer will permit different films to be used during a single mission (El-Baz and Ondrejka).[5]

The orbital inclination of the first shuttle missions to carry the LFC will be relatively low (28.5°). This low orbital inclination will be suitable for coverage of the Earth's equatorial regions, and thus for study of the desert belts north of the equator.

Data from manned space missions and Landsat satellites show that the reflectance properties of desert surfaces are indicative of the composition of the exposed rock and rock rubble. Examples have been given of distinct color zones in desert photographs that correlate with the amount of sand, desert-varnished pebbles, and clay minerals in the exposed soil (El-Baz).[7] Similarly, it has been shown that desert sands become redder as the distance from their source increases. This is due to the fact that individual sand grains become coated with an increased amount of iron oxides with the passage of time (Norris).[8] Color zones in orbital photographs of these areas can be used to delineate age zones within the sand fields. This would allow better determination of areas of the world that are in danger of desertification by shifting, relatively young (light-colored) sand accumulations.

Desertification, the spread of desert-like conditions in arid and semiarid regions due to overuse of land or to climatic change (Rapp)[9] has led to widespread starvation and to disruptions of landuse patterns along the perimeters of many of the world's deserts. Desertification results from local changes, such as dune migration and soil denudation as a result of slash-and-burn techniques, as well as more regional changes. The Agency for International Development (AID),[10] estimates that over one million square kilometers and about 60 million people are affected by desertification. The United Nations estimates that 20 million square kilometers have recently reverted to desert or desert-like conditions (UNEP).[11]

The features associated with desertification are most easily studied from orbital images, largely because the extensive areal coverage facilitates investigations of regional patterns and of large-scale landforms. Orbital images taken in different seasons or years can be compared to determine changes in desert areas. Desertification studies have been made using orbital images and photographs by MacLeod, *et al.*[12] and Brera and Shahrokhi,[13] among others.

From the above it is clear that orbital photography is an excellent tool for the study of world deserts. High resolution photographs taken from orbit can help in classifying desert surfaces and monitoring the changes to the desert environment. Planning for the use of Large Format Camera photographs in desert study should commence on both a local and regional basis.

METEOROLOGICAL OBSERVATIONS

In recent years, studies of the evolution of desert landscapes have been facilitated by the interpretation of orbital images. For example, McKee and Breed[14] and Breed[15] have used Landsat imagery to formulate a classification scheme for sand seas. As stated above, because of the large regional coverage and the ability to monitor distributions that vary with time, space photographs have provided a new way of observing the desert environment. However, the results of these studies arehindered by the lack of extensive ground-truth data, including details of dune morphology, and most importantly, information on wind speed and direction.

Space age technology has provided a wide variety of sensing devices that can, for example, measure air and water quality, monitor the flow within a pipeline, or detect earthquakes. It is here proposed that the same kind of technology be applied to monitoring climatic and meteorological conditions in remote desert regions. Automated stations can obtain the measurements, average them over a period of time, and relay them to orbiting satellites. The satellites, in turn, can beam these data to Earth for analysis and synthesis.

At the present time, the U.S. Geological Survey is operating a program of remote monitoring of hydrologic data for water resources assessment (Paulson).[16] Preliminary results of this program indicate both the cost-effectiveness of the method, and the advantages gained from realtime collection of the data. In at least one instance this program has resulted in a practical benefit. According to Carter and Paulson,[17] improved estimates of runoff were instrumental in avoiding flood damage and generating extra electrical energy during the runoff period.

There is a great number of satellites that can receive and transmit data collected on the ground. These include satellites of the Landsat type, Nimbus-F, GEOS, Tiros-N, and the communications satellites of Comsat, Intelsat, Anik, and conceivably Arabsat.

The proposed meteorological data collection scheme is a telemetry system that uses Earth orbiting satellites to relay data from a great number of stations to one or more receiving stations. There are three basic elements to such a scheme: 1) a field radio, usually called Data Collection Platform, that is connected to the sensor recorders; 2) a radio transponder with receive and transmit capability on board a satellite; and 3) a data receiving station for retrieval, processing and dissemination to investigators.

Remote meteorological stations have been developed with the capability of transmitting data to orbiting satellites. These stations consist of an aluminum free standing tripod with a mast that extends to 6.1 m for mounting the wind sensors. Swiveling tundra pads are provided with holes for securing to the ground, and one leg is adjustable for leveling. A lockable steel enclosure is provided to house the electronic equipment.

The standard complement of sensors is usually made up of seven components which measure: 1) cumulative precipitation; 2) wind speed with a time-weighed average; 3) wind direction; 4) temperature with vane aspirated radiation shield; 5) relative humidity with sky shield; 6) barometric pressure; and 7) battery power.

The only device that needs to be added to such an array of sensors for desert monitoring is a dust particle counter. This counter would help monitor dust storms which plague desert regions. Dust storm prediction and monitoring would be a useful addition because of benefits to aviation and other forms of transportation in desert regions.

Based on the above, it is recommended that testing of this method of meteorological monitoring via orbiting satellites be conducted in the Western Desert of Egypt. This desert is ideal for this type of research. In this desert "the free interplay of sand and wind has been allowed to continue for a vast period of time, and here, if anywhere, it should be possible in the future to discover the laws of sand movement, and the growth of dunes" (Bagnold).[18]

Present-day meteorological stations in Egypt are concentrated in populated areas, where their locations are chosen primarily on the basis of accessibility. However, these stations are usually located in oases within depressions where local topography exerts a great influence on both wind direction and velocity. Consequently, the local environment for the proposed stations should be carefully chosen to give representative wind data for the region. Although only 8 stations are proposed at present (Figure 4), it is anticipated that no less than 20 stations would be required to provide adequate coverage for both research and practical benefits.

The proposed program for remote monitoring of near-surface (up to 3 m) meteorological information in the Western Desert will help to close the gap between orbital and Earth-based studies in four ways. These monitors will aid the study of: 1) the relationship between dune form and the mean (or effective) dune-forming wind, 2) the relationship between macro- and micro-environment and location of extensive sand accumulations, 3) both long- and short-term meteorological effects on areas prone to desertification, and 4) orbital interpretations of dune orientation with respect to local and regional wind regimes.

The major scientific objectives of the proposed research can be accomplished through measurements of the magnitude, frequency and duration of winds. By continuous monitoring of these factors and by field checks (to determine amounts of ablation and deflation, and rates of dune migration), quantitative estimates of local and regional sand budgets can be made. Near populated areas, this analysis will aid immediately in plans to stop desert growth, while in more remote regions, it will provide the ground-truth information necessary for studies of orbital images.

Although these data could be recorded *in situ* with chart recording devices, the personnel, field support and time requirements would be prohibitively expensive. In addition, there would be an extensive time lag between acquisition and use of the data. Consequently, the proposed method of monitoring coupled with satellite relay is considered the only feasible way of gathering the required data in a timely manner.

Fig. 4: Landsat image mosaic of Egypt showing the proposed minimum number of locations of meteorological satellite monitoring stations at: 1) north of Qattara; 2) west of Nile Delta; 3) south of Faiyum; 4) north of Kharga; 5) Abu Tartur plateau; 6) South New Valley; 7) Oweinat area; and 8) Great Sand Sea.

SUMMARY

Deserts are among the least understood of Earth's features. Basic research of the world deserts has been hampered by inaccessibility, and harsh conditions. World-wide drought problems and increased desertification of semiarid lands has focused some attention on deserts, which occupy a large, unused part of the landmasses of the Earth.

This attention has been increased by the recent realization that space photographs can assist in the study of desert morphologies. The desert environment is particularly suited for orbital surveys because of the usually clear weather. Deserts also display different color tones which can be recorded on film, to aid in photointerpretation.

To satisfy the need for detailed studies of the desert environment it is necessary to obtain: 1) stereo photographs to allow photogeologic interpretations and the making of accurate topographic maps; 2) high-resolution photographs to allow the study of dune morphologies and sand movement; and 3) color photographs to depict color-tones that are indicative of the chemical makeup and relative-age of desert sands. These requirements can be met by presently planned use of the Large Format Camera on the Space Shuttle. The Shuttle provides the unique capability of carrying these cameras into Earth orbit, and returning them with the exposed film.

In addition to orbital observations and photographs of the deserts, space age technology allows the timely monitoring of their meteorological conditions. This can be accomplished by installing automated stations in remote localities to monitor the environment. These stations can monitor wind speed and direction, temperature, humidity, rainfall and barometric pressure. They also include dust particle counters to monitor dust storms and give early warnings of their development.

Data from these meteorological stations can be transmitted by radio to any number of Earth orbiting satellites with receiver capabilities. These satellites can then beam the data to Earth for analysis and dissemination. It is here proposed to test such a data collection system by installing eight automated monitoring stations in remote areas of the Western Desert of Egypt.

The space program has provided us with numerous new technologies. These are being utilized to a maximum extent in the developed countries. Utilization of such technologies in the understanding and monitoring of the desert environment will have far reaching benefits directly or indirectly to all mankind.

REFERENCES

1. H. Masursky, G.W. Colton, and F. El-Baz (1978) Apollo over the Moon: A view from orbit. NASA SP-362, 255 pp.
2. G.H. Grosvenor (1977) National Geographic, v. 151, n. 1, p. 1.
3. F. El-Baz (1977) Astronaut Observations from the Apollo-Soyuz mission Smithsonian Studies in Air and Space, Number 1. Smithsonian Inst. Press, Washington, D.C., 400 pp.
4. F.J. Doyle (1978a) The next decade of satellite remote sensing. Photogramm. Engin. and Remote Sens., v. 44, n. 2, p. 155-164.
5. F. El-Baz and R.J. Ondrejka (1978) Earth orbital photography by the Large Format Camera. Paper presented at the 12th Inter. Sym. on Remote Sens. of Environ., 20-26 April 1978, Manila, Philippines.
6. F.J. Doyle (1978b) A large format cartographic camera for Space Shuttle mission. Paper presented at the 44th Ann. Meet. of the Amer. Soc. Photogramm., 26 Feb. - 4 Mar., 1978, Washington, D.C.
7. F. El-Baz (1978) The meaning of desert color in Earth orbital photographs. Photogramm. Eng. and Remote Sens., v. 44, n. 1, p. 69-75.
8. R.M. Norris (1969) Dune reddening and time. Jour. Sed. Pet., v. 39, p. 7-11.
9. A. Rapp (1974) A review of desertification in Africa--with vegetation and man. Secretariat for International Ecology, Stockholm, Sweden.
10. A.I.D. (1972) Desert encroachment in arable lands: Significance, causes and control. Agency for International Development, Washington, D.C.

11. U.N.E.P. (1975) Review of the environmental situation and of activities relating to the environmental programme. Governing Council, Nairobi, 12 April - 2 May, 1975.

12. N.H. MacLeod, J.S. Schubert and P. Anaeijonu (1977) Report on the Skylab 4 AFrican drought and arid lands experiment. In: Skylab Explores the Earth. NASA SP-380, p. 263-286.

13. A. Brera, and F. Shahrokhi (1978) Application of Landsat data to monitor desert spreading in the Sahara region. In: Remote Sensing of Earth Resources; F. Shahrokhi, ed., Univ. of Tenn. Space Inst., Tullahoma, v. 3, 813 pp.

14. E.D. McKee, and C.S. Breed (1974) An investigation of major sand seas in desert areas throughout the world. Third Earth Res. Tech. Satel.-1 Symp., v. 1, Tech. Pres. Sec. A, Paper G6, p. 665-679.

15. C.S. Breed (1978) Morphology and distribution of dunes in sand seas observed by remote sensing. In: A global study of sand seas. U.S. Geol. Surv. Prof. Pap. (in press).

16. R.W. Paulson (1978) Use of Earth satellites for automation of hydrologic data collection. In: Collection, storage, retrieval, and publication of water resources data. U.S. Geol. Surv. Circ. 756, p. 8-14.

17. W.D. Carter, and R.W. Paulson (1978) Introduction to monitoring dynamic environmental phenomena of the world using satellite data collection systems. Comm. on Sp. Res. (COSPAR), Tech. Man. Ser. Man. No. 8.

18. R.A. Bagnold (1933) A further journey through the Libyan Desert. Geo. Jour., v. 82, p. 103-129.

THE USE OF REMOTE SENSING FOR MONITORING DESERTIFICATION PROCESSES AND NATURAL RESOURCES IN DESERT AND ARID AREAS

M. ABDEL HADY
Professor of Remote Sensing, School of Civil Engineering, Oklahoma State University, U.S.A., and Director, Remote Sensing Center, Academy of Scientific Research and Technology, Cairo.

E. M. EL SHAZLY
Chairman/President, Nuclear Materials Corp., Ministry of Electricity and Energy, Cairo.

V. MYERS
Director, Remote Sensing Institute, South Dakota State University, Brookings, U.S.A.

ABSTRACT

Facilitating the proper programming for halting desertification, and if possible its reversal in deserts and arid areas in the world, is the focal aim of the international effort for combating desertification. Desertification leads to the deterioration of an environment which has been suitable for the development of human communities into an environment which is getting progressively less suitable for man to carry out his activities.

Climate, geomorphology, rock substrata and human activities can lead to desertification when man tries to exploit groundwater and soil without prior knowledge of the impact of such exploitation. Sand dune movements, salinization, alkalization and drainage are but a few factors which play important roles in accelerating, slowing down or halting the desertification phenomena.

Desertification can be halted and reclamation procedures inititated only if there is a physical assessment of the processes involved, an estimate of the magnitudes of the problem, and the specific problem areas located along with areas of potential for reclamation. Proper assessment of realities of the situation is the only means of ensuring evaluation of any risks that might result from proposed programs against desertification in arid areas.

Assessment of the resources of arid and desertified areas, whether they be improvements or further deterioration, can best be made by using remote sensing procedures and indeed may be the only way where large areas are involved. Also, remote sensing may be the only practical means to define the areas of critical concern that justify more intensive monitoring and documentation expense.

Examples will be given in this paper illustrating work carried out by the Remote Sensing Center in Egypt in the application of remote sensing techniques for the monitoring of the processes of desertification and natural resources in various parts of the Western Desert, which are based on Landsat and aircraft imagery interpretation coupled with field investigations and laboratory analysis.

INTRODUCTION

The objectives of a monitoring program for desertification are: to assess existing ecological and resources conditions, to detect physical indicators of a desertification, and to observe changes representative of improvement or deterioration of natural resources.

Concern for environmental problems such as desertification appears to be increasing throughout the world. This, as anticipated, has led many people to expect a global solution to the problem. While this is a worthy goal to pursue, it is more likely that solutions must come on a regional basis, or even on a national level. During the past six years the Remote Sensing Center in Egypt took the lead, in cooperation with several other Egyptian and U.S. Agencies involved in combating desertification, in the use of the Remote Sensing Techniques for monitoring processes of desertification and natural resources over most of the Western Desert of Egypt. Results of these studies are published in a series of technical reports and some are presented in Proceedings of International Conferences and professional journals. The experience gained from these studies can be extended to other arid areas of the world with similar environmental conditions.

Assessment of natural resources and desertification indicators of the affected areas along with the monitoring of changes, whether they be improvements or further deterioration can best be made using remote sensing procedures, and indeed may be the only way where large areas are involved. Also, remote sensing represents a powerful analytical tool for eliminating the often high percentage of nonfragile and nonproductive areas of deserts that are therefore of no concern, leading to the definition of the areas of critical concern that justify monitoring and documentation expense.

JUSTIFICATION FOR MONITORING THE PROCESSES OF DESERTIFICATION

Desertification processes can be halted and reclamation procedures initiated only if there is a physical assessment of the processes involved, an estimate of the magnitude of the problem, and the specific problem areas located along with areas of potential for reclamation. Remote Sensing techniques can be a major tool for fulfilling this need.

In Egypt, as well as in many other countries in desertified areas of the world, plans are in progress towards completion of natural resources surveys in a methodical, conventional manner over a period of many years, and often in a noncoordinated manner.

A practical method, however, of completing such surveys in an accelerated and coordinated manner with priorities allocated to areas of high development potential, is by the utilization of remote sensing techniques, especially by satellite imagery.

Aerial photo coverage, in black and white, of most of the Western Desert of Egypt, except of the south Western part, is available. However, recent coverage in color, which is important for resources surveys, is generally not available. Such low altitude coverage is generally unmanageable for regional studies of such large areas as the Western Desert because of the large scale and great quantity of photos required to cover such a large area.

Also, such photos do not provide the spatial and temporal advantages inherent in spacecraft imagery. These photos may be necessary, however, for studies of limited sized calibration sites or for multistage studies.

Advantages of Landsat

As an advantage in monitoring desertification processes and natural resources, in recognizing boundaries and in classifying resources, Landsat satellite images have distinct advantages: the synoptic view, near-orthographic scenes, repetitive coverage and multispectral coverage. Also, the digital form of the data lends itself to computer handling classification and automatic identification of processes and features.

Monitoring of processes and changes, especially when repeating the coverage several times, over large desert areas, such as the Western Desert of Egypt, must be done at a very low cost to be feasible. Landsat satellite technology has reduced the costs in such a way that permits the accomplishment of tasks that were previously not feasible in a reasonable time frame. For the Western and Eastern Deserts of Egypt, as well as the Sinai Peninsula, for example, the Remote Sensing Center using Landsat image interpretations performed extensive surveys over 75% of these areas (about 750,000 km^2) in four years at a very low cost compared to conventional techniques.

In using Remote Sensing, especially Landsat, resource integration and monitoring program is feasible. Existing and new data related to geology, topography, soils, vegetation, present land use, and climate can be taken and placed in a uniform scale and degree of accuracy on reliable base maps for easy reference and use.

Qualitative or quantitative data concerning important desertification phenomena such as soil moisture, shifting and migrating sand dunes, changing vegetation patterns, degrees of salinity and alkalinity in cultivated areas whether periodic or permanent changes, and other phenomena require a relatively cheap method of periodic surveillance. Repetitive data over Egypt can be obtained rather quickly and cheaply from Landsat satellites now in orbit (every six days if desired for Landsat I, II and III) with good resolution for broad reconnaissance purposes. The temperature measuring capability of Landsat III and of the NOAA satellite (although at less resolution than Landsat) can be of great value in monitoring soil moisture.

MONITORING OF PROCESSES AND FEATURES RELATED TO DESERTIFICATION

Recognition of certain features and Processes are essential to a desertification study. Can such features and processes be monitored by Remote Sensing Techniques, using primarily satellite imagery?

The answer is definitely yes, especially when we consider that a study of desertification must be conducted initially on a reconnaissance level. Such features are:

Natural Vegetation

The objective here is to survey current vegetation cover, the successional changes that may occur, evidence of mismanagement, and seasonal changes.

Landsat satellite images as enhanced color composites and band 5 of the multispectral imagery at a scale of 1:250,000 can show general distribution and conditions. Desert vegetation must be observed through these satellite images at a time most favorable to plant growth to exert significant influence on the spectral signature. Different seasons coverage can be important.

This, followed by medium to low altitude aerial photographs for sample areas would allow verification of satellite interpretations and provide information on vegetation types within relatively homogeneous areas shown on satellite imagery. Finally, ground sampling must be conducted to determine species and other specific information desired. Natural vegatation having shallow root systems can be also used as soil indicators.

Soils

The objective here is to map soil associations and general potential land use and soil limitations on a regional scale.

Landsat satellite color composites and band 5 of the Landsat imagery, at a scale of 1:500,000 is utilized for this purpose. General pedological soil association maps over large desert areas can be prepared using Landsat imagery as a base and final maps can then be prepared by synthesizing background material with observations and data obtained from the field. Final descriptions of each mapping unit and complete soil properties are prepared on the basis of ground sampling and laboratory analysis of major mapping units.

The Remote Sensing Center in Egypt utilized this technique to prepare regional soil maps for the Central part of the Western Desert of Egypt, namely Bahriyya Oasis-South Siwa Oasis area and Farafra Oasis-Nile Valley area. The soil maps for these large areas have been constructed on a scale of 1:500,000, utilizing Landsat imagery interpretations in conjunction with field investigations of the soil units and subunits, measurement of soil profiles, and laboratory determination of the physical and chemical properties of the collected soil samples.

A land use reconnaissance map based on the previously mentioned soil map has been prepared for Bahriyya Oasis - South Siwa Oasis area and Farafra Oasis - Nile Valley area. The soil units have been classified in the former map into seven grades according to their suitability for reclamation purposes pending the fulfillment of other requirements, such as fresh water, etc.. Soils of Grade I include those of the Holocene flood plain and the surrounding flat stretches of the Holocene river terraces. Grade II soils incorporate those of the pediplains and depressions. Grade III soils are represented by those of the playas, bajadas and Holocene lacustrine clays. Grade IV soils comprise those of the sandy plains, while Grade V soils are represented by those of the old river terraces, dry valleys, fans, outwashed plains and locations surrounding salty marshes. Grade VI soils include those of the sand dune belts. Finally, Grade VII soils are represented by those of the desert pavements, talus slopes, wind blown depressions, dry sabkhas, hammadas and rocky exposures.

High Water Tables, Salinity and Alkalinity and Soil Moisture

The objective here is to map the extent of high water tables, and identify areas affected by salinity and alkalinity and water logged areas. Also, moisture in bare soils to depth of at least 3 inches, and moisture available in root zones can be estimated.

Where salinity or alkalinity are very severe, they become directly visible on satellite and aircraft images, with various degrees exhibiting different spectral reflectance characteristics especially in color composites, and bands 6 and 7 multispectral satellite images. In less extreme cases, analysis is generally performed indirectly through effects on vegetation. The timing of the imagery in this case is critical since the conditions are often seasonal or only seasonally visible.

Research in the Remote Sensing Center in Egypt has shown correlations between crop reflectance, especially in the near infrared, depth to water table, and concentration of salinity.

Regarding soil moisture, visible and near infrared remote sensing techniques proved valuable for detecting and mapping soil moisture parameters. Results of investigations carried out by the Remote Sensing Center in Egypt in this respect are available in published literature, for areas around the Suez Canal and in the Western Desert. Thermal infrared images obtained at night on bare soils in these areas provided temperature patterns that correlated well with soil moisture to a depth of several inches. Also, thermal infrared sensing of vegetation canopies which cover the soil can be used to relate vegetation temperature to soil moisture status.

The thermal sensor aboard Landsat III satellite is expected to provide excellent data on moisture detection all over Egypt.

Groundwater

The objective here is to detect and map shallow aquifers and detect indicators of deep aquifers.

For very shallow aquifers the techniques described above for water table and soil moisture are applicable. Also, shallow sand, gravel and deterious limestone aquifers can often be identified by remote sensing techniques on the basis of the geologics units, structure, landforms and other surface features that produced the aquifers.

Keys to detection of deeper aquifers and groundwater studies include the occurrence of fracture traces and lineaments on satellite imagery, and careful study of the geology, geomorphology, surface drainage pattern, and age and vegetation patterns from Landsat satellite images.

The Remote Sensing Center in Egypt has conducted over the past several years some extensive shallow and deep groundwater investigations in the Western Desert of Egypt.

Sand Dune Migration and Shifting

The objective here is to map major sand dunes, and detect their movements, pattern and direction, and encroachement on developed or potential development areas in the Western Desert.

Mapping of sand dunes may be carried out on a regional or on semidetailed or detailed scales through the application of remote sensing techniques. The regional mapping is done by the delineation of sand dune belts and sand accumulation over a period of several years through Landsat imagery interpretation. On the other hand, the semi-detailed and detailed investigations of sand dunes migration may be carried out in critical locations by airborne techniques over several years. In all cases, ground investigations are necessary for the substantiation of the spaceborne or airborne technologies applied.

Mineral and Energy Resources

These resources are becoming increasingly important in the development of deserts and the progress of human communities in general. In this part of the world the most important energy resources are petroleum and natural gas, while uranium represents a very important energy resource in immediate, medium term and long term senses. Mineral resources of particular importance include phosphates and iron ores.

The studies of the Remote Sensing Center have proved that the application of remote sensing techniques is particularly important in the development and exploration of mineral and energy resources. These studies are carried out on a regional basis through Landsat imagery interpretation on a semi-detailed and detailed basis, and through the use of airborne techniques. Mineral identification by remote sensing techniques may be direct when the tone and/or texture of the mineral occurrences are noted or it may be indirect, through recognizing favorable structures, lithology, geomorphology, etc., which host the mineral occurrences.

A notable example of the use of Landsat imagery in mineral development and exploration is the discovery of iron occurrences in Gebel Qalamun and environs in the Western Desert of Egypt under geological and structural conditions which are comparable to those where known deposits occur in Bahriyya Oasis to the south west of Gebel Qalamun. The bright thermal infrared anomalies exhibited by the maniferous shales in Gebel Qatrani area in the northern part of the Western Desert is a case illustrating the application of airborne remote sensing techniques in mineral development and exploration.

REFERENCES

1. M. Abdel Hady and R.V. Matalucci (1968). Infrared acarial survey in environmental engineering. Proc. Am. Soc. Civil Eng. 94, SA-6:1071-1084.

2. M. Abdel Hady and R.V. Matalucci (1969). Infrared acarial surveys in surface and subsurface explorations. Highway Research Board Special Report No. 102. National Research Council, Washington, D.C. pp. 1-13.

3. M. Abdel Hady, H. Karbs and M. Abdel Hafez (1970). Subsurface drainage mapping by infrared imagery techniques. Proc. Oklahoma Acad. Sci. 50:10-18.

4. M. Abdel Hady and H. Karbs (1971). Depth to groundwater by remote sensing techniques. Proc. Am. Soc. Civil Eng. Irrigation and Drainage Div. Journal 97:IR 3.

5. E.M. El Shazly, M.A. Abdel Hady, I.A. El Kassas and M.A. El Ghawaby. Geologic interpretation of ERTS-1 satellite images for West Aswan area, Egypt. Proc. NASA Third Earth Resources Technology Satellite-1 Symposium. Vol. 1 Sec. A:921-942 (1973). Also, in Proceedings, Ninth Int. Symp. on Remote Sensing of Environment, Michigan. (1974). pp. 119-131.

6. E.M. El Shazly, M.A. Abdel Hady, I.A. El Kassas and M.A. El Ghawaby (1974). Geologic interpretation of ERTS-1 satellite images for East Aswan area, Egypt. Proc., Ninth Int. Symp. on Remote Sensing of Environment, Michigan. pp. 105-107.

7. E.M. El Shazly, M.A. Abdel Hady and M.A. Morsy (1974). Geologic interpretation of infrared thermal images in East Qatrani area, Western Desert, Egypt. Proc., Ninth Int. Symp. on Remote Sensing of Environment, Michigan. pp. 1877-1889.

8. E.M. El Shazly, M.A. Abdel Hady, M.A. El Ghawaby, I.A. El Kassas and M.M. El Shazly (1974). Geology of Sinai Peninsula from ERTS-1 satellite images. Proc., United Nations/FAO Regional Symp. on Remote Sensing of Earth Resources and Environment, Cairo. pp. 335-354.

9. E.M. El Shazly, M.A. Abdel Hady and I.A. El Kassas (1977). Delineation of land features in Egypt by Landsat satellite images. Proc., Sixth Annual Remote Sensing of Earth Resources Conference, Tulahoma, Tenn.

10. E.M. El Shazly, M.A. Abdel Hady and M.M. El Shazly (1977). Groundwater studies in arid areas in Egypt using Landsat satellite images. Proc., Eleventh Int. Symp. on Remote Sensing of Environment, Michigan.

11. E.M. El Shazly, M.A. Abdel Hady, M.A. Abdel Hafez, A.B. Soliman, M.A. Morsy, M.A. El Rakaiby, I.E. El Aassy and A.F. Kamel (1977). Interpretation of multispectral and infrared thermal surveys of the Suez Canal Zone, Egypt. Proc., Eleventh Int. Symp. on Remote Sensing of Environment, Michigan.

12. E.M. El Shazly, M.A. Abdel Hady, M.A. El Ghawaby and S.M. Khawasik (1977). Application of Landsat satellite imagery for iron ore prospecting in the Western Desert of Egypt. Proc., Eleventh Int. Symp. on Remote Sensing of Environemnt, Michigan.

13. E.M. El Shazly, M.A. Abdel Hady, M.A. El Ghawaby, S.M. Khawasik and M.M. El Shazly (1977). Application of Landsat satellite imagery in assessing the regional geological, structural, environmental and groundwater conditions in the Qattara Depression area, Egypt. 28th Int. Astronautical Congress, Prague.

14. E.M. El Shazly and M.A. Abdel Hady (1978). Landsat satellite mapping in Egypt and its possible applications in petroleum and natural gas exploration. Tenth Arab Petroleum Congress, Tripoli.

15. E.M. El Shazly, M.A. Abdel Hady, M.M. El Shazly, M.A. El Ghawaby, S.M. Khawasik, A.A. Haraga, S. Sanad and S.H. Attia (1978). Application of Landsat imagery in the geological and soil investigations in the Central Western Desert, Egypt. Proc., 12th Int. Symp. on Remote Sensing of Environment, Manila.

16. E.M. El Shazly, M.A. Abdel Hady and A. Shata (1977). Feasibility study, transnational project: the management of the major regional aquifers in Northeast Africa and the Arabian Peninsula. UN World Conference on Desertification, Nairobi, Kenya.

17. V.I. Myers, H.S. Mann, D. Moore, M. DeVries and M. Abdel Hady (1978). Remote sensing for monitoring resources for development and conservation of desert and semi-desert areas. Int. Symp. on Arid and Semi-arid lands, Jodhpur, India.

18. A.M. Aly and M.A. Abdel Hady (1968). Annotated bibliography on IR photography and remote sensing thermal techniques for mapping saline soils and vegetation. Oklahoma State University, School of Civil Engineering, Research Report No. 33.

19. H. Karbs, A. Aly and M.A. Abdel Hady (1969). Special relectance qualities of soils with various salts and moisture levels in the ultraviolet, visible and near IR regions of the electromagnetic spectrum. Oklahoma State University, School of Engineering, Research Publication No. R(S)-2.

20. E.M. El Shazly, M.A. Abdel Hady, I.A. El Kassas, H. El Amin, A.B. Salman, M.M. El Shazly and A.A. Abdel Meguid (1976). Geology of Kharga Dakhla Oases area, Western Desert, Egypt, from Landsat-1 satellite images. Academy of Scientific Research and Technology, Remote Sensing Center, Cairo.

21. E.M. El Shazly, M.A. Abdel Hady, I.A. El Kassas, H. El Amin, A.A. Abdel Meguid, S.I. Mansour, M.M. El Shazly and M.A. Tamer (1977). Geology and groundwater conditions of Tushka Basin area, Egypt, utilizing Landsat satellite images. Academy of Scientific Research and Technology, Remote Sensing Center, Cairo.

22. E.M. El Shazly, M.A. Abdel Hady, M.M. El Shazly, S. Sanad, S.H. Attia, M.A. El Ghawaby and S.M. Khawasik (1977). Soil investigations at Bahriyya Oasis - South Siwa Oasis area, Western Desert, Egypt. Academy of Scientific Research and Technology, Remote Sensing Center, Cairo.

23. E.M. El Shazly, M.A. Abdel Hady, M.M. El Shazly, A.A. Haraga, S. Sanad, S.H. Attia, M.A. El Ghawaby and S.M. Khawasik (1977). Regional geological and soil investigations at Farafra Oasis - Nile Valley area, Western Desert, Egypt, using Landsat satellite imagery. Academy of Scien. Res. and Tech., Remote Sensing Center, Cairo.

24. E.M. El Shazly, M.A. Abdel Hady, M.M. El Shazly, S. Sanad, M. El Ghazzawy and S. Abdel Mogheith (1978). Subsurface geology and geochemistry of Pliocene-Quaternary aquifers in northwest Nile Delta area. Acad. Scien. Res. and Tech., Remote Sensing Center, Cairo.

GUAYULE AS A POTENTIAL SOURCE OF RUBBER FOR EGYPT

WILLIAM G. McGINNIES
Director Emeritus, Office of Arid Lands Studies,
University of Arizona, Tucson, Arizona

ABSTRACT

Guayule, a native of the North American deserts, is a known producer of natural rubber equal in quality to Hevea rubber. It has been produced under cultivation in North America under conditions similar to those found in Egypt. The purpose of this paper is to explore the needs and possibilities of production in Egypt and to outline a program for production, including (1) the availability of suitable climate and soils, (2) cultural practices, (3) rubber extraction facilities, and (4) a technological assessment for production in Egypt.

Guayule, *Parthenium argentatum,* is a small shrub native to western Texas in the United States and the central drylands of northern Mexico. It is the only plant, with the exception of *Hevea brasiliensis,* that has produced rubber in significant commercial quantities.

Because it is a known rubber producer and has been cultivated successfully in arid regions, it deserves consideration for possible production in Egypt. I think this consideration might well be based on the answers to four questions:
(1) Is guayule capable of producing a high quality rubber?
(2) Will there be a market for additional natural rubber?
(3) Can guayule be produced under present economic conditions?
(4) Is there a place for guayule in the Egyptian socio-economic structure?

The answer to the first question is definitely yes. Speaking on the quality of guayule rubber at a recent U.S. Congressional hearing on guayule, Noel D. Vietmeyer, Study Director, National Academy of Sciences, Panel on Guayule, reported that the National Academy of Sciences panel subjected guayule rubber samples to the most modern and sensitive tool of chemical structure elucidation, and that the results showed conclusively that guayule rubber and hevea rubber have chemical and physical properties that are virtually identical. Both are polymers isoprene, having double bonds with cis configuration (this gives rubber its "bounce"). They have identical microstructure and approximately the same molecular length and weight.

In a statement made at the same Congressional hearing, J.J. Riedl[2], General Manager, Plantation Operations, Goodyear Tire and Rubber Company, stated, "We believe the technical qualities of purified guayule rubber make it acceptable for use as a replacement for hevea natural rubber and synthetic natural rubber in certain formulations .... The property differences that do exist can probably be corrected by changes used in compounding the raw polymer".

Scientists from Firestone Tire and Rubber Company have reached essentially the same conclusions and both of these companies are actively engaged in studies leading toward guayule rubber production.

As further proof that a rubber of commercial quality can be produced from guayule, the Centro de Investigacion en Quimica Aplicada supported by funds from the Mexican government has operated a pilot plant in Saltillo, Mexico for the successful production of a high quality rubber and as a result plans are underway to construct a rubber extraction mill with an annual capacity of 50,000 metric tons of native shrub.

Experimental passenger tires with guayule replacing hevea rubber in all components where natural rubber was required have been produced by the Goodyear Tire and Rubber Company, and small truck tires were built and tested in Mexico. These passed high speed performance, tire endurance, and other tests.

The importance of these tire tests can be fully appreciated by reference to the fact that tire manufacturing represents about 72 percent of all natural rubber used worldwide (Pierson).[3] Bias and biasply automobile tires are 13 percent natural rubber and radial tires twice as much. Truck and bus tires are 30 to 75 percent natural rubber and aircraft tires 90 percent, and it does not appear that these percentages will decrease, because the characteristics of natural rubber are essential for all parts of tires except the tread.

As there seems to be little if any concern among knowledgable persons of the capability of guayule rubber to meet market demands, the next step is to take a look at the rubber market.

During the congressional hearings previously mentioned, Vietmeyer stated, "The world'd demand for natural rubber is increasing so rapidly that economists are predicting that the rubber tree plantations cannot produce enough, and that natural rubber shortages will soon occur. From model calculations the Malyasian Rubber Research and Development Board concludes that the annual demand for natural rubber will outstrip supply by about 1 million tons in 1980 and by about 2 million tons in 1985. As a result, the World Bank expects prices to increase (in current U.S. dollar terms)...to about 60¢ per pound in 1980 and 78% per pound in 1985."[1]

The reports from many sources indicate that the chances for increasing the supply of hevea rubber are limited. The Malaysian production is approaching its maximum and while there has been a recent effort on the part of major rubber users to increase hevea rubber production in South America and Africa, the production potentials are not great in terms of worldwide needs. Furthermore, there is always the chance of disease or other catastrophes that may reduce supplies. It therefore seems that the market for guayule rubber should be very good.

On the basis of this evidence and much more that could be cited, I think it is fair to say that guayule can produce a high quality rubber, that can be used in place of hevea rubber and that there appears to be a market for additional rubber.

Before discussing questions 3 and 4 regarding economic production and the adaptation of guayule to Egyptian conditions, I would like to talk about guayule, its native habitat, its history, its growth characteristics, and some experiences in growing, harvesting and milling as these may have a bearing on the answer to these questions.

THE GUAYULE PLANT AND ITS ENVIRONMENTAL RELATIONSHIPS

Guayule is a low growing much branched gray-green shrub usually less than 75 centimeters tall. The narrow leaves, two to five centimeters in length, are covered on both surfaces by hairs (or trichomes) which give the gray-green sheen to the plant. The woody branches terminate in flowering heads, containing sterile disk-florets and five fertile ray florets each of which may produce seeds.

The root system of guayule consists of a tap root that may reach a depth of 6 meters under cultivation and shallow fibrous roots with a spread of 3 meters in the upper 15 centimeters of the soil. In its native habitat plants may live 30 years or more.

Guayule occurs at elevations between 600 and 2,000 meters (1800 to 6600 ft) in the semidesert drylands of north central Mexico and Big Bend area of Texas in the United States which have a rainfall of 200 to 375 millimeters (8 to 15 inches) occurring mostly in summer. Within its native habitat it grows in scattered patches from one half to several hundred hectares in size. It is largely restricted to outwash slopes and fans with calcareous soils. Daytime temperatures seldom exceed a maximum of 49°C (102°F) and a minimum of -18°C (0.0°F). Under cultivation guayule has shown a greater adaptibility to soil and moisture conditions and less tolerance of low temperatures, the minimum being in the neighborhood of -9°C (15°F).

Under cultivation guayule has been grown on noncalcareous well drained soils but it has very little acid tolerance, and limited alkali tolerance. It also was grown very successfully in regions of winter rainfall of 300 millimeters (12 inches). Usually supplemental irrigation is needed where precipitation is below 350 millimeters (14 inches). When annual moisture exceeds 600 millimeters (24 inches), excessive growth rather than rubber formation may occur and on all but very well drained soils, death may result.

HISTORY OF GUAYULE RUBBER PRODUCTION

Guayule was "discovered" in 1852 by Dr. J.M. Bigelow in Texas, but its rubber producing characteristics were known by Indians several centuries earlier (McGinnies and Haase).[4] The first commercial production of rubber was in 1888 by the New York Belting and Packing Co., of New York USA from 45,000 Kg (100,000 lbs.) of shrub imported from Mexico. In 1904 the pebble mill extraction method was developed in Mexico and by 1910 rubber production from native Mexican shrub was a booming industry, and from 1910 to 1946 the United States imported more than 68 million kg of Mexican guayule rubber.

In 1910 the Intercontinental Rubber Company initiated studies in Mexico to grow guayule. In 1912 their activities were moved to Valley Center near San Diego in the United States, and in 1916 to southern Arizona 40 Kilometers (25 miles) south of Tucson, where guayule was grown under irrigation. In 1925 operations were moved to Salinas, California where about 3240 hectares (8000 acres) were planted mostly without irrigation. About half of this acreage was plowed up during the period of low rubber prices in the early thirties, but the company did mill shrub from about 1750 hectares (4400 acres) producing 1.4 million kg of rubber prior to the purchase of the holding by the U.S. Government in 1942.

Because of the shortage of natural rubber and the non-availability of hevea rubber from southeast Asia during World War II, the United States purchased the holdings of the Intercontinental Rubber Company and planted 13,000 hectares in California, Arizona, and Texas before the program was curtailed at the end of the war owing to the development of synthetic rubbers and the reduced need for a domestic natural rubber supply.

During the three and half years the Emergency Rubber Project existed, 1,400,000 kilos of rubber were produced. At the end of the period the remaining 12,000 hectares of shrub with an estimated 10 million kilos (22,000,000 pounds) of rubber were destroyed.

During the decade following World War II some research, especially relating to rubber extraction, was carried on but in the early fifties all guayule research was terminated in the United States. During the 1970's there has been a renewal of processing research in Mexico leading to the development of improved rubber quality.

POTENTIALITIES OF GUAYULE AS A RUBBER PRODUCER

The rubber in guayule is contained in single, thin walled cells, mainly in the outer layers of the stem, but the old cells of the wood and pith produce rubber for several years. Two thirds of the rubber is in the stems and branches and the remainder in the roots. There is no rubber in the leaves (Hammond and Polhmaus).[5]

Genetically guayule is a rather unusual plant. It belongs to a genus with 16 species and it can be crossed with all of them. They vary from small annual plants to one 5 to 7 meters in height. The chromosome number of guayule may vary from 2n=36 to 2n-144. The 36 Chromosome plants produce seeds by sexual reproduction while plants with higher chromosome numbers are usually apomictic -- that is, they produce seeds without the union of male and female elements and hence the progeny inherit the characteristics of female only. However, plant breeders have been able to make crosses with the higher chromosome number plants and later to return progeny to the apomictic stage, thus allowing plants with desirable characteristics to breed true.

As a result of natural and man made crosses, guayule exists in a wide variety of strains, some with much rubber and others with almost none. Strains containing up to 26 percent rubber have been found, and cultivated strains have produced up to 20 percent rubber after 4 years growth. The rubber content depends not only on genetic characteristics but also environmental conditions. Actively growing quayule produces little or no rubber, but when growth is slowed by cool weather or reduced moisture supply, the rubber content increases.

GUAYULE AS A CULTIVATED CROP

There is a great amount of information available on the culture, harvesting and milling of guayule. Over 900 references were listed in a bibliography[6] published in 1975, and in addition the accomplishments of the Emergency Rubber Project during World War II are well documented. The following discussion is largely based on the experiences of the Emergency Rubber Project.

SEED AND SEED TREATMENT

Guayule is a prolific seed producer. By the end of the first year abundant seed is produced in plantations. Under cultivation a seed harvest of 300 kg (660 lbs.) per hectare was common and collections as high as 1100 kg (2400 lbs.) per hectare were made.

The seed consists of an embryo and some attached parts which tend to inhibit germination. There may also be some inherent dormancy but all dormancy may be overcome by treatment with hypochlorite, customarily in the form of household bleach -- sodium hypochlorite (chlorox). Seed exposed to air lost its dormancy in a year or two but soon after deteriorated. On the other hand, seed in sealed containers with a moisture content of 5 percent remained viable for 30 years or more.

While seeds germinate readily, it has been found that they have low emergence energy and hence are unable to compete with weeds. Moisture control is also essential. Because of these limitations it has been customary to grow guayule seedlings in nurseries, and later transplant them to the field.

NURSERIES

During the operation of the Emergency Rubber Project seed was sown in standard nursery beds 1.2 by 122 meters (4 x 400 feet) at a rate to produce 200 useable seedlings per square meter (20 per sq. ft.). The aim was to produce 2,000,000 to 2,500,000 useable plants per hectare (800,000 to 1,000,000 per acre) of nursery space. This was sufficient to plant 12 to 16 hectares (30 to 40 acres) depending upon spacing used in field planting. Irrigation was supplied by oscillating overhead sprinklers. Weeds were controlled by especially designed cultivators and oil sprays (present day herbicides were not available at that time). Seedlings were transplanted at 4 months or later depending on size of plants and planting needs. If necessary, seedlings could be held up to two years in nursery beds.

It was found that to insure good survival in field plantations seedlings should be hardened by cool temperatures or low soil moisture content prior to transplanting. Seedlings to be transplanted were mowed to a residual height of 5 centimeters (2 inches) and the roots cut at a depth of 12.5 to 17.5 centimeters (5 to 7 inches). It was found that guayule nursery seedlings could be transplanted at any time in the year. However, best results were obtained when minimum temperatures were above 10^oC and maximum temperature below 35^oC (50^oF and 95^oF).

FIELD PLANTATIONS

Planting was done with a machine planter such as a modified 4-row Holland or Lindeman planter. Seedlings were transplanted to the field on irrigated land in rows 70 centimeters (28 inches) apart at intervals of 30 to 60 centimeters (12 to 24 inches) within the row. The closer spacing was aimed toward harvesting at the end of two years, the wider spacing for harvest at the end of four years. In non-irrigated fields the standard spacing was 70 by 60 centimeters (28 by 24 inches) with the expected harvest in 5 years or more.

Weed control is essential for guayule production especially in the first two years. During World War II this was accomplished by cultivation and the use of oil sprays. Since then many herbicides have been developed, some of which may be superior to oil.

It was found that proper irrigation practices were very important in the production of quayule. Soils that were too heavy or too light made irrigation control difficult and shallow soils resulted in excessive disease losses and contributed to poor moisture holding capacities.

RUBBER FORMATION

In its native habitat guayule shows a buildup of rubber percent during the winter semi-dormant period and it appears that this buildup is a result of cool temperatures and moisture stress. Observations in warm moist areas substantiated the belief, but experimental studies showed that plants could be harvested at any time of the year with a fair rubber content, provided the plants had been subjected to moistur stress. Experience also indicated that growing the crop with limited water supplies would produce somewhat less tonnage but with considerably higher rubber content. In general water requirements are less than those for cotton and small grains.

Stressing is very important in rubber formation and in areas where there is no cool dormant season the control of soil moisture is essential to provide stress conditions. Best results were obtained on loamy soils that could be dried slowly. On light and shallow soils withholding moisture often resulted in shock because of the rapid removal of soil moisture.

INSECTS AND DISEASES

Insect problems have not been great. Defoliation by grasshoppers and seed destruction by Lygus were encountered, but both could be controlled. Damage from other insects was minimal.

Diseases were important sources of loss, and in some situations were sufficient to necessitate plowing up of depleted stands and in other cases resulted in reduced production because of fewer plants and unsatisfactory stress conditions. For moisture stress 80 percent survival was considered optimal as gaps between surviving plants were usually small and stress by withholding moisture was uniformly obtained.

HARVESTING

Because about one third of the rubber content is in the roots the usual custom was to undercut the plants at about 20 centimeters (8 inches). Mowing the plants at least once and harvesting the tops was tried to prolong the life of the plants between plantings. However, based on limited trails this practice is still open to question.

Prior to and during World War II it was thought that drying the shrub was necessary for coagulation of the rubber, but post war research showed that drying was not necessary. However, under field conditions it may be desirable to dry and bale the shrub to decrease transportation and handling costs.

RUBBER EXTRACTION

Up to the end of World War II rubber was extracted by means of pebble mills. The basic equipment consisted of a horizontal cylinder lined with flint bricks in which a charge was placed of one third volume of flint pebbles, and one half of water together with the shrub. The cylinders were slowly rotated, resulting in the crushing of the shrub. After discharge from the pebble mill the rubber worms were floated out, cleaned and dried.

Under the improved method recently developed in Saltillo, Mexico, rubber is extracted and processed as follows: The shrubs are first dipped in hot water for 10 minutes at 75°C (167°F). This coagulates the rubber and removes foreign material and leaves, which contain no rubber, but do contain copper, manganese and resinous compounds that contaminate the rubber. The plants are then passed through a hammer mill and a Bauer mill (a device used in paper making) which break open the rubber filled cells. The pulping is done in water with caustic soda added. The material is transferred to a slurry tank where the water-logged bagasse sinks, the rubber worms float and are skimmed from the surface. They are then run through a second tank where they are rinsed and then treated with a detergent. Guayule worms contain 17 to 25 percent resins which are removed with acetone. The acetone is then distilled and recycled. During this process the rubber may be treated to modify the product.

GUAYULE AS A POSSIBLE CROP FOR EGYPT

The third question, "Can guayule be produced under present economic conditions?", cannot be answered positively as we have no recent studies on which to base conclusions. The best we can do is to project experiences gained during World War II. This is not too satisfactory because the operation was conducted by the government under war time pressure to produce rubber. There is also the problem of rising prices for labor and materials, complicated by the changes in the dollar value.

The greatest expense in guayule culture is in the establishment of plantations. During World War II seedlings were produced in nurseries at a cost of $2.50 per thousand, and

and probably could be produced now at a cost of $5.00 to $10.00 a thousand. The cost of seedlings produced in a greenhouse nursery may vary from $10.00 to $100.00 a thousand. At a spacing of 70 x 60 centimeters, 28,000 seedlings per hectare would be needed (11,200 per acre). Direct seeding has been tired, but with limited success and when the cost of weed control and extra irrigation are considered it is about as costly as the transplanting methods and poorer stands are obtained. Low cost labor would naturally decrease the cost in either case. The overall costs of growing guayule after the plants are established would be about the same as for other row crops.

In field plantations herbicides can be used to control weeds, but some hand labor is necessary and might be increased to largely replace the use of herbicides if ample low cost labor is available. After the second growing season, the in-row and between-row spaces are largely closed in and only minimum weeding is necessary. Irrigation costs are similar to those for cotton except that guayule can be grown successfully with about a third less water.

The fourth and last question, "Is there a place for quayule in the Egyptian socio-economic structure?", is what this workshop is all about. With my very limited knowledge of Egypt, I would not attempt to answer this question. I can, however, mention some of the items I consider important in reaching a reasonably sound conclusion.

I have attempted to point out some of the salient items regarding guayule as a rubber product. If we accept these, it can be concluded that guayule is a plant that can produce rubber under conditions in Egypt. The basic requirements are well drained soils with a pH within the range 6.0 to 8.5, and adequate water supply (about 50 centimeters, 20 acre-inches). The optimum amount would be dependent upon evapotranspiration rates and soil texture. The quality requirements are not definitely known but there are indications of only moderate salt tolerance. When harvesting is done with heavy machinery for digging and baling, the cost is relatively low. However, transportation is an item to be considered as even when baled, guayule is a bulky commodity.

Milling costs are unknown at present. Projection from World War II operations are unreliable, because it is probably that new or improved methods of rubber extraction will be developed. Also to be considered are the compensation returns from resins and other recoverable products including leaves and bagasse. The last two might be used for several purposes, including livestock food, soil amendments and fuel. It is even considered likely that these by-products may cover the cost of milling. Bagasse has been used to supply the boiler heat needed to mill guayule.

In the final analysis the economy of rubber production depends directly on the percentage of rubber in the harvested plants. The costs are the same for planting and care of plantations whether the rubber content of the plants is low or high. The milling costs are almost entirely dependent on the weight and volume of shrub milled. On the average during World War II about 25 metric tons (27.5 short tons) of shrub were produced per hectare at the age of 4 years. With a 10 percent rubber content the rubber production would be 2.5 metric tons (2.75 short tons) per hectare.

The minimum size of a factory unit has been set at 4000 hectares (10,000 acres). On a four year rotation 1,000 hectares (2,500 acres) would be harvested each year. Total shrub would be about 25,000 metric tons, producing 2,500 metric tons (2,750 short tons) of rubber that would be worth $2,915,000 in 1983 at 53¢ per pound and $3,190,000 in 1986 at 58¢ per pound at the end of the 4 year cycle. This would amount to $2,915.00 to $3,190.00 per hectare or $729.00 to $797.00 per hectare per year (approximately $290.00 to $320.00 per acre per year). With 20 percent rubber content, a realistic figure for the future based on plant breeding and bioinduction techniques, this figure might be double which would mean returns of more than $1500 per hectare per year.

Assuming it would cost $1.00 a kilo to produce rubber, the cost for 2,500 metric tons would be $2,500,000 leaving a net profit of $415,000 in 1983 and $690,000 in 1986 or $104 per hectare for year 1983 and $172 per hectar per year in 1986.

REFERENCES

1. Noel D. Vietmeyer. Hearing, Subcommittees on Agriculture and Science, Research and Technology. U.S. House of Representatives, Washington, D.C. June 19, 1978.

2. J.J. Riedl. Hearing, Subcommittees on Agriculture and Science, Research and Technology. U.S. House of Representatives, Washington, D.C. June 19, 1978.

3. R.M. Pierson (1975). In, W.G. McGinnies and E.F. Haase, An International Conference on the Utilization of Guayule. Office of Arid Lands Studies, University of Arizona, Tucson.

4. W.G. McGinnies and E.F. Haase (1975). Guayule, a rubber-producing shrub for arid and semiarid regions. Arid Lands Resource Information Paper No. 7. University of Arizona, Tucson.

5. B.L. Hammond and L.G. Polhamus (1965). Research on guayule (*Parthenium argentatum*): 1942-1959. USDA Tech. Bull. 1327.

6. W.G. McGinnies and E.F. Haase (1975). An International Conference on the Utilization of Guayule. Office of Arid Lands Studies, University of Arizona, Tucson.

ECONOMIC POTENTIALITY OF *THYMELAEA HIRSUTA* (ENDLE.), A PERENNIAL SHRUB IN EGYPTIAN DESERT LANDS

A.A. EL GHONEMY
Faculty of Science, University of Tanta, Egypt

A. WALLACE
Laboratory of Nuclear Medicine and Radiation Biology, University of California, Los Angeles

H.K. El Din
National Paper Company, Alexandria, Egypt

ABSTRACT

Paper pulp is in great shortage not only in Egypt but also in other nations in arid regions of the world. A perennial desert shrub, Thymelaea hirsuta, *which grows abundantly along the Mediterranean coast of Egypt, has characteristics which make it a candidate for a source of paper pulp. The pericycle of this plant is composed of numerous large groups of long fibers measuring 3000 to 10,000 microns in length and 14 to 22 microns in diameter. Xylem tissues have lignited fibers varying in length from 600 to 2500 microns with diameters of from 15 to 23 microns. The plant can be partially harvested on a rotational basis from the 2.5 million acres in Egypt where it is a dominant species without harm to the plant and to the ecosystem. Plants can be so handled about every fourth year. Studies have been made of the technical suitability of the species for paper pulp production and these will be reported. The results are very encouraging. With planning and care most of the need for paper pulp in Egypt could be met from this source.*

INTRODUCTION

One of the major problems facing many nations is the shortage in raw materials used in paper industries. In Egypt, it is estimated (Fouad and El Din)[1] that by 1985 the gap between the production and consumption of paper and board will amount to about 322,000 tons annually. This amount requires for its production about 150,000 tons of long-fiber chemical pulp.

The present high (and rising) cost of imported long-fiber chemical pulp has led the authors to search for local sources of cellulosic raw materials that might replace, partially or wholly, the imported long-fiber chemical pulp for the supply of locally produced paper.

Thymelaea hirsuta, a perennial desert shrub covering vast plains along the Mediterranean coast of Egypt, has been recommended for this purpose (El-Ghonemy *et al.*).[2] This is the first attempt in Egypt in which a desert plant is examined for its suitability for the production of paper pulp.

PART I

THYMELAEA HIRSUTA FROM THE BOTANICAL POINT OF VIEW, TAXONOMIC POSITION AND ECOLOGICAL BEHAVIOR

Thymelaea hirsuta is the only species of the genus *Thymelaea* in Egypt. It belongs to the Thymelaeceae, order Myrtiflorae, class Dicotyledonae of the subdivision Angiosperm. Its vernacular name in Egypt is Mithnan.

Thymelaea hirsuta is circum Mediterranean in its distribution but with only minor importance along the European coastal belt. Previous workers have contributed considerable information and data about the general distribution, predominance and socioecological relations of *Thymelaea*.

According to Tadros and Atta,[3] the community dominated by *T. hirsuta* sprreads nearly all over the uncultivated desert areas of the Mediterranean coast. El-Ghonemy and Tadros[4] have shown that the southern limit of *T. hirsuta* in Egypt is approximately at 70 km from the coast.

Details of soil properties, climatic features and nature of vegetational groupings in regions supporting *T. hirsuta* are published elsewhere (El-Ghonemy *et al.*)[2,8].

MORPHOLOGICAL AND ANATOMICAL STRUCTURE

Thymelaea hirsuta is a perennial shrub, stem erect and densely branched, 40-80 cm high, attaining one meter or more under favorable conditions.

From the anatomical point of view, the phelloderm, which occupies a relatively narrow area, contains among its cells many cellulonic fibers, scattered in the form of small groups of 2-4 fibers each. The pericycle is composed of numerous large groups of long fibers forming a ring interrupted by parenchyma. These fibers have thick cellulosic walls, narrow lumen and measure 3000 to 10,000 microns in length, and 14-22 microns in diameter.

The vascular tissue forms a continuous ring, the phloem is relatively wide and consists mainly of long cellulosic fibers similar to those of the pericycle. The xylem tissue, on the other hand, is formed mostly of a mixture of short and long lignified fibers. They measure 600-2500 microns in length and 15-23 microns in diameter. The percentages of the different plant fractions in the disintegrated stem tissues of *T. hirsuta* are given in Part II.

THE PRODUCTIVITY RATE OF *T. HIRSUTA* IN RELATION TO SOIL TYPES

The vigor and productivity of *T. hirsuta* vary in different habitats, in apparent relation with the prevailing environmental conditions. The results obtained from studying the effects of four distinctly different habitat types on growth habit and standing crops of *T. hirsuta* are summarized in Table 1. The two fertile Ermonsol sites are characterized by greater harvest rates ranging between 1500-1700 Kg per hectare. The harvest rate on the Rhegosol site is also reasonable (900 Kg per hectare) and for industrial purposes could be economically sound.

It should be mentioned, however, that although the ecological amplitude of *T. hirsuta* in Egypt may cover the extent of 2.4 million hectares (rocky ridges, salt marshes, sand dunes not included), only 40% of such area (1.0 million hectares) may be considered as covered by *T. hirsuta* community. It may be added that the figures given for harvest rates in Table 1 represent data for stands typical for this community. Stands from other allied communities may give much lower harvest rates.

FLOWER TYPES, SEX RATIO AND SEED CHARACTERISTICS

Observations on the flowers of *T. hirsuta* in two distinctly different habitats, *viz*, calcareous sand dune and an inland wadi bed, indicate that the plant is mostly dioecious. Hermaphrodite and sterile flowers have been recorded. The number of individuals carrying male flowers exceeds significantly that of individuals carrying female flowers (Table 2). The average number of flowers (male or female) per plant of average size (height 70 cm, width 100 cm) is about 20,000, occurring generally on terminal branches, of a mean length of about 20 cm.

TABLE 1. Variations in some characters and productivity rate of *Thymelaea hirsuta* in localities of different climate and edaphic features.

						Plant Characters				
		Height (cm)		Diam. (cm)		Weight per plant (oven dry Kg)		Density No. per ha		Yield (Kg oven-dry wt. Per hectare
Locality	Soil unit	Mean	Range	Mean	Range	Mean	Range	Mature Plants	Seed-lings	
Ei-Kingi	Lithosols	50	25-80	60	35-100	0.1	.05-.55	1500	900	370
Burg El-Arab	Ermosols	80	55-115	95	50-110	1.7	.35-515	360	140	1500
K. 56 South of Alex.	Ermosols	70	45-100	83	65-130	0.8	.30-1.7	910	620	1800
K. 72 South of Alex	Silic Regonsols	88	50-150	87	40-140	0.59	.30-3.6	300	110	950

TABLE 1, cont.

ENVIRONMENTAL INDICES

Oven-dry weight of leaves (% of plant)	Ratio of bark to xylem	Rainfall (mm)	Aridity index (Q of Emberger)	Conductance of soil	Soil Phosphorus (pts per million)		Soil texture
					Total	Available	
16	40:50	170	25	2.0	175	5	Med.
20	42:58	150	20	2.0	360	12	Fine
16	41:59	100	15	2.4	200	5	Fine-Med
18	44:56	70	12	1.0	55	Nil	Coarse

The seed of *T. hirsuta* is harvested in the hull, is brownish in color, about 3 mm long and 2 mm wide. There are between 250,000 and 300,000 seeds in a kilogram. A seeding rate of one kilogram of these seeds per hectare, if evenly distributed, would result in some 70 to 75 seeds per square meter of surface.

In a given habitat the average number of reserve seeds under a given *T. hirsuta* shrub varies from one individual to another as well as from one site to the other. The average number of reserve seeds at the dune habitat exceeds significantly those at the habitat of the inland wadi beds (Table 3). A number of damaged seeds (on the average 2% of the total number of seeds) were found in both habitats. The damage is presumably caused by some insects and animals, which feed on cotyledons and endosperm.

GERMINATION AND PROPAGATION

Seeds of *T. hirsuta*, as the seeds of many desert species, exhibit some sort of dormancy. Preliminary tests under laboratory conditions have shown that germination of untreated seeds is generally less than 1%. Treatments with sulphuric acid followed by gibberellic acid hormone raised seed germination under laboratory conditions to about 50%.

The average number of seedlings of *T. hirsuta* which emerge naturally per unit area during the entire period of emergence (December to February) has been estimated. The number of seedlings below the shrub (female individuals) canopy (1 m^2) and within its surrounding area (10 m^2) varies from one individual to another as well as from site or habitat to the other. In the dune habitat the average number of seedlings below the canopy is 7 compared with 0.25 in the inland wadi bed. The corresponding values for the surroundings outside of the canopy is 1 and 0.2 respectively.

The mortality rate of seedlings also varies from one habitat to the other. It is generally high (up to 100%) in the dune habitat and low (25%) in the inland depressions. In the dune habitat, it is assumed that the rapid drying of the soil surface is the most important factor causing such high death rates of *T. hirsuta* seedlings. However, the presence of distinct size groupings of *T. hirsuta*, representing different ages, indicate that successful establishment of seedlings of this plant may occur periodically every 5 to 10 years.

TABLE 2. Sex ratio in *T. hirsuta* growing in two habitats of different environmental conditions.

Sex Percentage	LOCALITY & HABITAT Littoral calcareous dunes at Gharbaniat	Inland Depression at Omayed
Male	61	50
Female	29	42
Predominantly Male	6	4
Hermaphrodite	2	2
Predominantly Female	1	1
Male + Sterile	1	1

TABLE 3. Variations in numbers of reserve seeds of *T. hirsuta* in relation to habitat and position relative to the shrub.

Number of Seeds per shrub	LOCALITY & HABITAT Littoral calcareous dunes	Inland Depression
Leeward side of shrub		
Mean	2500	12
Range	300-5000	5-30
Windward side of shrub		
Mean	100	5
Range	20-200	0-15
Total		
Mean	2600	17
Range	320-5200	5-45

During the first year of establishment most of the seedlings show poor development and it is estimated that such seedlings may reach their full size within five or six years.

Regrowth after cutting the top of mature plants varies according to how cutting is done. When the top green foliage is removed entirely, the remaining part of the stem soon dies. On the other hand, when plants are partly cut, even if the remaining parts are but small little branches, they continue their growth at quite a high rate. These pre-treated plants reach their average full size within four years after cutting.

MINERAL COMPOSITION

An analysis is presented of the inorganic element relations of five populations of *T. hirsuta* growing in distinctly different habitats. This study is considered basic for any future application of the plant in industrial purposes. At the same time it is recognized that any overall evaluation of plant conditions, fertilizer requirements and recommended soil management practices will require considerable soil and management information beside what is furnished by the analysis of the plant.

The technique adopted in plant analysis is that of Wallace *et al.*[6] From data given in Table 4 we may arrive at the following conclusions:

1.(A) Leaf tissues (Table 4): For certain elements (N, P, Na, Ca, Mg, Mn, Ti, and Sr) the overall variations in concentrations are significant, while for other elements (K, Zn, Fe, B, Al and Si) this variation is insignificant.
1.(B) Stem tissues: Elements that show significant overall variations include P, Na, K, Ca, Mg, B, Al, Si, Ti and Sr, while those with insignificant overall variations include P, Zn, Fe and Mn.
2. The pattern of uptake and accumulation of the different elements varies from one plant organ to another.
3. Correlations between different nutrient elements in plant tissues are undoubtedly affected by various combinations of biological factors, but in view of the variety of sites included in this analysis, the similarity in the biochemical functions of pairs or groups of elements seems to be the most important factor contributing to the observed correlations between most of them (Table 5).

TABLE 4a. Ordered means for nutrient concentrations that showed various levels of variations in leaf samples of *T. hirsuta* collected from different habitts (dry weight basis). The analyses of variance is according to the unequal cell analysis technique. The means bracketed are not significantly different from each other at the 0.05 probability level, while those over different brackets are.*

						P
Nitrogen,%	2.08(S)	1.92(Si)	1.83(D)	1.83(A)	1.66(R)	<0.05
P, μg/g	2928(S)	1417(A)	1141(R)	956(Si)	937(D)	<0.005
Na, μg/g	2945(D)	1455(R)	905(S)	803(A)	698(Si)	<0.005
K, %	1.22(S)	1.18(D)	1.12(A)	1.09(R)	0.99(Si)	<0.1
Ca,%	1.39(S)	1.36(D)	1.20(R)	1.14(A)	1.11(Si)	<0.005
Mg, %	0.35(R)	0.34(S)	0.33(D)	0.32(A)	0.28(Si)	<0.005
Zn, μg/g	71(Si)	58(D)	29(A)	24(R)	21(S)	<0.1
Fe, μg/g	1336(A)	1264(Si)	1047(R)	1000(S)	915(D)	<0.1
Mn, μg/g	140(Si)	135(A)	132(R)	107(S)	42(D)	<0.05
B, μg/g	68(D)	67(S)	56(Si)	55(R)	50(A)	<0.1
Al, μg/g	933(S)	909(A)	789(R)	640(Si)	497(D)	0.1
Si, μg/g	5066(S)	3708(A)	2115(R)	2096(Si)	2040(D)	0.1
Ti, μg/g	110(S)	97(A)	95(R)	84(Si)	52(D)	0.05
Sr, μg/g	858(D)	462(R)	382(A)	363(S)	254(Si)	0.005

*D=Calcareous sand dune; S=Saline; R=Rocky; A=Alluvial; Si=Silicious sand

TABLE 4b. Ordered means of nutrient concentrations that showed various levels of variations in stem samples of *T. hirsuta* collected from different habitats. The analysis of variance is according to the unequal cell size technique. The means bracketed are not different significantly from each other at the 0.05 probability level, while those over different brackets are.*

						P
Nitrogen, %	0.86(S)	0.85(A)	0.84(D)	0.82(R)	0.74(Si)	<0.1
P, µg/g	5092(S)	2170(R)	2095(A)	1750(D)	259(Si)	<0.05
Na, µg/g	2145(D)	1372(R)	983(S)	746(A)	491(Si)	<0.005
K, %	1.38(S)	1.29(Si)	1.22(A)	1.09(R)	0.94(D)	<0.005
Ca, %	1.03(D)	0.88(R)	0.76(A)	0.70(S)	0.68(Si)	<0.025
Mg, %	0.28(D)	0.27(R)	0.23(S)	0.23(A)	0.16(Si)	<0.005
Zn, µg/g	62(A)	61(S)	54(R)	52(Si)	48(D)	<0.1
Fe, µg/g	2120(S)	2023(R)	1972(A)	1577(D)	798(Si)	<0.1
Mn, µg/g	68(R)	62(S)	57(A)	56(Si)	38(D)	<0.1
B, µg/g	32(D)	30(R)	29(Si)	26(S)	25(A)	<0.025
Al, µg/g	1448(R)	1368(A)	1241(S)	1030(D)	361(Si)	<0.005
Si, µg/g	7173(R)	5688(A)	4856(S)	3590(D)	1873(Si)	<0.005
Ti, µg/g	150(R)	132(S)	129(A)	103(D)	39(Si)	<0.005
Sr, µg/g	545(D)	421(R)	378(Si)	316(A)	226(S)	<0.01

*D=Calcareous sand dune; S=Saline; R=Rocky; A=Alluvial; Si=Silicious sand

TABLE 5. Complete simple correlation coefficient matrix for 14 elements showing the levels of relationships present between them in both stem and leaf tissues of *T. hirsuta*.

	Element	STEM TISSUES P	Na	K	Ca	Mg	Zn	Cu	Fe	Mn	B	Al	Si
LEAF TISSUES	P	-	0.80†	0.03	0.03	0.37	0.35	0.03	0.70	0.39	0.17	0.46	0.41
	Na	-0.28	-	-0.60	0.67†	0.69†	0.16	-0.16	-0.16	0.18	-0.23	0.65§	0.10
	K	0.01	0.18	-	-0.36	-0.30	0.22	0.15	0.09	0.40	-0.43*	-0.01	0.01
	Ca	0.02	0.18	0.41	-	0.65§	-0.06	-0.04	0.23	0.06	0.46	0.36	0.22
	Mg	0.15	0.17	0.19	0.45*	-	0.15	0.06	0.57§	0.20	0.47*	0.60§	0.60§
	Zn	-0.22	0.19	-0.09	-0.13	-0.19	-	0.59§	0.45*	0.43*	0.10	0.07	0.27
	Cu	-0.05	0.19	-0.22	-0.23	-0.12	0.81	-	0.35	0.21	-0.24	0.31	0.24
	Fe	0.24	-0.49*	-0.01	-0.42	-0.05	0.10	0.14	-	0.59§	-0.03	0.73†	0.71
	Mn	-0.15	-0.57§	-0.16	0.22	0.08	-0.10	-0.04	0.33	-	-0.04	0.23	0.46*
	B	0.09	0.43*	0.24	0.39	0.28	0.01	0.13	-0.48*	-0.28	-	-0.11	0.04
	Al	0.75†	-0.52	0.04	-0.08	0.18	-0.21	-0.15	0.66†	0.17	-0.23	-	0.85†
	Si	0.12	-0.18	-0.13	-0.40	-0.17	0.01	0.18	0.25	0.28	-0.17	0.30	-
	Ti	0.63§	-0.52§	-0.07	-0.00	0.20	-0.28	-0.28	0.66†	0.29	-0.23	0.91†	0.23
	Sr	-0.39	0.46§	0.24	0.54	0.19	0.10	0.10	-0.54	-0.19	0.15	-0.40	-0.33

* = P 0.05; § = P 0.01; † = P 0.001

4. The mineral contents of *T. hirsuta* are generally of intermediate values compared with those of other associated species (El Ghonemy *et al.*[5]

PART II

THYMELAEA HIRSUTA FROM TECHNICAL AND INDUSTRIAL VIEW POINTS
MATERIALS, METHODS AND TECHNIQUES

MECHANICAL SEPARATION FOR CORK, PHLOEM AND XYLEM FRACTIONS

Air dry branches and stems of *T. hirsuta* were sawed into pieces 4-7 cm long. These pieces were further mechanically crushed into little parts of about 3x1x0.05 cm. A suspension of the crushed plant material in water having a consistency of 6% was prepared in a laboratory disintegrator with 350 rotations per minute for 45 minutes. The suspension was then diluted with water to a consistency of 2.0% and left for one hour. The floating cork parts, which were found to be fiber free, were removed through repeated decantation processes. The remaining parts of the crushed plant material were screened over a 20 mesh screen and washed with a strong jet of water. The soft fiber of the phloem passes through the screen, while the woody parts of the xylem are retained on the screen. Some traces of the cork may pass through the screen and become incorporated with the phloem fraction. These three fractions of the plant (phloem, xylem and cork) were collected, oven dried and weighed.

COOKING TECHNIQUES FOR THE SEPARATE FRACTIONS OF *T. HIRSUTA*

The xylem fraction, separated by the above mentioned technique was cooked using the Kraft method in a laboratory digestor of 8 liters capacity. The sulphidity of 25% and pressure of 6 atmosphere were kept constant for all cooks.

The phloem fraction was cooked by caustic soda method in the digestor with pressure adjusted at 6 atmosphere and for a cooking period of only 2 hours. The rather low lignin content (6.7%) of the phloem fraction justifies the use of soda cooking (with active alkali from 4.65-9.3%) instead of the Kraft method.

EVALUATION OF THE PHYSICAL PROPERTIES OF THE PUMPS

The pulps produced from the two separate fractions of *T. hirsuta*, the xylem and phloem, were tested for their physical properties according to Tappi. Grade index (GI) for the pulp was calculated according to the formula:

$$GI = \frac{\frac{A+B}{2} \quad C+D}{3}$$

where B, B, C and D are the breaking length, burst factor, tear factor and double fold respectively, expressed as percentages of the corresponding values for the softwood kraft pulp.

RESULTS AND DISCUSSION

MECHANICAL SEPARATION OF CORK, PHLOEM AND XYLEM

The xylem fraction represented the highest percent (57.7) followed by that of the phloem (18.3) and cork (15.6) (Table 6). The chemical composition of the phloem and xylem fractions determined according to Tappi[7] is given in Table 7. The phloem fraction is more rich in alcohol-benzene extractives, ash and cellulose. The xylem fraction is rich in hot water extractives, lignin, and pentosans. The alcohol-benzene fraction for the cork was also determined, and being as high as 24.8%, deserves special attention.

CHARACTERISTICS OF THE XYLEM PULP

CHEMICAL PROPERTIES

Table 8 presents chemical characteristics of *T. hirsuta* xylem pulp produced by the Kraft method under different cooking conditions, except for sulphidity and pressure. These two parameters were kept constant for all cooks at 25% and 6 atmosphere respectively. From these data it is clear that if the cooking time were kept constant at 4 hours, while increasing the amount of active alkali (as Na_2O) relative to plant material (cooks 1, 2, 4, 6 and 7) the percentage of the chemicals consumed did not appreciably increase. A notable decrease in the amount of rejects took place also with increasing cooking chemicals. Above 18% alkali the amount of rejects became negligible.

TABLE 6. Percentages of the different plant fractions in disintegrated stem tissues of *T. hirsuta*.

Fraction	Mean %	Range %
Cork	15.6	12.0 - 18.9
Phloem	18.3	13.9 - 21.4
Xylem	57.7	51.7 - 63.3
Wastes	8.4	6.9 - 11.6

TABLE 7. Chemical characteristics of the phloem and xylem fractions of *T. hirsuta*.

Property	Phloem %	Xylem %
Hot water extractives	5.8	14.2
Alcohol-benzene extractives	11.6	3.7
Lignin	6.7	16.1
Ash	3.8	1.8
Pentosans	23.2	27.3
Cellulose	52.6	39.5

TABLE 8. Chemical properties of *T. hirsuta* xylem pulps produced by Kraft method, with sulphidity of 25%, pressure 6 atm, and other various conditions.

Cook No.	1	2	3	4	5	6	7
Active alkali, % of plant material	14	16	18	18	18	20	22
Chemical concentration as Na_2O gm/l	28	32	36	36	36	40	44
Cooking time, hours	4	4	2	4	6	4	4
Chemical consumed:							
% of plant material	12.6	13.3	13.1	13.3	15.7	13.5	13.9
% of total chemical	90.3	82.2	72.8	73.7	86.9	67.5	63.4
Yield							
Total	59.3	53.3	53.1	48.0	46.9	45.1	43.2
Screened	53.8	52.7	47.7	47.2	46.1	45.1	43.1
Chemical constituents calculated as % of xylem fraction used							
Holocellulose	52.8	48.8	50.1	45.3	41.9	43.0	41.3
Lignin	3.9	2.2	2.3	1.3	1.0	0.3	0.2
Pentosan	13.5	12.3	12.2	10.9	9.9	10.2	9.4
Cellulose	33.4	35.7	37.3	33.6	32.5	33.5	32.3
Ash	1.1	-	1.1	1.0	1.0	1.2	0.8
Chemical constituents calculated as % of the xylem pulp produced							
Holocellulose	88.2	91.2	94.4	97.3	89.3	95.3	95.6
Lignin	6.6	4.1	3.4	2.8	2.1	0.4	0.4
Pentosan	22.7	23.0	23.0	22.8	21.1	22.6	21.8
Cellulose	62.5	66.7	70.3	69.9	69.3	74.2	74.8
Ash	1.8	2.3	2.0	2.1	2.1	2.6	1.8
Permanganate No.							
Unscreened	31.8	16.0	20.3	14.7	12.4	11.9	11.5
Screened	24.7	15.3	16.0	12.5	12.4	11.9	11.5
Brightness	24.2	33.5	35.2	39.2	35.0	41.2	43.9

At 20 and 22% active alkali the holocellulose and cellulose contents of the produced pulps retained constant values of about 95 and 74% respectively; the lignin contents, on the other hand, dropped to 0.4%, and the permanganate number reached a constant value of 11.9.

The pulp produced from cook 5 (with 20% active alkali and under the above mentioned cooking conditions) with a lignin content of 0.4% can, therefore, be considered as the bleachable chemical pulp of the xylem fraction of *T. hirsuta*, while the pulp produced from cook 4 (with 18% active alkali) with a lignin content of 2.8% may be taken for the unbleached chemical pulp of the xylem fraction.

Table 8 (cooks 3, 4 and 5) shows also the chemical characteristics of the xylem pulp produced under a constant percent of active alkali (18%) while changing the time of cooking (2, 4 and 6 hours). These data indicate that the increase in the cooking time from 2 to 6 hours has resulted in a corresponding increase in the amount of the chemicals consumed from 13.3 to 15.7%. The cellulose of the pulp produced, however, has remained constant at about 70% irrespective of the cooking time. The lignin and pentosan contents, on the other hand, showed a remarkable drop with increasing time of cooking.

PHYSICAL CHARACTERISTICS

The strength properties of the xylem pulps produced under different cooking conditions are given in Table 9. Obviously the grade index of the pulp produced increased gradually till a maximum value of 53.4 at an active alkali of 18% (cook 4). Above this concentration the strength properties started to decrease particularly due to a remarkable drop in the number of double fold.

CHARACTERISTICS OF THE PHLOEM PULP
CHEMICAL PROPERTIES

Table 10 shows the chemical characteristics of the phloem fraction pulp of *T. hirsuta* produced under different cooking conditions, except for cooking time which was kept constant in all cases at only two hours. These data show that, although the lignin content of the phloem material is originally low (6.7%), the pulp produced from cook 4 (with an active alkali of 9.3%) contains still a remarkably high lignin content of 4.7%. This was thought to be due to the high (11.6%) alcohol

TABLE 9. Physical characteristics of *T. hirsuta* xylem pulps produced with sulphidity of 25%, pressure of 6 atm and other various conditions

Cook No.	Active Alkali %	Chemical Concentration as Na_2O gm/l	Cooking Time hrs	Initial Freeness °Sr	STRENGTH PROPERTIES Breaking Length m	Burst Factor	Tear Factor	Double Fold	Grade Index %	Brightness
1	14	28	4	30	5333	37.5	83.3	577	42.1	24.2
2	16	32	4	27	6111	57.4	83.3	1230	53.2	33.5
3	18	36	2	20	6444	51.6	8.0	1218	51.9	35.5
4	18	36	4	30	6555	50.4	90.0	1128	53.4	39.2
5	18	36	6	30	5555	51.6	75.0	1289	49.8	35.0
6	20	40	4	30	6000	49.2	80.0	1325	51.5	41.2
7	22	44	4	30	5889	49.2	88.3	720	47.9	43.9

TABLE 10. Chemical characteristics of *T. hirsuta* phloem pulps produced at two hours cooking time, pressure 6 atm, and other various conditions

Cook No.	1	2	3	4	5
Active alkali, % of phloem fraction	4.65	6.2	7.75	9.3	6.2
Chemical concentration as Na_2O,gm/l	9.3	12.4	15.5	18.6	12.4
pH of black liquor	7.8	8.3	8.8	8.9	9.1
Yield					
Total, %	65.7	63.8	60.3	58.6	66.7
Screened, %	59.2	56.8	56.0	55.2	61.3
Chemical constituents calculated as % of phloem fraction used					
Lignin	4.7	3.8	3.1	2.8	4.3
Pentosan	9.1	10.2	10.0	9.9	11.7
Cellulose	41.5	41.8	39.6	40.3	46.8
Ash	1.5	1.7	1.8	1.8	1.9
Chemical constituents calculated as % of phloem pulp produced					
Lignin	7.1	5.9	5.5	4.7	6.4
Pentosan	13.8	16.0	16.6	16.9	17.5
Cellulose	13.1	65.5	65.6	68.7	70.1
Ash	2.3	2.6	3.0	3.1	2.9
Solubility in alcohol-benzene	2.7	1.2	0.3	0.1	0.1
Permanganate No.					
Unscreened	19.5	18.8	18.2	17.4	18.3
Screened	18.2	17.3	17.1	16.0	14.3
Brightness	31.5	30.2	31.8	32.9	37.5

benzene extractives of the phloem fraction which might have preferentially reacted with cooking active alkali.

Cook number 5 (Table 10), however, was carried out after extracting the phloem material with an alcohol-benzene solvent, and under cooking conditions similar to that of cook 2. It is clear that the lignin content in the pulp produced did not drop but rather showed a slight increase (6.4%). Consequently the alcohol-benzene extractives associated with the phloem material have no effect on the solubility of the phloem lignin in the cooking alkali.

It is obvious also that the ash content of the phloem has dropped from 3.8% before cooking to an almost constant value of 1.8% in cook 4. This might indicate that this part of ash is resistant to active alkali.

Correlation of the mineral composition of *T. hirsuta* stem (Table 4b) with the ash content of the original plant material and the pulps produced is of some interest. The silicon content (as si%) of the *T. hirsuta* stem is generally less than 0.3%. This very low level is considered an advantage of *T. hirsuta* over the Egyptian rice straw which contains about 13% silicon oxide (SiO_2), rendering a recovery system for the cooking chemical not feasible.

Elements giving soluble compounds in caustic soda (alkali metals and aluminium) form a ratio of the total mineral content (Table 4b and El Ghonemy *et al.*[8], ranging from 42.8% to 64.5%) with an average of 51.5%. This may be correlated with the results of the ash content of xylem and phloem fractions before and after cooking (Tables 8 and 10). The ash content of these two fractions respectively were 1.8% and 3.8% before cooking and dropped to 1.1% and 1.8% after cooking. This means that the alkali soluble portions of the ash were about 39 and 53% of the xylem and phloem fractions respectively. The average of these two figures (46%) is obviously comparable to that calculated from the mineral content composition (51.5%).

PHYSICAL CHARACTERISTICS

Table 11 gives the strength properties of the phloem pulp produced under different cooking conditions. It is clear that there is a significant drop in the grade index (taken as a measure for the strength properties) from 151% for cook 2 to 95% for cook 5. These two cooks were, however, carried out more or less under the same cooking conditions except for a

TABLE 11. Physical characteristics of *T. hirsuta* phloem pulps produced at two hours cooking time, pressure 6 atm, and under other different conditions

Cook No.	Active Alkali %	Chemical Concentration as Na_2O gm/l	Cooking Time Hours	Initial Freeness °Sr	Breaking Length m	Burst Factor	Tear Factor	Double Fold	Grade Index %	Brightness
1	4.65	9.30	2	27	8778	89	270	5039	142	31.5
2	6.20	12.40	2	24	8667	97	253	6479	151	30.2
3	7.75	15.50	2	24	9556	109	150	5915	126	31.8
4	9.30	18.60	2	22	9111	100	130	5265	112.8	32.9
5	6.20	12.4	2	26	9000	101	127	3275	95.6	37.5

pretreatment of the phloem material in cook 5 with an alcohol-benzene mixture to remove waxes and resins naturally associated with the phloem fibers. Accordingly it can be concluded that resins and waxes when kept in the phloem pulp would significantly increase its strength properties.

The pulp produced in cook 4 can be considered as the chemical pulp of the phloem fraction of *T. hirsuta*. This pulp has a grade index of 112% compared with softwood kraft pulp.

The strength properties of the pulps produced from the two fractions of the *T. hirsuta* stem (the xylem and phloem) compared with other types of pulps are given in Table 12. Obviously, the pulp produced from *T. hirsuta* phloem has the highest grade index (112.8), followed by that of softwood pulp (100%), and then by *Juncus rigidus* depithed pulp (71%), *T. hirsuta* xylem pulp (53.4), bagasse (sugar cane) pulp (34.9), and finally by rice straw pulp (27.2).

Dale[9] has pointed out that the phloem fibers currently used in making paper of high quality include that obtained from *Edgeworthia papyrifera* (Thymelaea). This latter plant with pulp properties comparable to that of *T. hirsuta* is now propagated in Japan as a source for extremely strong and high quality paper.

COMMENTS AND CONCLUSIONS

The preceding pages show that *T. hirsuta*, one of the most common and widespread species in the Egyptian coastal desert, could be considered as one of the most promising economic wild fibrous plants, and it may be used as a raw material in good quality paper industry because of its attributes: (a) The ash content of its raw material is extremely low, being 1.8% and 3.6% for the phloem and xylem fractions respectively. (b) *T. hirsuta* contains a rather low percentage of lignin (6.7% in phloem and 16.1% in xylem) and a high percentage of cellulose (52.6% in phloem and 39.5% in xylem). (c) *T. hirsuta* gives a rather high yield of unbleached pulp (47.2% for xylem and 56.8% for phloem). (d) Strength properties of the phloem fraction pulp are much higher than any of the local or imported raw materials and give a grade of 112% compared to imported softwood long-fibered unbleached kraft pulp (GI = 100%). The xylem fraction pulp has still higher strength properties (GI = 53.4) compared with those of rice straw, bagasse or undepithed *Juncus rigidus* pulps. (e) *T. hirsuta* pulp may be used alone to give good quality paper or mixed with local raw

TABLE 12. Physical characteristics of *T. hirsuta* phloem and xylem pulps compared to those of other materials at 50°Sr

STRENGTH PROPERTY	*T. hirsuta* Phloem Pulp	*T. hirsuta* Xylem Pulp	Softwood Kraft pulp	Bagasse Pulp	Rice straw Pulp	*Juncus rigidus** Not Depithed	*Juncus rigidus** Depithed
Breaking lenght(m)	9111	6555	9500	6722	4444	7000	9667
% Kraft pulp	95	69	100	70.1	48.8	70	96.7
Burst factor	99.6	50.4	75	44	32.8	5.5	73
% Kraft pulp	132.8	67.2	100	59.3	43.7	68.6	91.2
Tear factor	130	90	140	46.7	44.7	43	80
% Kraft pulp	92.8	64.2	100	33.3	31.9	39	72.7
Double fold no.	5265	1128	4000	270	134	300	1570
% Kraft pulp	131.6	28	100	6.8	45	10	52.7
Grade Index	112.8	53.4	100	34.9	27.2	40	73

* Zahran *et al.*, 1972.

material to produce paper for common uses. (f) The bleached chemical pulp of the phloem fraction could be used for the production of speciality papers, particularly cigarette papers. Such a project, which requires detailed economic and technical feasibility studies, might not use more than 50 tons of the plant material daily, a quantity which may be provided with the present availability of the plant in the Egyptian desert. Such amount is estimated at the present time to be about one million over-dry tons.

ACKNOWLEDGMENTS

This study was supported in part by grant GB 38937 (PL480) between the National Science Foundation and the University of California, Los Angeles, and grant GF 39244 between the National Science Foundation and the University of Alexandria, Egypt, and by the National Paper Company, Alexandria.

REFERENCES

1. Y. Fouad and H.K. El-Din (1972). Report submitted to the National Paper Company (Alexandria, Egypt).

2. A.A. El-Ghonemy, H.K. El-Din, and H.A. Razzak (1976). *Thymelaea hirsuta L.*, a possible cellulosic raw material for paper pulp industry in Egypt. Proc. J. Egypt. Acad. Sci. 27:51-66.

3. T.M. Tadros and B.A. Atta (1958). The plant communities of barley field and uncultivated desert areas of Mareotis, Egypt. Vegetatio 8:161-175.

4. A.A. El-Ghonemy and T.M. Tadros (1970). Socioecological studies of natural plant communities along a transect 200 km long between Alexandria and Cairo. Bull. Fac. Sci. Univ. Alex. 10:392-407.

5. A.A. El-Ghonemy, A. El-Gazzar, A. Wallace and E. Romney (1977). Mineral element composition of the natural vegetation along a transect at Mareotis, Egypt. Soil Sci. USA 124(1):16-26.

6. A. Wallace, E. Romney and collaborators (1972). Radioecology and ecophysiology of desert plants at Nevada Test Site. USAEC Monograph TID-25956.

7. TAPPI (1960). Tech. Assoc. Paper Pulp Industry, USA.

8. A.A. El-Ghonemy, E. Shaltout, W. Valentine, and A. Wallace (1977). Distributional pattern of *Thymelaea hirsuta (L.)* Endle. and associated species along the Mediterranean coast of Egypt. Bot. Gaz. USA 138(4):479-489.

9. J.D. Dale (1969). Structural and physical properties of pulpwood. In, Pulp and Paper Manufacture, 2nd ed., Vol. 1. McGraw-Hill Book Co., New York.

<u>See also</u>

M. Zahran, H.K. El-Din and T.S. Boulos (1972). Potentialities of the fiber plants of the Egyptian flora in national economy. I. *Juncus rigidus* and paper industry. Bull. Inst. Des. d'Egy.

JOJOBA - A PROMISING PLANT FOR THE SEMI-ARID REGIONS

D.M. YERMANOS
Professor of Agronomy, University of California,
Riverside, U.S.A.

ABSTRACT

As a result of the world-wide concern for conservation of resources and for the development of new renewable resources, a great deal of attention has been directed towards the development of Jojoba as a potential new crop for semi-arid regions. Jojoba grows wild in the Southwestern U.S. and N. Mexico, in areas where the annual precipitation is 4"-15". Under cultivation it appears to grow satisfactorily in areas of marginal soil fertility, high salinity and high atmospheric temperatures. It appears to have no major natural enemies and to be tolerant to chemical treatments, if they were to become necessary. It is perennial with an assumed life span in excess of 150 years. Jojoba seed has an average weight of about 0.5 grams per seed, and it contains a unique liquid wax which is a superior lubricant and a potential replacement of whale oil, obtained from the sperm whales, an endangered species. Thus, Jojoba has a double appeal, first as a potential crop of semi-arid regions requiring low cultural and energy inputs and second as a source of a valuable commodity. It should be pointed out that a cheap and abundant source of lubricants will disappear when our stock of fossil fuels are exhausted and that none of our new sources of energy have lubricants as by-products. Finally, Jojoba could be grown with other companion crops such as guayule or mesquite for more diversified farming.

Jojoba (pronounced "hohoba") has long been known as a unique plant of the Sonoran desert and a potential industrial crop. Its scientific name, Simmondsia chinensis (Link) Schneider, is somewhat misleading as to its area of distribution. Natural populations of jojoba exist only inside the quadrangle formed by drawing lines from Riverside, California, to Globe, Arizona, to Guaymas, Mexico, to Cabo San Lucas, Baja

California (Mexico). Jojoba has not been found growing naturally in any other part of the world. The area mentioned comprises about 100,000 square miles between latitudes 25° and 31° North, and includes many disjunct populations, some having few plants and others having several thousand. The populations occupy elevations from sea level to 1500 meters (4500 ft) and occur on coarse, sandy or gravelly soils with good water penetration and a pH range of 5-8.

Jojoba is a dioecious, perennial, bushy shrub with a natural life span of at least 100 years and possibly over 200 years. In desert areas with about 75 mm (5 in.) of precipitation jojoba grows 90-120 cm (3-4 ft) in height. Where rainfall reaches 250-400 mm (10-15 in.) jojoba grows to 5 meters (15 feet) in height. Most jojoba plants have a spherical shape, while in windswept coastal areas, a low prostrate growth habit prevails much like ground cover.

Jojoba has recently attracted worldwide attention for several reasons: (a) its seed contains a liquid wax which, in addition to several miscellaneous uses, can serve as replacement of sperm whale oil, a product derived from an endangered species; (b) it is an extremely drought resistant species; (c) it can grow in areas of marginal soil fertility, high atmospheric temperatures, high soil salinity and low humidity; (d) it has low fertilizer and energy requirements; (e) it can be grown and processed with commercially available equipment; (f) it appears to have potential in several semi-arid areas around the world with high population density and limited employment opportunities, where little else can be grown.

Jojoba has an extensive and deep root system, it is non-poisonous, it does not accumulate large quantities of flammable dry material under the canopy, it is perennial and evergreen and requires little care. These attributes make it a valuable soil conservation and landscape plant for highway shoulder plantings, rest stops, city parks and greenbelts around cities suffering from sand or dust pollution.

Jojoba plants are either male or female. The flowers, male and female, start to appear late in the summer; they are fertilized in early winter and the seed matures in August or September. Although various insects visit jojoba plants it is believed that pollination is done entirely by the wind. At maturity the seed weighs 0.2-1.5 grams, has a diameter of 3-15 mm and contains 44-58% liquid wax. Jojoba wax can be extracted from the seed with similar equipment and techniques commonly used for other oil seeds. The wax is not a "fat" or triglyceride but a "liquid wax," or wax ester of high molecular weight. A wax molecule consists of one molecule of a long chain alcohol esterified with one molecule of a long chain fatty acid. Jojoba wax consists of two such esters, mostly, one with 40 carbon atoms (30%) and one with 42 carbon atoms (50%). Jojoba wax has several advantages over sperm whale oil: it possesses no fish odor and, indeed, has a mild, pleasant "nutty" odor; the crude oil contains no fats and little besides the liquid wax and requires no refining for use in most industrial processes; it has a very high viscosity index and very high flash and fire points which are important industrial parameters; it does not darken to the same extent as other oils on sulfurization and the highly sulfurized jojoba oil remains liquid, whereas highly sulfurized sperm oil requires the addition of mineral oils to remain liquid. One of the most important properties of jojoba wax is that it is undamaged by repeated heating to high temperatures and does not change viscosity with temperature change.

The meal of jojoba contains about 30% protein; it also contains a monoglucoside not found in any other plant species, called "simmondsin," which acts as an appetite depressant when the meal is fed to animals. Simmondsin may find significant uses in the pharmaceuticals industry; at this time, however, it constitutes a problem as it makes the meal unsuitable as feed. Research is underway to develop methods of removing or neutralizing simmondsin.

The only sources of jojoba seed at present are the natural populations. At maturity the outer tissues of the jojoba fruit, the hulls, turn from green to brown, shrivel, and separate, releasing the seed which drops to the ground. Thus, there is no need to shake the plants in the course of mechanical harvesting. Each fruit produces 1-3 seeds. Harvesting data indicate that shrubs in natural populations may produce as much as 4 kg (10 lb) of clean dry seed per plant. A few plants, 40-50 years old, growing in botanical gardens, have exceeded 13.5 kg (30 lbs) of clean dry seed per plant. The ratio of male to female plants varies in different areas and is either close to 1:1 or males outnumber females.

A major effort is now underway in the U.S., Mexico and Israel to domesticate jojoba. During the last 3 years several commercial plantings were established and processing facilities were built. All indications are that jojoba stands an excellent chance of soon becoming a new commercial crop; it responds favorably to cultural practices such as pruning, fertilization and irrigation. One and a half acre feet of water per year appear to be sufficient for optimum jojoba development. The greatest need for water occurs during the late winter and spring. Thus, when grown under irrigation jojoba does not compete with other irrigated crops, most of which have their peak requirements for water during the summer. Jojoba withstands heavy pruning and can easily be trained to grow in a variety of shapes. Male branches may be grafted onto female ones and vice versa. A number of herbicides may be used for weed control. No major pests or diseases have been observed as yet. Temperatures of up to 49°C (120°F) do not seem to have adverse effects on plant growth or seed production. Flowers are damaged, however, at -4 to -5°C (22-24°F) and seedling damage may be heavy at temperatures below -6°C (20°F). Dry seed may be stored for over 20 years without any deterioration in the quality of wax it contains.

Present trends are to grow jojoba in rows spaced 10-20 ft (3-7 m) apart, with plants spaced 3-5 ft (1-1.5 m) on the row. Plant propagation can be done from seed, cuttings, or through tissue culture. With asexual propagation the sex ratio of the seedlings is known in advance and a plantation may be laid out in accordance with a pre-planned pattern and frequency of male and female plants. When seed is used for planting, the sex of seedlings cannot be determined until 2-3 years after planting, when flowers appear. To circumvent the problem of sex identification, seeds are planted closer together (50-75 cm apart) on the row. After blooming, the desired ratio and arrangement of male and female plants are maintained in the field by roguing out the extra plants. Until more data become available on soil type requirements, field plantings of jojoba should be made preferably on light permeable soils. Seedlings should be irrigated during the first few months of growth until they establish their root system.

Seed may be harvested off the plants by hand or may be swept and picked or vacuumed after they drop to the ground with equipment and techniques comparable to those available for almonds or walnuts. It can also be caught in plastic nets placed under the bushes as is done with macademia nuts.

The establishment of a jojoba plantation should not cost any more than that of an almond grove. With lower expenditures for irrigation and fertilization, the maintenance of a producing plantation should be lower than that of most tree crops. Although jojoba starts producing seed in the second or third year, with currently available strains, it appears that commercial harvest of seed might not be profitable or even practical, until the fifth year after planting. The disadvantage is more than offset, however, by the long life span of the plant. Since there are no producing commercial plantations of jojoba available, no detailed data on production and returns are available.

Jojoba is not thought of as a crop which will compete for acreage with our traditional ones, but rather as one which will fit in regions where traditional crops cannot be grown profitably, due to various limiting environmental factors. Extrapolating from data obtained from the harvest of wild plants and from limited information from research plots it appears that seed yields of 8 year or older plantations might range from 1-2 tons of dry, clean seed per acre. Development of higher yielding strains and better cultural techniques should lead to higher yields and lower costs of production per acre. Current expectations are that the small quantities of wax that will be available initinally from cultivated plantations will be absorbed by the more lucrative markets such as the cosmetics, waxes, and possibly pharmaceuticals, at price levels of \$2-4/kg (\$1-2/lb) of wax. The real challenge for jojoba will be to penetrate the larger but less lucrative market of lubricants. This may not be too difficult however, because with the disappearance of fossil oils, cheap sources of lubricants will also be lost. Lubricants will be needed in the near future, as badly as fuels. None of the new sources of energy now contemplated (solar, atomic, geothermic) have lubricants as by-products.

POTENTIAL USES OF JOJOBA WAX

Lubrication: superior lubricant for high-speed, high-temperature machinery, cutting, grinding and transformer oil; substitute for sperm whale oil

Cosmetics: hair oil, shampoo, soap; face creams; lipstick, tanning lotions, insect repellents

Pharmaceuticals: carrier or coating for medicinal preparations, antifoaming agent; stabilizer of penicillin products; potential treatment for skin ailments

"Factice": potential in linoleum manufacture, printing ink, varnishes, chewing gum

Polishing waxes: for automobiles, floors, furniture, vinyl

Solvent: of polyethylene

LIQUID WAX CHARACTERISTICS

Analysis		Published Values
Acids	Oleic	0.66%
	Palmitoleic	0.24%
	Eicosenoic (C_{20})	30.30%
	Docosenoic (C_{22})	14.20%
Alcohols	Eicosenol (C_{20})	14.60%
	Docosenol (C_{22})	33.70%
	Nexaconsenol (C_{26})	2.00%
Melting Point		11.2-11.8°C
Solidifying Point		6.7, 7°C
Boiling Point at 757 mm under N_2		398°C
Flash Point (C.O.C.)		555°F
Fire Point (C.O.C.)		640°F
Viscosity, Centistrokes at 25°C		48.4
Viscosity, Centistrokes at 100°F		27.0
Viscosity S.U. at 100°F, Sec.		127
Viscosity S.U. at 210°F, Sec.		48
Viscosity Index (Dean and Davis)		173
Color Number (A.S.T.M.)		2
Corrosion at 212°F, copper strip		Nil
Pour Point		10°C
Carbon Residue		0.01%
Refractive Index at 25°C		1.4648, 1.4650
Density at 25°C		0.8642, 0.8990
Specific Gravity 25°/25°C		0.8635, 0.8640
Saponification Number		92.2, 92.6, 95, 165.7
Acid Number		0.23, 0.32, 0.57
Iodine Number (Hanus)		81.7, 82.1, 88.4
Acetyl Number		6.8
Reichert Meissl Number		0.70
Polenske Number		0.31
Unsaponifiable Matter		37.62%, 48.3%, 51.1%
Iodine Number of Unsaponifiables		77.2, 79.3-80.2
Acetyl Number of Unsaponifiables		171.8, 172
Total Acids		51.9%, 53.5%
Soluble Acids (as butyric)		2.43%

LIQUID WAX CHARACTERISTICS (Cont.)

Analysis		Published Values
Saturated Acids		1.54% - 1.64%
Insoluble Acids		59.43%
Iodine Number of Total Fatty Acids		76.1
Neutralization Value of Fatty Acids		172.0
Average Wax Ester Molecular Weight		606, 610
Hydrogenated Wax	Melting point	65-68°C, 73-74°C
	Hardness (trionic gauge)	90, 92

THE POTENTIAL OF CONTROLLED-ENVIRONMENT AGRICULTURE IN DESERT REGIONS

MIGUEL R. FONTES
Research Horticulturist, Environmental Research Laboratory, University of Arizona, U.S.A.

ABSTRACT

The Environmental Research Laboratory (ERL), a research unit of the University of Arizona, has been devleoping controlled-environment agriculture (CEA) systems in desert regions for several years. ERL has established CEA systems in Abu Dhabi, Kharg Island, Iran, Puerto Penasco, Mexico, and Yuma and Tucson, Arizona. These installations produce vegetables and high-priced animal protein.

Controlled-environment agriculture is the total concept of modifying the natural environment for optimum plant and animal growth. For plants, this means controlling both the aerial and root environments for optimum growth.

Controlled-environment agricultural systems have some inherent characteristics that could have very positive impacts on food production in desert regions. Large quantities of food can be produced using limited amounts of land and water. In outdoor areas, production would not be possible due to the severity of the climate. The systems to date have been used only to produce high-value animal protein and vegetables. The main reasons for this limited use are high capital investment and energy requirements. Research and development are needed to develop low-cost systems with reduced energy inputs.

INTRODUCTION

Controlled-environment agriculture (CEA) is the total concept of modifying the natural environment for optimum plant and animal growth. For plants, this means controlling both the aerial and root environments. Since the control of the environment is a major factor in these systems, production takes place inside enclosures. This allows for control of air temperature (heating and cooling), light, water and nutrient, and protection from adverse climatic conditions.

A major part of the material presented in this paper is from a technological assessment of CEA prepared for the National Science Foundation by International Research and Technology Corporation (IRT). The Environmental Research Laboratory (ERL) acted as a subcontractor to IRT in the assessment.

Historically, CEA developed in northern climates for out-of-season production of fruits and vegetables. The largest acreages occur in Europe; however, Japan has the largest acreage of a single country with 10,800 hectares (Figure 1). In most northern climates, light limits production in winter months. Therefore, one major reason for considering arid lands for CEA is the abundant light for maximum crop growth.

Some of the inherent characteristics of CEA which could lead to further development in desert regions are maximum yield per unit growing area, efficient water usage, and protection from adverse climatic conditions. Yield is increased because of an enhanced growth rate, increased number of crops per year, and diminution of losses due to weather.

CEA IN ABU DHABI

The Environmental Research Laboratory has been instrumental in establishing CEA systems in several desert locations. The initial research and development activities were conducted in Puerto Penasco, Mexico, on a desert seacoast. The first large-scale project was constructed in Abu Dhabi and became operational in 1972.

Air-inflated polyethylene structures comprise half the two hectares of horticultural enclosures; structured greenhouses cover the remaining one hectare. Low growing crops that need no overhead support, such as lettuce, turnips, peppers, eggplant, radishes, etc., are cultivated in the air-inflated greenhouses. Cucumbers, tomatoes and string beans, because they are trained vertically on support wires, are grown in the structured houses.

Seawater is used to evaporatively cool the air upon entrance into the greenhouse. Because of the high humidity in summer, it is difficult to maintain optimum temperatures inside. Waste heat from engine-driven electric generators is used to desalt seawater.

Plants are grown in the indigenous fine calcium carbonate sand, which has a pH of about 8.3. Because the sand is low in nutrients and cation exchange capacity, fertilizer is provided the plants at regular intervals by dissolving selected commercial-grade fertilizers in the distilled seawater. The nutrient solutions include all the elements required for plant growth except those taken from the atmosphere.

Since the solution passes through small openings in the irrigation systems, all the nutrients must be water soluble to avoid plugging. They also must be in the form most readily absorbed by plants, and thus available for uptake once they are applied. The only cations and anions applied are those the plant uses; no carrier ions are added. Thus, no salt residue builds up in the beds once the plants have removed the nutrients. The only residue left is that of the unused fertilizers.

All desalted water is costly. It is important, then, to use it as efficiently as possible. A plastic liner is placed beneath crops that are grown in double rows. This liner prevents water loss through percolation and lateral movement, and fertilizer loss through leaching. The water is applied near the roots by different systems. For double-row crops, a 1.3 cm diameter polyethylene pipe runs down between the double rows, and small capillary tubes distribute the solution to the base of each individual plant within the rows. On the other hand, such closely-spaced crops as okra, cabbage and lettuce are watered with a plastic hose that is perforated at intervals equal to the spacing of the plants. The plants are irrigated 3 to 4 times daily, rather than once, in order to optimize the availability of nutrients around the roots.

The facility produces a variety of vegetable crops which are sold in the local markets. Yields for some of these crops are given in Table I. The most profitable crops are tomatoes and cucumbers.

KHARG ISLAND

The success of the Abu Dhabi installation led to the construction in 1974 of a one-hectare unit on Iran's Kharg Island in the gulf between Iran and the Arabian Peninsula. Kharg Environmental Farms was a gift from the consortium of oil companies to the people of Kharg Island.

The facility produces a variety of high-quality vegetables for distribution and consumption throughout the Kharg communities. Typical products grown are tomatoes, cucumbers, lettuce, cabbage, eggplant, okra, turnips and radishes, and at any one time, a variety of vegetables are in various stages of growth.

The irrigation and fertilization systems are similar to those used in Abu Dhabi. The ambient temperatures are lower in winter than in Abu Dhabi; thus, heating systems are used in the nursery area.

A simple but massive cooling system is used to control summer heat. Large exhaust fans, roof-mounted in the center of the house, pull air from both ends of the house. The end walls are fitted with special evaporative cooling wall media which use seawater to cool the air. This system is capable of completely changing the air in the greenhouse in less than one minute, and provides relatively cool, moist air around and over the plants. This also helps reduce their requirements for fresh water by reducing transpiration rates.

The roof of the greenhouse is a series of large replaceable air filled bags made of clear, light-stabilized polyethylene, 150 microns thick. These bags can be shaded in the summer and easily replaced or cleaned in the fall to maintain optimum light and heat levels in the house. The air pressure in the bags can be controlled to provide proper structural rigidity during periods of high wind. Table 2 summarizes the yields obtained from the Kharg installation.

TUCSON, ARIZONA

ERL was also involved in designing a four-hectare complex in Tucson, Arizona, owned by Superior Farming Company. This facility produces European-type cucumbers and high-quality tomatoes. Although this facility is located in a desert region, the winter temperature can go as low as -7°C.

In the winter, two central heating stations supply hot water to conventional fan coil units located adjacent to the exhaust fans in the greenhouses. Heat is distributed uniformly within the greenhouse through the use of polyethylene duct tubes. Horizontal aspen pads and centrally-located exhaust fans are used to cool in summer.

The plant growing system is the sand culture method of hydroponics. Precise amounts of mineral nutrients are mixed with irrigation water by proportioners. Watering cycles are fully automated by timeclock controls. Water and nutrients are conserved through the use of trickle irrigation techniques and a polyethylene barrier beneath the 0.3 m deep sand growing medium. Drain pipes installed above the polyethylene barrier insure that adequate root aeration conditions are maintained.

Table I. Abu Dhabi controlled-environment agriculture yields

	Yield per Crop Mt/ha	Crops per Year Number	Annual Yield Mt/ha
Cucumbers	179	3	538
Tomatoes	135	2	270
Radishes	20	9	182
Cabbage	52	3 to 4	155 to 206
Chinese Cabbage	45	4+	179+
Beans	10	4	40

Table II. Kharg Island controlled-environment agriculture yields

	Yield per Crop Mt/ha	Crops per Year Number	Annual Yield Mt/ha
Cucumbers	190	3	570
Tomatoes	159	2	318
Cabbage	57	3	172
Lettuce	56	3	168
Peppers	49	2	97
Eggplant	28	2	56

High yields of high-quality produce are consistently obtained in this system due to the optimization of the growing environment. The location of the complex in a region of high solar radiation has eliminated the constraint of poor winter light conditions, which often limit yields of greenhouse crops in the midwestern United States. Also, the winter solar load helps reduce the costs of heating the greenhouses.

CEA systems in use today have rather high capital costs, ranging from \$20-70/m^2. This is one of the major reasons use has been limited to high-value crops. Another significant cost area is energy for environmental control. This is especially important today with the rising cost of energy and limited supplies in certain areas of the world. Labor input is high, as the roof-supporting structures limit the use of large agricultural machinery.

As part of a technological assessment of CEA for the National Science Foundation, a CEA system was designed conceptually which addressed these factors limiting its diffusion. Figure 2 is a schematic drawing of a passively cooled greenhouse.

Fresh air is drawn through an opening along the south end, moves through and across the crop growing area, and is vented up a solar chimney along the opposite side. After moving through the enclosure, the exhaust air is heated by passing over a solar absorbing strip 2.6 m wide, and then convectively rises through the chimney causing a draft pressure to maintain air circulation across the crop. The rectangular chimney runs the length of a long side of the enclosure and is 3 m wide and 10 m high.

The chimney structure, of steel columns, braces and beams supporting sheet plastic or glass-based film walls and covers, also acts as a support tower for a suspension cable system that supports the roof assembly above the crop area. The second support tower is the chimney structure for another parallel module. Thus, one additional tower that is not a chimney is required for each set of parallel adjoining modules. The space between modules, down through which the inlet air flows, is five meters.

The roof system is suspended in a peak and valley configuration to facilitate rainwater run-off and drainage of foam used for insulation in cold weather. Drainage gutters, normally structural members of substantial weight and cost, are made of plastic sheeting.

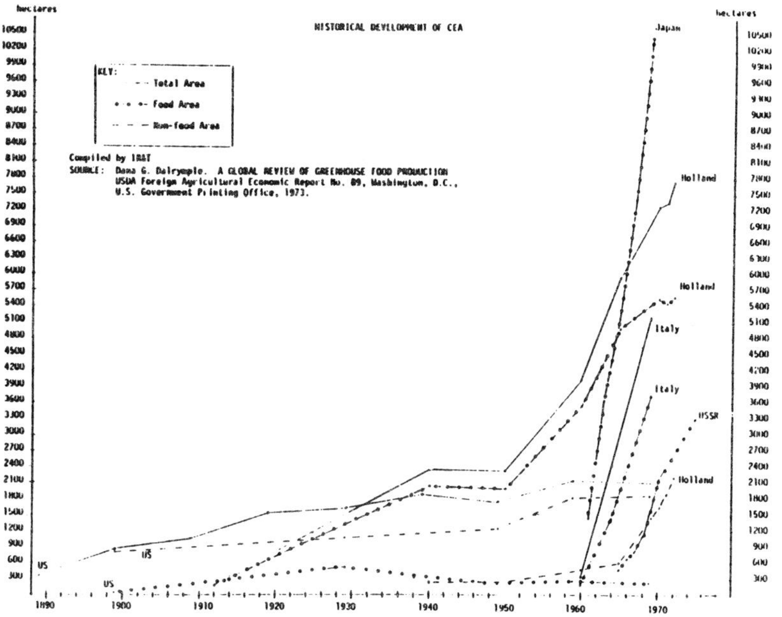

Fig. 1: Historical development of controlled-environment agriculture.

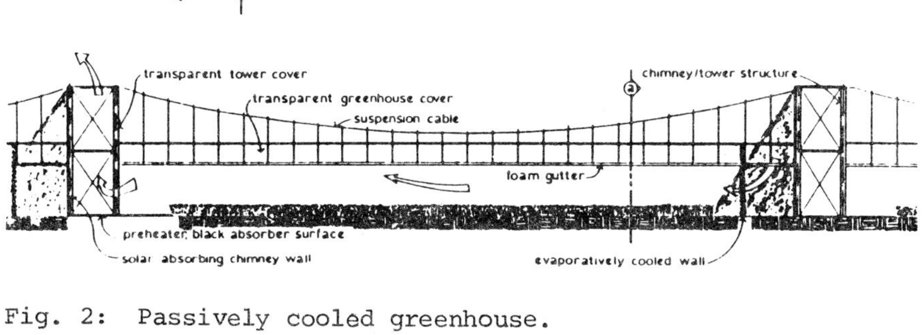

Fig. 2: Passively cooled greenhouse.

The enclosure cover is made of double-walled, air-inflated "pillows" of glazing material supported by steel cables or wires that take up the tension caused by inflation. The inflation pressure of several tenths of an inch of water is maintained against leakage by several blower fans for each module. The maximum thickness of the air-inflated roof is about 30 cm. The pressurized roof also is the container for a layer of detergent-based foam that is used for insulation of the roof at night or during periods of heavily overcast, unusually cold weather during the day. This foam distribution system is inside the air-inflated roof; the foam and liquid associated with it are contained in a recycling system.

The solar heat load on the enclosure is dissipated primarily by evaporation of water transpired by the plant surfaces or, during the early stages of plant growth, by evaporation of water from the surface of the growth medium between the plant canopies. The evaporated water is transported out of the enclosure as water vapor in the exhaust that rises through the solar chimney.

When the outside air temperature (dry bulb) is above about 32°C, the air is evaporatively cooled by passing through a moistened pad of aspen tree cellulose or a plastic water-retaining material such as CELdek. The 15 cm thick pad is vertically placed parallel to the air inlet area along one of the long sides of the enclosure. The evaporation rate in the pad, and therefore the temperature and humidity of the air that has passed through it, are controlled by the airflow rate and the rate of supply of feedwater to the cooling pad. The airflow rate can be controlled by adjustment of the size of the inlet or outlet areas.

All needs for energy for heating the air in the enclosure during cold weather are met by a combination of solar energy absorbed by the plant canopies and transferred to the air and, at night or during periods when the outside daytime air temperature is unusually low, by using the local crop wastes as fuel for heating the enclosure atmosphere.

For agronomic crops, the types of equipment and procedures used for plowing and cultivating the soil, seeding, irrigation and harvesting are similar to those used for open-field agriculture, with adjustments required to manage crops with higher plant densities than are typical for open-field cultivation. The object of this suspended type structure is to allow for use of conventional agricultural equipment, thus decreasing labor requirements. The actual labor requirement will depend directly on crop type.

Capital cost was estimated at $162,000/ha at 12% annual interest. Assuming the average life of a CEA system is 15 years, the annual capital cost per hectare would be $23,785. This is still a very high annual capital cost, but the design does decrease dramatically the energy and labor operational costs. Break-even analysis conducted on several crops shows that only the high-priced vegetable crops, such as lettuce, cucumbers and tomatoes, looked profitable. The design is promising enough to warrant building of test models to check the performance of a solar chimney.

CEA AQUACULTURE

Research and development of controlled-environment aquaculture of crustaceans by ERL began in 1972 as an evolutionary product of the controlled-environment agriculture technology developed by the Laboratory. The research facility, located in Puerto Penasco, Mexico, is a cooperative project of ERL and its counterpart, CICTUS, at the University of Sonora.

There are several contrasts between the high-density culture possible in controlled environments and other systems most frequently encountered. For example, in open-environment culture, the crustaceans are in cages or net enclosures. In pond culture, large bodies of captive water enclose the cultured populations. In the latter system, there is greater human involvement in the construction and maintenance of ponds and some supplemental feeding is required, while in the former, intensive feeding may be required. By contrast, the intensive system of ERL encloses the complete environment. Water use is programmed, nutrition is almost 100% by supplementation, and there is continuous population and biomass monitoring.

Intensive CEA offers several advantages over its counterparts. CEA allows maximal growth rate, survival and stocking densities through optimization of water quality, feeding regimes and disease management. The high densities possible in CEA systems allow higher productivity on less acreage. For example, it has been demonstrated that CEA can produce 45,000-67,000 kg tails/ha compared to a maximum of 2,750 kg live weight per hectare in an extensive system. In a CEA system, predation is eliminated, and easy visibility of the animals permits rapid recognition and therapy of disease conditions. In CEA, populations and growth rates can be readily followed, thereby permitting programmable, easily-anticipated year-round harvest. In addition, the high visibility of CEA uniquely permits manipulation of brood stock to meet the spawning requirements.

Wild adult shrimp are collected from the Gulf of California and are held in the raceways for maturation. Spawnable females are then removed and transferred to the hatchery as needed. The postlarvae are transferred to a fiberglass flow-through nursery system until they reach 10 to 50 mg. The postlarvae may be transferred to a post-nursery system, a miniature raceway where they are stocked at 500 to 600/m^2 and placed on supplemental feeding. When the animals have reached 0.6 to 1.2 g, they may then be transferred to a grow-out system where they are stocked at a density of approximately 170/m^2. They will spend 20 to 25 weeks in this grow-out system and will usually be harvested at approximately 21 g (35 count).

Water pumped from a seawater well is introduced into the raceway by headers at the end or along the long axis of each raceway. Both makeup and recirculation water are introduced in this manner and are sprayed directly onto the surface of the raceway to increase efficiency of aeration. Both overflow and bottom drains have been provided in the raceways.

This type of aquaculture system uses rather large quantities of seawater. On the average, water in the raceways is changed 5 to 7 times daily. This water carries away waste products of the shrimp and contains some nitrogen and phosphorus. A program was initiated in 1977 to use this water for irrigation of halophytes. The approach used is to search for plants which grow in seawater, rather than attempting to alter conventional food plants.

Plant material and seeds of several species of plants have been collected and planted in test plots in Puerto Penasco. A cooperative research program has been established with Dr. Fred Somers, University of Delaware, who has been conducting similar studies for several years. The plants tested so far have included several species of *Distichlis*, *Spartina* and *Atriplex*, and also *Batis*, *Salicornia*, *Sporobolus*, *Nitraria* and *Eucalyptus*.

The major problem to date has been germination in seawater. So far, all seeds here had to be germinated in fresh water and adjusted to full strength seawater through a salt gradient. It is hoped to establish growth rates and yields for several of these species in 1979.

SUMMARY

For controlled-environment agriculture systems to have a significant impact in desert regions, additional research and development are needed to reduce capital costs and energy inputs. Studies to date indicate that significant reductions are probable. The major advantages of CEA applications in desert regions will be high productivity and water use efficiency, and protection from adverse climatic conditions.

Controlled-environment aquaculture systems, producing animal protein on desert seacoasts, will become a reality in the near future. This development will have a positive impact on desert seacoast areas. Development of a seawater-based agriculture system using halophytes is an area of vast potential and should be given high priority in research and development program planning and funding.

ORDINATION ANALYSES IN SURVEYS OF DESERT VEGETATION

WILLIAM H. BROOKS
Environmental Programs, Office of Arid Land Studies,
University of Arizona, Tucson

ROBERT L. BURGESS
Environmental Sciences Division, Oak Ridge National
Laboratory, Oak Ridge, Tennessee

ABSTRACT

Using ordination analysis to determine general plant distribution patterns in perennial desert vegetation and the phytosociological characteristics of the dominant species is discussed. Ordination analysis provides a three-dimension spatial ordering of stands based on expression of vegetational similarity. Constructing an ordination from a matrix of similarity values makes it possible to infer phytosociological relationships from which elevational, topographic, edaphic, and species correlations can be interpreted. These correlations can be applied to survey techniques such as remote sensing, in the development of ground control stations or in any instance where a permanent record of perennial vegetation is required. Ordination analysis permits delineation of "association-segregates" from a large group of "core" species that may predominate over most desert habitats. It is the interrelation of phytosociology and physical environment within the "association-segregate" that produces vegetational patterns characterizing various desert regions which should be quantified in natural resource surveys of desert areas.

INTRODUCTION

Phytosociology is the study of the structure of vegetation. Its aim is to describe plant species in quantitative terms and to explain or predict plant distributional patterns. Phytosociological concepts generally have been linked with studies of the structure, composition, and dynamics of plant communities based on ecological responses to certain environmental conditions. Included in these concepts are community development, interrelationships, and distribution, as well as all causal factors. However, in this paper, phytosociology is interpreted to mean vegetation analysis.

One approach to this aspect of plant ecology involves the study of vegetation along environmental gradients in which the pattern of overlapping populations has been called a "continuum." Curtis[1] based the term on the observation that plant species occupying similar habitats have evolved different requirements and limits and are competing rather than cooperating in community functioning. This results in gradual population change along a gradient. The "individualistic" concept first proposed in 1926 by Gleason[2] forms the basis upon which the continuum theory was later developed. It is based on the fact that all plant communities are composed of individual species and that genetic make-up controls or governs their ecological amplitudes. This means that every environment has its own biotic potential to support pland and animal life. The biotic potential of a community often can be partially illustrated through the ordination of "association-segregates."

The term "association-segregate" was first introduced by Braun[3] to denote vegetation climax units after being advanced by Clements and Clements[4] in North America to include climatic groupings. The association-segregate does not imply number of species dominants nor area occupied, but does connote origin and relationship. Since no two plant species behave in exactly the same manner, the members of complex communities segregate in various ways to produce a plethora of community types. It is these which we call "association-segregates." Associations are not true plant communities; they are conceptual units.

Vegetation is dynamic. Its change often is described as a flowing stream governed and directed by the interaction of factors determined by the scale in which the factors work and in the speed of change they induce. Cooper[5] considered species associations as forks in the "main stream" and plant communities as the termini of "streamlet" elements.

Some vegetation types indigenous to extreme environments, such as deserts, are separated from the rest of the vegetation by relative discontinuities. This zonation is distinguished gradation as a tendency toward partial segmentation of the vegetation continuum. The distributional patterns of a species are a reflection of its own physiology and genetics. Vegetation is a complex and largely continuous population pattern; association-segregates throughout a particular vegetation matrix reflect the reproductive abilities of vascular plant species in the face of competition and the physical variables of the environment, particularly in arid zones. Refinement of methods for obtaining quantitative data long has been necessary so that systems of community descriptions could be supported and changes in development patterns and reproductive potential could be documented.

SOME INFLUENCES IN THE DEVELOPMENT OF DESERT VEGETATION

Desert vegetation is distinct as to species. In the southern deserts of Western North America, these species have evolved from warm-adapted, essentially tropical and subtropical stocks, the genera of which resemble genera in adjacent regions. In the Middle East, derivatives of the xeric Indo-Malaysian flora prevail. The line which marks the extreme latitudinal limit of frost is the most important climatic boundary in restricting extension of these perennial species.

If desert vegetation is destroyed there is no ordered succession. The xerophytic perennials are the dominants in desert vegetation and represent examples of a direct adaptation to environmental rather than successional factors. Although there is no ordered plant succession in the classical sense, there is direct development and change. This produces a vegetation mosaic which varies with the independent variables of topography or other factors. As Forest Shreve[6] stated, "Desert vegetation varies over distance according to the manner in which species respond to the presence or absence of certain environmental gradients such as edaphic and elevational factors." The latter is a great influence on distributional patterns of desert species due to differences in climate encountered within short vertical distances. The distributional limitations of plant species are determined by their physiological behavior in relation to environment. Some species show a departure from classic types of respiration, transpiration, and water intake. These balances are difficult to analyze because the actual quantity of water available is not adquately indicated by precipitation data alone. Low moisture content of the soil often prevents successful germination and establishment of many new individuals of perennial species. This factor is partially responsible for the openness of desert vegetation.

Another important physical factor affecting species distribution long recognized by desert plant ecologists is cold air drainage. It plays a major role in determining not only the level of minimum temperatures, but also their duration. The thin layer of cold air near the ground at many topographic sites during temperature inversion nights seems to be a potent factor in seedling establishment. Latitudinal limits of many desert perennials are determined by these diurnal fluctuations as well as more constant seasonal extremes.

In summary, most desert regions are distinctive as to species composition and resemble adjacent regions as to genera. Aspect dominance is mostly by desert shrubs. Within this life form group, two or three genera usually dominate the vegetational structure in a manner determined largely by the physical factors of the environment which influence the establishment and distribution or perennial species. Among these factors are climatic and edaphic differences encountered within short vertical distances. The duration of minimum temperature, transpiration-absorption water balance, and soil moisture, especially during periods of germination, are among the important physical influences which analytical methods, used in conjunction with remote sensing techniques, help to quantitatively substantiate in terms of their expression through existing phytosociological patterns.

ORDINATION ANALYSES

A number of objective methods for comparative analyses of large samples of vegetation are available. The importance value, first introduced in 1951 by Curtis and McIntosh[7] and applied to a grassland-forest border region, is an index which is derived through the summation of relative values expressing the vegetational importance of a species within a stand. Field data (frequency, density, and cover) are converted to relative terms for each stand. By adding the relative values of density (the proportion of the density of one species to that of the stand as a whole), frequency, and cover (the proportion of crown cover of one species to that of the stand as a whole), the importance value (IV) of each species is determined. This index is useful in experimental studies of communities, for comparing one community with another, for showing changes in a plant community over time, and as an analytical tool in interpreting aerial photographs when it is applied in conjunction with data supplied by ground control stations.

The importance value can be applied further in constructing an ordination which implies phytosociological patterns in the vegetation. This provides a spatial ordering of stands in three dimensions based on expression of vegetational similarity. The formula for this expression is:

$$C = \frac{2w}{a + b} \times 100$$

"C" is the Index of Similarity, and "a" and "b" are the sums of the quantitative species data in stands "A" and "B" respectively. "w" is the sum of the values common to stands "A" and "B." In this paper, importance values are used for "a" and "b" and results in a matrix of "C" values.

Constructing an ordination from the matrix of similarity indices basically follows the principles outlined by Bray and Curtis,[8] Burgess,[9] and Loucks.[10] The stand with the lowest total of similarity values is the stand having the least in common with all other stands. This stand constitutes one end of the first axis. The stand showing least similarity with this first stand is recognized by the lowest "C" value. It represents the opposite limit of spatial distribution in the first dimension. The positions of other stands along the first axis are calculated directly from the Index of Similarity and expanded to 100 units.[9]

The closer stands are clustered toward the center of an axis, the greater the probability that they will show spatial separation in another dimension. Stands which are located toward one end of the axis are more likely to be similar since by sharing a correlatively high relationship with an end point of the axis, they are more likely to be related to each other. Therefore, two stands proximate to the center of the first axis often will form the ends of the second axis. All other stands on the second axis are positioned between them. The end points for the third axis are derived in the same manner.

Constructing the ordination shows the distribution of stands in three dimensions. From the ordination it is possible not only to infer phytosociological relationships, but elevational, topographic, and species correlations also may be interpreted.

PHYTOSOCIOLOGICAL RELATIONSHIPS

In a study to delineate the interrelationships of species, Sonoran Desert stands in central Arizona (Southwestern United States) were ordinated. Importance values for all species were ordinated to study phytosociological relationships. Figure 1 illustrates the utility of an ordinational procedure in the study of interrelationships. In the upper figure, the distribution of stand elevations (to use one environmental parameter) is shown when ordinated. Progressively larger circles imply higher elevations and higher species importance values. In the views below, total cactus and thrub components of the ecosystem are plotted. It is obvious that the "Y-axis" of the ordination reflects, at least in part, the elevational gradient. To further test this assumption, a regression analysis was run (Figure 2) which indicates the presence of a highly significant P value correlation (P=0.01). Slope aspect also contributes to species and life-form segregation.

Within the elevational range studied, there is also an ordinational correspondence between total cactus distribution and elevation. By contrast, the ubiquitous shrub component occupies a different portion of the ecological sphere and shows no such relationship (Figure 1).

Graphically, the species separate into semidistinct biological groups. Burgess[9] reported a similar biological segregation from an intrastand ordination of an oak forest in Wisconsin. The association-segregate concept advanced by Braun[3] mentioned earlier is a convenient term to describe the component parts of this desert community. From the ordinational distribution of the importance values, it can be seen that certain species appear to cluster more in some stands than in others. If a number of these species patterns are studied, it is noted that stands having some environmental commonality reflect similar species patterns. From this observation it is then possible to categorize species showing similar patterns into "community groups" containing species which show maximum development in certain areas of the ordination.

Seven community groups in the central Arizona desert have been recognized subjectively from among the myriad objective patterns of the ordination. Because each of these groups will have at least one of the community dominants, they are not to be construed as discrete communities, but rather community-variants or association-segregates.

The first of these seven groups is represented in this study by a large core of species in which all of the community dominants are found. In addition to these dominant species, secondary species may be represented in this nuclear group. This grouping would seem to make possible a listing of species

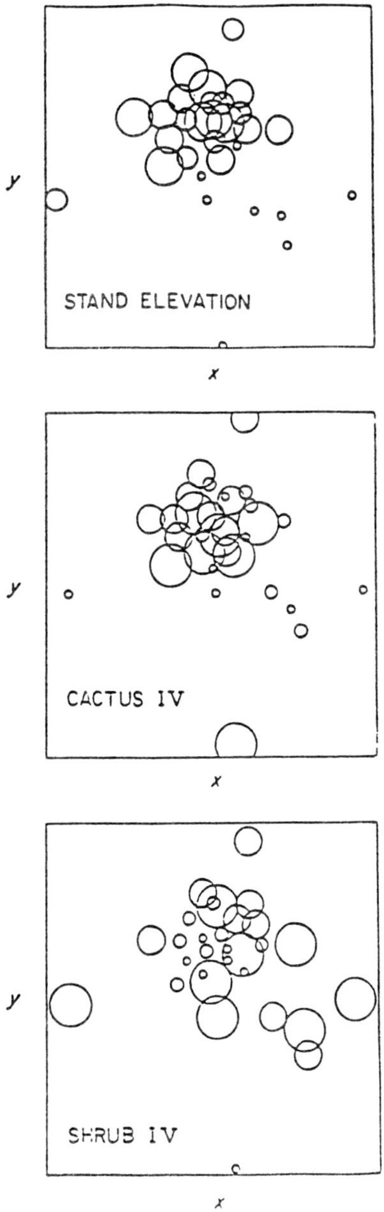

FIGURE 1

Distribution of stands within the ordination to illustrate the influence of elevation

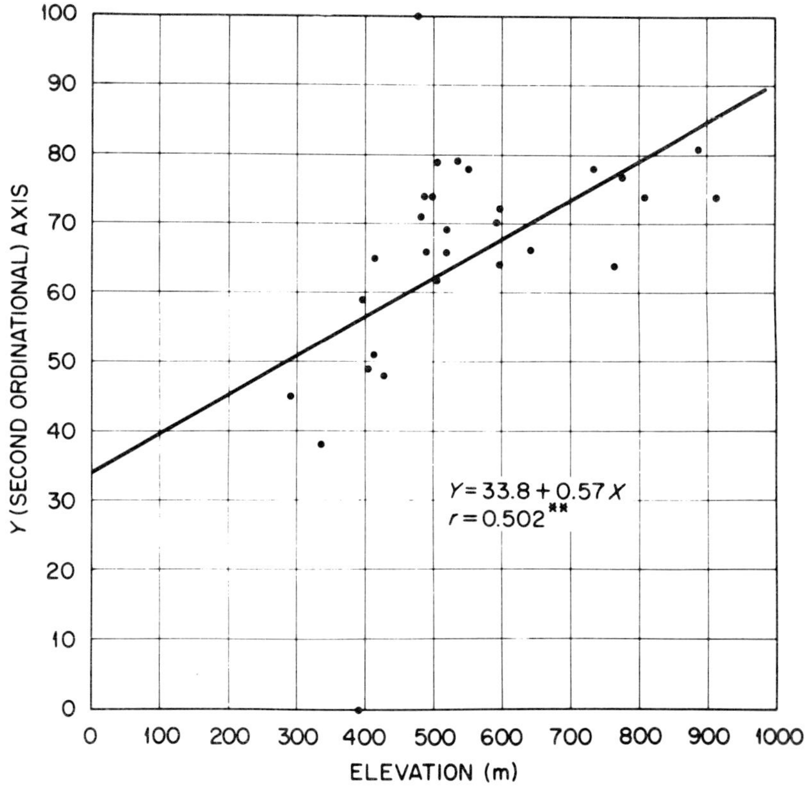

FIGURE 2

Regression analyses of stand elevational gradient

to be generally expected within the borders of the central Arizona upland desert as elevational, topographic, edaphic, moisture, and temperature realms are explored.

Within this core, *Cercidium microphyllum, Encelia farinosa, Franseria deltoidea, Larrea tridentata, Opuntia acantho carpa*, and *O. bigelovii* are dominant. Their two-dimensional ordination patterns are shown in Figure 3. The larger circles represent those stands where the species exhibit higher importance values. While similarities are apparent, each species occupies its own region within the ordination, i.e., no two species show identical patterns. In part, this illustrates theat desert vegetation is also "individualistic." This grouping predominates in the middle to higher elevation stands in the range established, as well as on slopes.

Bajadas encircle all isolated mountains in central Arizona. Approximately one-half the area of intermontane plain in the Arizona upland is designated as upper bajada (Shreve).[10] The vegetation of these upper bajadas is richer in composition than vegetation in any part of the Sonoran Desert north of the international boundary and outside the moist floodplains.

Franseria deltoidea is the most important species in the core group. Except in isolated physiographic situations and in lower elevation stands, *Franseria deltoidea* is the single most important species in the central Arizona uplands. It is uniformly distributed and usually contributes the most to cover and absolute density. The plant is relatively short-lived. It occurs on coarse, rocky soils, but rarely in sandy areas. In some areas, *Larrea tridentata* is an important co-dominant. There are instances, however, where elevation and soil characteristics will influence the separation of these two species in an ordination.

Community Groups II-VII are illustrated in Figure 4. The composite patterns of the species comprising each group occupy slightly different modal regions within the ordination. While overlap is readily apparent, the relative distinction of the high values in each group permit the recognition of these association-segregates.

Community Group II includes species which reach maximum development at the lower to middle elevations (290 to 600m), on floodplains and on gently rolling foothills. The group is composed almost exclusively of shrubs. Since stem succulents are associated more with slope and relief, they are not important within this group. *Larrea tridentata* is the leading

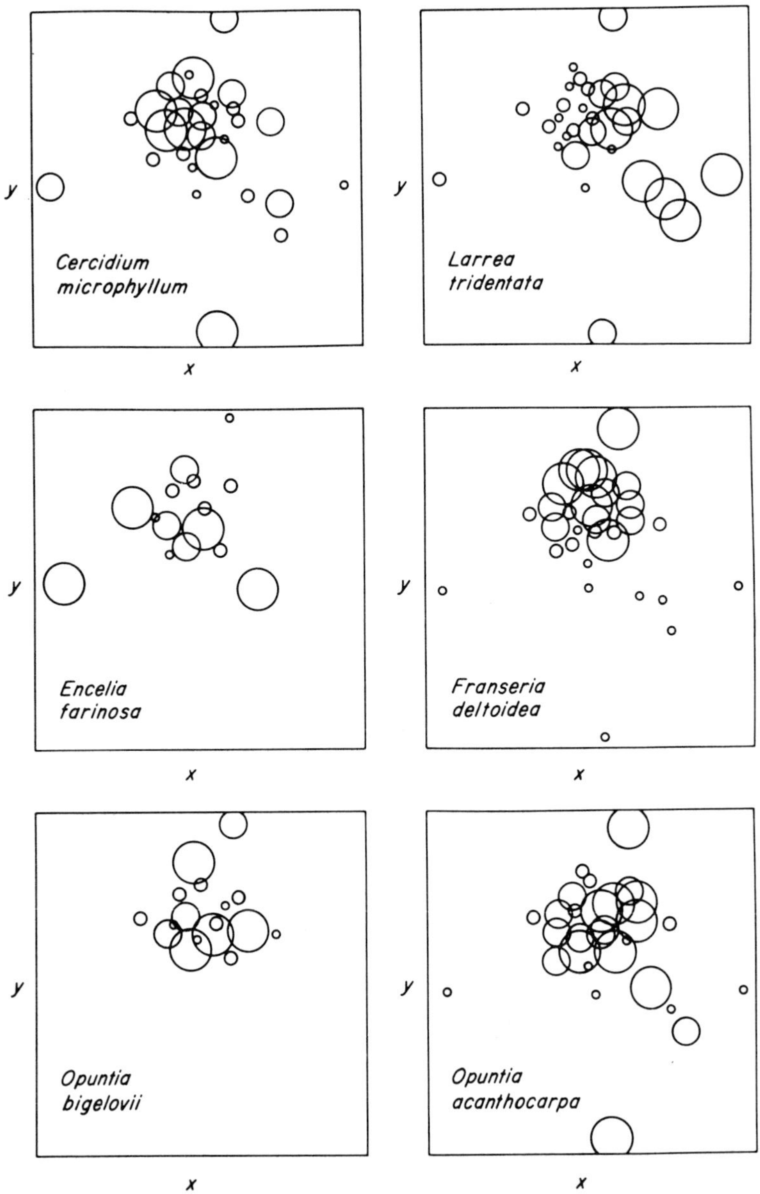

FIGURE 3

Ordinational distribution of species importance values in Community Group 1 (core)

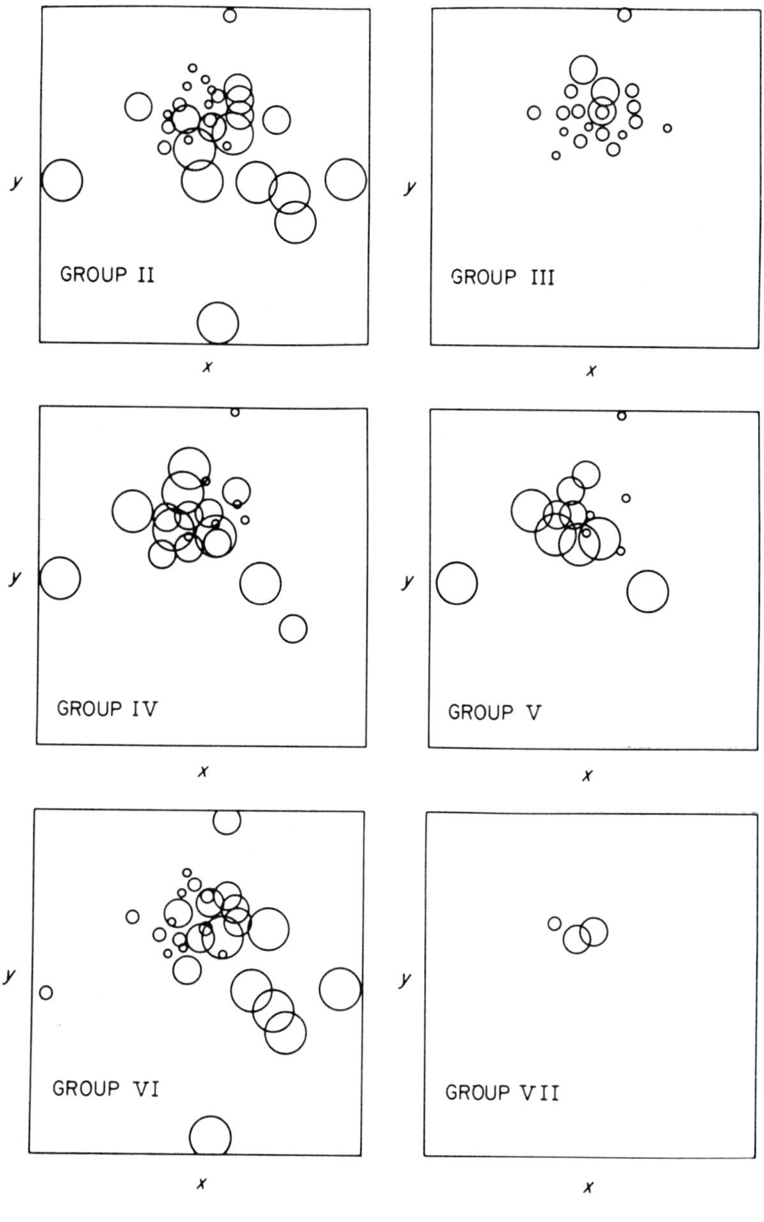

FIGURE 4

Ordinational distribution of species importance values in Community Groups II-VII

dominant where *Franseria deltoidea* is not strongly represented. *Larrea* is able to persist in regions where rainfall is often minimal.

Community Group III includes species in stands on rolling topography from middle to higher elevations. *Franseria deltoidea* is the group dominant and is uniformly distributed in all stands which are characteristic of this group. *Echinocactus wislizenii* and *Opuntia bigelovii* are found on more steeply sloping topography. *Opuntia bigelovii* is generally distributed over the slope areas, while *Echinocactus wislizenii* tends to be present in greater density on the lower slopes and at the bases of foothills. In this group, *Opuntia bieglovii* is the most important associate. *Simmondsia chinensis* (jojoba) tends to be present where stand diversity is high. *Simmondsia* occurs on rocky, well-drained slopes in the higher desert and chaparral from 500 m to 1600 m in elevation.

Community Groups IV and V include *Ereus giganteus, Encelia farinosa, Eriogonum fasciculatum,* and *Eriogonum inflatum*. These groups reach high importance primarily on slopes. With the exception of *Encelia farinosa,* other species are generally distributed, but are included in the lower quartile of community importance of those species studied. *Cereus giganteus* expands the geographic range of this group by virtue of its ability to grow on rolling foothills.

The results of the ordination show the vegetation to be a community dominated by seven woody perennials. These are distributed widely over central Arizona and tend to determine the physiognomic aspect in all areas. In some instances, such as rolling topography, most of the seven species will be present while in others, such as streamways, one or two species may be predominant. Other species are more or less localized and may become more important under the influence of physiography or elevation. It is not to be expected that associate species will be limited necessarily to one species group any more than the community cominants are so limited. What is illustrated is the expression of optimal conditions for establishing these associate species under certain environmental influences. This expression, and thereby the indication of maximum development within certain community areas, is based on the sum total of community importance values of all species in a particular group.

In some areas species are able to persist in the least favorable habitats, in stands with simple rather than complex floristic composition. Nearby, in other habitats of identical climate, but perhaps different soil, grow numerous other species whose establishment is made possible by more favorable edaphic conditions. It is readily apparent that in such communities the presence of a particular habitat is recognized by the repetition of its characteristic species composition.

The methods used make it possible to distinguish certain species which are important in each life-form class. Importance values aid in determining those species which influence aspect to the greatest extent in certain habitats and in the community as a whole.

From the ordination of stands it was possible to determine the relationship of the dominant species to each other and to the other members of the community. From the phytosociological patterns of species within the ordination, groupings of species were evidently representative of various habitats. There are seven groups in this study, one of which is a large core group which contains all of the major trees, shrubs and cacti indicative of most of the vegetation in the central Arizona upland desert. All seven of the community dominants are in this group.

Community Group II is composed primarily of those species growing on the alluvial plains at lower elevations. The group dominant is *Larrea tridentata,* which assumes a greater importance value in the lower elevations in the western part of the study area. In Community Group VI, *Larrea tridentata* and *Franseria deltoidea*, although found together as dominant species on the middle to higher elevations, separate at lower elevations where *Larrea tridentata* grows principally on floodplains.

The remaining groups are composed of species which are found in association in various areas. Their importance within these groups is influenced by physical factors of the environment and group composition.

A major conclusion drawn from the ordination is that association of species can be differentiated by its use, and that these associations are due mainly to the interrelation of the component individuals and the physical environment. This interrelation of phytosociology and environment within a group association-segregate results in vegetational patterns.

As a final note, one additional index may be used in conjunction with the methods thus far discussed. Burgess[12] used a behavior index (relative cover X frequency) to provide an indication of the degree to which a species may increase or decrease in dominance within association-segregates, depending upon such factors as slope aspect, elevation, and edaphic differences. Brooks[13] employed these methods in a study within the Great Basin Desert of North America and, more recently, in a study which embraced a portion of the juniper woodland of the Asir Province in southwestern Saudi Arabia.

The Government of Saudi Arabia, recongizing the potential as well as the continuing threat to plant and animal populations in the area, had proposed that a combined national park/ wildlife preserve be developed in the western portion of the Asir Province. During the summer of 1974, steps were taken to evaluate the proposal in terms of wildlife protection and possible long-term reestablishment of species known to have existed in the region during earlier years. As a result of that survey and the attendant recommendations, a quantitative vegetational survey of the existing tree and shrub cover was undertaken during the fall of 1974 so that additional information relative to cover and food could be evaluated and some reference point established for monitoring successional patterns once the area was defined and fully protected. In addition, it was thought to be essential to accumulate information regarding the present state of the juniper woodland in the region since it had fallen prey to increased pressure from woodgathering in the recent past. Methods which have been discussed in this paper were employed with the result that vegetational patterns were rapidly delineated, changes in degree of dominance of the major perennial species documented (Figures 5, 6), and ground control stations were established for possible aerial surveys in the future.

As arid zones continue to be assessed in terms of their natural resources and surveyed for potential utilization, or in searching for reasons for their decline, we believe the methods discussed are useful analytical tools.

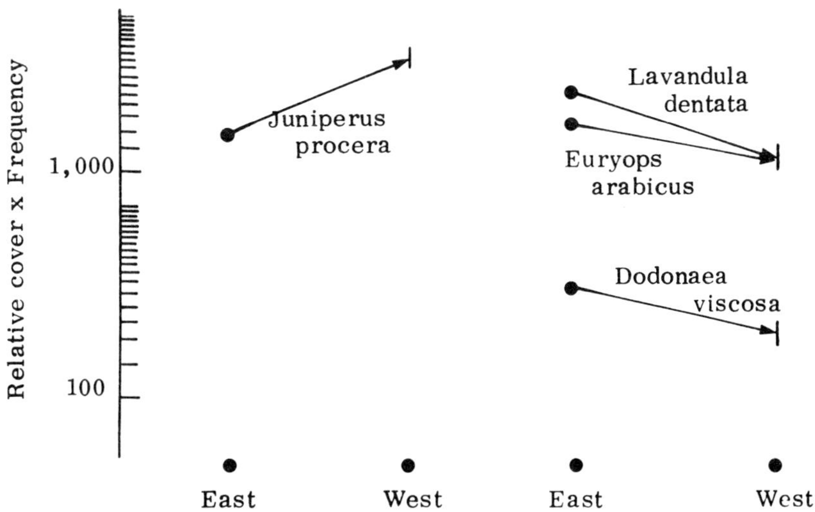

Figure 5 Behavior Indices of dominant perennial plant species near Sudah, Saudi Arabia as influenced by east and west facing slopes in the lower elevational limits of the juniper forest.

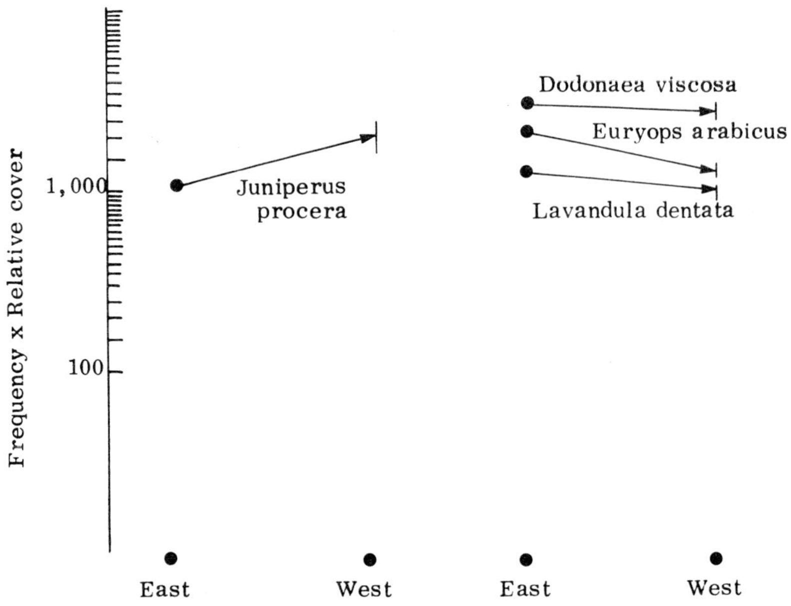

Figure 6 Behavior Indices of dominant perennial plant species near Sudah, Saudi Arabia as influenced by east and west facing slopes in the higher elevational limits of the juniper forest.

REFERENCES

1. J.T. Curtis (1955). A prairie continuum in Wisconsin. Ecology 36:558-566.

2. H.A. Gleason (1926). The individualistic concept of the plant association. Bull. Torr. Bot. Club 53:7-26.

3. E.L. Braun (1935). The undifferentiated deciduous forest climax and the association-segregate. Ecology 16:514-519.

4. F.E. Clements and E.S. Clements (1930). Climax and Climate. Carnegie Inst. Yearbook 29. Washington, D.C.

5. W.S. Cooper (1926). The fundamentals of vegetational change. Ecology 7:391-413.

6. F. Shreve (1942). The desert vegetation of North America. Bot. Rev. 8:195-244.

7. J.T. Curtis and R.P. McIntosh (1951). An upland forest continuum in the prairie-forest border region of Wisconsin. Ecology 32:476-496.

8. J.R. Bray and J.T. Curtis (1957). An ordination of the upland forest communities of southern Wisconsin. Ecol. Monogr. 27:325-349.

9. R.L. Burgess (1959). An intrastand ordination of an oak forest community in southern Wisconsin. MS Thesis, University of Wisconsin.

10. O.L. Loucks (1962). Ordinating forest communities by means of environmental scalars and phytosociological indices. Ecol. Monogr. 32:137-166.

11. F. Shreve (1951). The Vegetation of the Sonoran Desert. Carnegie Inst. Publ. 591, Washington, D.C.

12. R.L. Burgess (1965). A study of plant succession in the sandhills of southeastern North Dakota. Proc. North Dakota Acad. Science 19:62-80.

13. W.H. Brooks (1969). Some quantitative aspects of the present shrub community in Ginkgo State Park, Washington. Northwest Science 43(4):185-190.

PLANNING FOR TECHNOLOGICAL DEVELOPMENT IN DESERT REGIONS

STAHRL W. EDMUNDS
Director, Dry Lands Research Institute, University of California, Riverside, California U.S.A.

ABSTRACT

The development of desert regions is an exercise in technological planning which evolves from simple to complex technologies, from small to large investments, from public to private sector projects, and from product to integrated systems management methods. This process is illustrated by the development of the U.S. Sonoran Desert in the Southwestern United States, in which a desolate arid region moved through the stages of simple livestock production to highly developed agriculture, and more recently to light manufacturing and energy production. Water and energy are key constraints to relieve in desert development. Energy enables a desert region to pump water over longer distances and to reduce water consumption by increasing the efficiency of irrigation technology. Energy also improved habitability and employment in desert regions. Possible areas for energy development in Egypt may be in passive or absorptive solar systems, wind energy, nuclear, and perhaps geothermal energy.

The desert regions of the world are often rich in selected resources, such as sunlight, minerals, arable soils, fossil fuel, geothermal energy, land-space, and natural beauty. They are deficient for man's survival mainly in water, vegetation, and shelter or shade. Technologies can frequently remedy the deficiencies of desert environs, at least enough to make the human habitat tolerable, and then the rich resources of the desert become available. The trade-off is whether the technological state of the art at the time can remedy the desert deficiencies at a reasonable enough cost to realize some return on the resources, or at least a survival wage.

Seen in this perspective, the problem of desert utilization is one of technological planning, that is, assessing what technologies are becoming available, and at what cost, to make the resource utilization of the desert economically feasible. Simple technologies require only simple planning; and in this sense, simple technologies, like well-digging and invention of the goatskin waterbag, have made deserts usable since antiquity for nomadic or oasis life by simple prudence and planning to regulate human usage within bounds of the resource. However, as the technologies become complex, as in the case of large irrigation or energy systems, the planning also becomes complex, and the public investment becomes large. Then it becomes necessary to invent, design, interrelate, and assess the technologies to see whether the investment will yield an economic return, or at least survival wages for an overpopulated region, sufficient to warrant development. To make such assessments requires complex technological planning. Since planning is a form of management, the management aspects of increasing planning complexity are: (1) product management, (2) project management, (3) system program management, and (4) strategic policy or integrated planning. The production of goatskin waterbags or trucks, to enable humans to cross deserts, would be an example of product management. The development of a local irrigation network would be an example of project management. The development of an entire river basin would be an example of system program management. The development of an energy plan to utilize various alternate energy sources for a region would be an example of an integrated plan or a strategic policy.

This hierarchy of management techniques is of some interest to those interested in desert development because it can be shown that deserts, or any developing regions, go through stages of development from the simple to the complex. This means that the first adaptations are simple technologies. As higher technologies are applied, planning concurrently becomes increasingly interactive, the public investment increasingly large, and with large investments the management techniques become more systematic and comprehensive. I propose to illustrate this idea with the case of the development of the U.S. Sonoran Desert in Southwestern United States. Then I will try to show its applicability to technological planning for desert regions generally, such as in the case of Egypt. Finally, I shall hope the illustrations demonstrate that regional development is best achieved when the early applications of product and project technologies are introduced with some conception as to what the ultimate regional development or systems design might be. Without an ultimate design, errors or oversight in total system concept may occur, such as in the U.S. Sonoran Desert where there has not been sufficient forethought to what the total regional system development might be to prevent urban congestion, recreational damage to the desert, or air and water pollution in some areas.

THE CASE OF DEVELOPMENT IN THE U.S. SONORAN DESERT

The U.S. Sonoran Desert is a region of perhaps 60,000 square miles bounded by Phoenix and Tucson, Arizona, on the east, Mexico on the south, Riverside and San Diego, California, on the west. Part of this desert may be called the California desert, the Mohave, the southern end of the Great Basin, or the northern end of Mexico's Sonoran Desert. The region is arid, generally from zero to five inches of rainfall per year. Parts of the region are mountainous, particularly on the east and west. The higher eastern desert is vegetated by cacti and succulent community plants. The vegetation of the lower western desert is termed the creosote bush community, or else it is in sand dunes and salt flats. The U.S. Sonoran Desert is divided by the Colorado River which rises in mountains a thousand miles away. The Colorado, like the Nile in Egypt, provides for localized irrigation to several agricultural basins in the regions, particularly at Yuma and at Imperial County, which is one of the richest agricultural production areas in the United States. The Gila and Salt Rivers also provide extensive irrigated agriculture in Arizona.

The U.S. Sonoran Desert was a desolate, sparcely populated region at the turn of the century. Now it supports a population of perhaps five million people, mainly concentrated in a few urban and agricultural centers, with a reasonably balanced economy. The interesting part of this conversion from a highly underdeveloped region to its present partially developed state is that most of the growth in the economy during the first half century can be accounted for by public investment. Only recently, in the last two or three decades, have private development and industrial diversification become significant factors in augmenting the regions growth. Therefore, a chronology of the technologies, planning, and management means for development is of some interest to illustrate how desert regions evolve from simple to complex states.

The first technologies which opened the region around the turn of the century were the stagecoach and livestock grazing in the high plateau grasslands, together with the discovery of copper at Jerome and Tucson, Arizona. The Santa Fe Trail and the Butterfield Stage provided a simple transportation infrastructure, and the new immigrants employed themselves in product management of producing livestock and copper.

The second phase of development was initiated by the government, through the Bureau of Reclamation, building the irrigation networks for the Imperial Valley and Yuma from Colorado River water, and the Gila and Salt River irrigation projects of Arizona. These projects provided subsidized water and electricity at low rates to local farmers, who over the years have become very prosperous large agriculturalists. This project management phase of development brought into being an agriculturally based economy.

A third stage of development took place from perhaps 1925 to 1960 when the federal and state governments began locating large facilities in the region, such as schools, colleges, universities, agricultural experiment facilities, wartime supply depots, military ranges and reservations, and government offices. The government offices brought programs and services, especially for serving the elderly and welfare clients. This was an era, then, of program management. The region burgeoned as a retirement center for those seeking a warm, sunny climate.

The statistics of this period show the phenomenal impact of government on the economy. Personal incomes in current dollars increased 35 fold between 1929 and 1972, which reflects both increased population and inflation as well as higher incomes. But the contributions to the 35 fold increase were these: government expenditures increased 98 times; transfer payments (also government) 236 times; private sector 35 times; and farm income 8 times. Clearly the government expenditures and transfer payments were the engine driving the expansion.

In economic terms, the expansion of a primary income source has a multiplier effect in the form of derived demand and expansion in supporting services. Thus a new government facility will require new supporting services in terms of electric power, maintenance, repair, and materials. Also, new retirees settling in the region will require supporting services of grocery stores, home building, drug stores, restaurants, etc. The multiplier in economic literature is usually estimated to be in the range of a 2 to 2 1/2 expansion in derived demand based upon increase in a primary income source.

If the increase in government expenditures in the U.S. Sonoran Desert between 1929 and 1972 is looked upon as a primary income source (which it was, since the tax revenues were mostly imported from eastern states), then with a multiplier of 2, government expenditures account for two-thirds of the entire economic growth of the region, and private sector expansion accounts for only one-third of the growth and development.

The private sector expansion, which accounts for only one-third of the growth, has occurred mainly in the past twenty years. The urban centers have experienced substantial private investments in construction, electronics manufacturing, aluminum fabrication, light manufacturing, food processing, finance, trade, resorts, recreation, and services.

A fourth stage of development is now underway in the U.S. Sonoran Desert, which is based upon large scale technological development for geothermal energy production in Imperial Valley, plus the large new Central Arizona water project. Both of these involve huge investments, one private and one public. The expansion of private investment, which has formerly lagged behind public investment, is perhaps of greater interest in examining the future of desert development, because investments in energy are examples of high technologies pushing the state of the art, and of complex planning, which introduces the stage of integrated planning and strategic policy making.

GEOTHERMAL ENERGY DEVELOPMENT IN THE DESERT

The highly productive agricultural area of Imperial County has also been discovered to be the site of substantial geothermal energy resources, perhaps as much as 4,000 to 10,000 megawatts of electric power. Such electric power output would be equivalent to from four to ten large nuclear plants, or from 10 to 25 percent of California's present output. Clearly then, geothermal energy represents a potentially large resource.

Geologically, the geothermal resources occur at the juncture of the North American plate and the Pacific plate, where tectonic action produces a spreading, strike-slip fault in the Salton Sea Trough. The main fault block is a trapezoidal area between the Brawley and Imperial faults, which rotates to the right or counterclockwise as micro-earthquakes spread and fracture the overlaying strata from depths of basement rock at about 20,000 feet. Hot magma intrudes into the basement rock providing the heat source for water held in aquifers within the strata of impermeable clays separated by loose aggregates and silica. Due to the desert environs and drainage from salt laden soil, the hot geothermal fluids are highly mineralized and saline, from 10,000 ppm to 250,000 ppm of dissolved solids in some wells. The geothermal energy source, then, is a liquid dominated system caused by a leaky transform with highly corrosive fluids. Not a very desirable characterization of the resource, perhaps, except that the fluids in some wells are very hot, up to 250 degrees Celsius. The technological problem, then, is how to handle these hot, corrosive fluids to produce electric energy or direct heat energy conversion. This is where the complex technological planning begins to emerge.

Geothermal wells can be drilled by conventional oil well techniques; but new technologies are needed to develop corrosion resistant well-casings, high temperature down-hole pumps, and corrosion resistant reinjection pumps. The conversion of hot geothermal fluid to electric energy is practicable with either flash-steam or binary cycle technology. The flash-steam is more economical, at about $350 per kilowatt of installed capacity (in 1976 prices), but presents more problems in corrosion and environmental control. Binary cycle plants avoid some environmental problems by transferring the heat to a secondary fluid and keeping the geothermal liquids in a closed cycle, but the cost is perhaps $500 per kilowatt of installed capacity. These plant costs, added to the well costs of about $150 kilowatt of capacity, yield total investments of $500 to $650 per kilowatt of capacity, which equates to about 3.5¢ per kilowatt hour at the busbar. These costs make geothermal energy competitive with other forms of electric energy conversion. Electric generation from fossil fuels also costs about 3.5¢ per kilowatt hour, of which 2¢ is the added fuel cost caused by raising the price of oil from $2.50 to $12.00 per barrel. In other words, the rising price of oil is what has made geothermal, and other alternate energy sources feasible. The cost of nuclear energy is also in the 3.5¢ per kilowatt hour range, and coal generation about 4¢ per kwh.

In these terms, the development of geothermal energy in Imperial County appears feasible given a technological planning cycle which includes technical assessment, new design, development of new equipment, geological exploration, land use planning, production well drilling, pilot plant construction, transmission corridor planning, and perhaps direct or waste heat usage. These phases of technological planning are well on their way in Imperial County, and three 50 megawatt pilot plants are under construction at costs of about thirty to forty million dollars each. The large production plants are expected to be about 200 megawatts in capacity, costing about $125 million with field development costs included. Thus the total geothermal investment in Imperial County for full development to 4,000 to 10,000 megawatts of power might come to $2.5 to $10 billion of private capital.

The benefits to the community in Imperial County would be mainly increased tax revenues of about $5,000 per megawatt of capacity annually (or eventually twenty to fifty million dollars per year), plus some increase in employment amounting to perhaps five percent of the labor force. The disadvantages to the community would be mainly environmental, perhaps some increase in air and water pollution as well as more electric transmission corridors; however, these environmental impacts could be abated by careful planning and control. On the whole, the community is going ahead with development since the benefits are perceived to outweigh the disadvantages by a considerable margin.

TECHNOLOGICAL DEVELOPMENT IN NEW DESERT PROJECTS, SUCH AS EGYPT

The transfer of technological planning experience from one country, such as the United States, to another country with a different resource and cultural background, such as Egypt, is problematical. Still the purpose of this workshop is to examine analogues which might contribute to desert development in Egypt; and hence, some speculations about the course of development might be appropriate.

First, desert development is likely to proceed by the transference of simple technologies, now used in Egypt, to the desert regions, such as the adaptation of present agricultural techniques to more arid conditions, or the extension of irrigation systems to the degree that water can be pumped or transported over longer distances.

Secondly, irrigations efficiencies will need to be reexamined, just as we are having to reexamine them in California as a result of last year's drought. Water evaporation, and hence water consumption per crop, can be greatly reduced by moving successively from flood irrigation to furrow, sprinkler, or drip irrigation. But each savings in water consumption among these irrigation methods occurs by moving to a higher technology, a higher capital investment per crop, and a higher energy consumption. Hence the real problem becomes the trade-off of water costs with energy and capital costs for any specific crop. Also, the change to higher irrigation efficiencies necessitates more complex technological planning.

Thirdly, the development of desert regions beyond their agricultural dimensions will most probably require heavy public investment in government facilities and services which make desert settlements viable as places of employment or habitability.

Fourthly, the private investment in new desert regions (beyond the supportive economy established by government investments and programs) is likely to occur through either (1) exploitation of some new mineral or energy resources, or (2) establishment of small light manufacturing facilities to utilize a low cost labor base. The need to encourage private investment along these lines suggests the desirability of incentives to industry for mineral or energy production, or incentives to locate new plants in desert regions.

Fifthly, the ultimate development of new desert regions will depend to a considerable extent upon the availability of energy in the region. Water is certainly the first constraint, but to some extent energy can substitute for water by providing pumping capacity. In any case, energy availability will dictate the extent to which a desert region may be made habitable or provide employment for settlements. The need to develop energy resources will require complex technological planning, substantial private and public investment, as well as system program management methods.

From this brief sketch of the developmental process for a desert region, let me go on to hazard some guesses about the feasibility and timing of energy technologies for the development of new desert regions, such as in Egypt.

The most plausible source of energy additions in Egypt are probably, beyond drilling for oil, fairly simple applications of solar energy, especially passive solar systems of absorptive-convective solar equipment. Passive solar homes are being built in California with large window spaces, water columns, massive walls, and high ceilings to absorb and convect the solar radiation (or cooling) within the house at costs only slightly greater than conventional homes. While passive solar construction is a novelty in California, it was characteristic of the monumental buildings of ancient Egypt;

and with this venerable tradition of passive solar construction, Egypt might well find it a useful approach to heating and cooling, especially since some of the techniques are capable of do-it-yourself installation.

Another form of do-it-yourself solar energy conversion is through absorptive panels, either glass, pipe arrays, or black-body plates with convection of fluids to storage or heat transfer equipment. These absorptive-convection solar methods are suitable for hot water heating or space heating and cooling. Their investment costs are greater than natural gas units, by a factor of eight or ten, which means their amortization may take ten years. This long amortization of capital cost has not made such solar installations attractive as yet in the United States; but if labor rates are low or much of the equipment can be self-installed, the concepts may be close to being practicable in Egypt.

Beyond the simple solar technologies are more complex solar systems, such as the focused array high temperature towers, or photovoltaic cells. However, the investment costs of these technologies are still perhaps two orders of magnitude beyond the cost of fossil fuel or geothermal energy conversion; and hence they are rather unlikely, except perhaps for very remote regions.

Wind energy is closer to feasibility than the large solar systems. Small wind turbines are practicable for remote individual homes or installations. Large wind turbines, with blades as long as three hundred feet, are also being tested experimentally. Such turbines can convert energy with several hundred kilowatts of capacity at a cost about one order of magnitude beyond fossil fuel or geothermal energy.

The main large energy sources, then, which would seem practicable for desert regions are nuclear energy, which is fairly price competitive, if environmental protection can be handled or geothermal energy if it is available. While I am not an expert in this field, my reading of the geology literature suggests that Egypt may well have geothermal reservoirs in the vicinity of the Red Sea, where high temperature fluids and heat sources have been detected, or at the juncture of continental plates offshore from Egypt in the Mediterrean between Egypt and the Middle East. If these geological possibilities are corroborated by testing, i.e., magnetic or gravity anomaly measures or bore holes to measure heat gradients, then some government-funded exploratory drilling or incentive to private companies, would seem appropriate measures to prove the resources.

CONCLUSIONS

The development process in desert regions moves from simple to complex technological planning as the management shifts from the extension of present crafts and techniques to larger public-facility investments and then on to large private investment in industrial or energy programs. As investments become larger, they are accompanied by more intensive management methods, moving from product management to program management, systems management, and integrated strategic policy and planning. A plausible course of development in new desert settlements in Egypt might well be the extension of existing irrigation networks, an increase in the efficiency and technology of irrigation, the establishment of government facilities and programs in desert settlements, and ultimately private investment in light manufacturing or energy development. All of these stages will require technological planning and new technologies.

RURAL TECHNOLOGICAL PROBLEMS OF NEW MEXICO ARID AND DESERT REGIONS AND COMPARABLE PROBLEMS IN THE ANDEAN HIGHLANDS

WILLIAM A. GROSS
Dean, College of Engineering, The University of New Mexico, U.S.A.

ABSTRACT

Rural aridity problems in Africa, Asia, Australia and the Americas are comparable in many respects. There tends to be much sunlight and little rain. The ancients developed techniques, some of which we can use and try to improve upon, and others that we may supplement. The problems inviting better solutions are categorized by size of agricultural effort, and number of people involved. The focus here is upon labor-intensive micro-agriculture, and tiny business-related problems as differentiated from problems of capital-intensive and macro-agriculture. Proper solutions for one may cause failure if applied to the other.

The sun drives the environmental systems and also, because of its availability, offers opportunities for solutions. A problem with solar solutions shared by New Mexico and Andean areas is water distillation, for example. Increased production may result in more local self-sufficiency, better nutrition, time for handcrafts or small business, and the need for improved intermediate management. Solutions, however, need to be appropriate both for the short and long term, contributing to enrichment of the land and its potential, thus increasing the people-carrying power of the land.

The focus of this paper is upon problems faced by small scale farms, businesses and housing efforts in desert and arid regions of the southwestern United States and Latin America, with particular emphasis on solar energy related problems. Of particular concern to this author are the problems faced by small scale businesses (which are defined as those with 1-5 employees) and micro-agriculture (which is defined as farming involving about 1-5 persons and up to about 2 hectares of land). Like large scale enterprises, small scale enterprises face problems relative to food generation, food preparation, shelter, clothing, transportation, commerce, and the impact of large water reservoirs. Unique or "appropriate technology" solutions to these problems are essential for small scale enterprises since they lack the capital and the necessary experience and size to employ the capital intensive solutions which are appropriate for large scale enterprises. This paper will explore solar energy problems for small scale enterprises and will describe the status of current solutions to those problems.

The material for this paper is drawn from many sources, including various research efforts at the University of New Mexico and solar energy applications and appropriate technology efforts in the U.S. and Latin America. The author has been involved in the solar energy planning effort of the 13 Western States of the U.S.A., in the effort to develop a village technology program at the University of New Mexico, and has been a visiting investigator in the Andean countries.

I. A CONTEXT FOR PROBLEM IDENTIFICATION

Since the environmental and social context largely determines the types of problems which will be encountered in arid areas, it seems important to describe briefly some of the environmental and social conditions in New Mexico and the Andean countries. With its 4200 meter mountains and minimum elevation of about 100 meters, New Mexico has a moderately high desert environment. The State experiences large daily and annual temperature changes with summer temperatures typically ranging from 15C^{o} to 35C^{o} and winter temperatures from -15C^{o} to 25C^{o}. The State's heating requirements are comparable to those of the northern portions of the United States and its cooling requirements exceed those of most areas of the country. Like New Mexico, the Latin America high mountains area experiences extensive changes in elevation, substantial daily and moderate annual temperature changes. The temperatures are generally more clement than in New Mexico, thus lessening the need for heating and cooling systems. Rainfall is sparce over a considerable expanse, and there is substantial insolation. Technological developments for dispersed rural areas are of particular importance both in New Mexico and in the Andean area since a majority of the population in these areas lives in small communities or isolated rural settlements. Low cost technological developments are of importance to these areas because of their low per capita income (shown below in 1976 U.S. dollars)[1,2]

Navajo Indians in New Mexico	$ 900
Colombia	$ 630
Ecuador	$ 640
Peru	$ 800
Bolivia	$ 390

Actual per capita income in many rural areas is much less than what is shown above.

As previously indicated, this paper addresses the problems of small scale businesses, farms, and housing efforts in arid areas. The importance of scale of operation is frequently overlooked. Each increase by a factor of ten in the number of employees or persons involved or in the area cultivated appears to require a completely different structure to operate efficiently. Thus, businesses and agricultural efforts may be categorized as follows:

		Approximate Number of Employees/ Persons Involved	Approximate Area Cultivated
	Size 0	1 - 5 employees	Not applicable
Businessess	Size 1	5 - 50 employees	"
	Size 2	50 -500 employees	"
	Size 0	1 - 5 persons	0 - 2 hectares
Agriculture	Size 1	5 - 10 persons	2 - 10 hectares
	Size 2	10 - 20 persons	10 -100 hectares

Clearly these categorizations could be expanded to include larger scale operations. Note that the classification for businesses is determined more by the number of employees and for agriculture by the number of hectares because of the increasing mechanization likely to be used on larger farms. This paper focuses on the size 0 operation and to some extent the size 1 farm, which are identified as small scale business or agricultural efforts, and on the housing needs of people associated with these types of operation. Such small scale efforts generally have very low capital/labor ratios, and machinery may be neither feasible nor affordable.

Small scale efforts are particularly relevant to arid and desert areas because such areas can only support a small population density completely and directly. Population density is limited by specific local conditions and the technology used. The technologies used in such environments must begin to meet long term - as well as short term considerations, such that the quality and carrying power of the land will increase -- obviously a reversal of a widespread situation today: overgrazing and over-use which impoverish the soil. Any population greater than the carrying power of the land must be supported by import/export activities involving material, hardware, and/or software.

It is becoming increasingly recognized and accepted that "appropriate technology" (or "intermediate technology") is the most viable development approach in areas where small scale agricultural, business and housing efforts predominate. Appropriate technology can be defined as decentralized, regionally adaptive technology which is mindful of employment, the level of technology, the level of dependence, and the necessity of low initial investment, operating and maintenance costs. Development of appropriate technologies requires sophisticated analysis of regional environmental and social characteristics. Appropriate technology offers an alternative to the capital intensive approaches used most commonly in developed countries and is a means of breaking the dependency relationship which occurs when underdeveloped countries rely heavily on goods produced by developed countries.

II. IDENTIFICATION OF DEVELOPMENT PROBLEMS IN ARID LANDS

Development of arid lands requires that those responsible for such efforts become problem solvers. The first step toward problem resolution is to be certain that all the important problem conditions are identified and understood. Otherwise, the wrong problem will be solved. One needs to be especially sensitive to the particular characteristics in different places, including the unique environmental characteristics of very small land areas. An understanding of the problem dimensions may require considerable technical and sociological sophistication. Since inappropriate results are likely to emerge if only the narrow immediate problems are addressed, it is essential to consider each problem in a comprehensive context and to predict long term impacts of proposed solutions.

A number of difficult problems relating to food generation, food production, shelter, clothing, transportation, commerce, and, where applicable, the impact of large water reservoirs on the area served, confront small scale development efforts in arid lands. Although high cost capital intensive solutions have been found to solve most of those problems, suitable lower cost, low capital, labor intensive solutions, need to be developed. Such solutions generally invovle the utilization of low cost locally available resources. Only low cost, practical, locally adapted technologies are financially feasible for the small enterprise. A list of specific problems of interest to U.S. and Latin American small scale development efforts is found in the Appendix.

In the interest of brevity, this paper will focus on identification of problems and solutions involving both direct and indirect use of solar energy. Solar energy related problems are particularly noteworthy since the sun is an important resource in arid lands, and small scale enterprises often cannot afford more expensive fossil fuels. The principal solar energy related problems for small scale farms, businesses and housing efforts in arid lands are described below.

Food Drying: Traditional food drying practices involve simply spreading the crop, fish, or meat in the sun. Dirt, insects, rain, the time required, and putrification reduce the efficiency of this traditional approach. As examples, about 20% of Peru's staple food, the dried potato, is lost due to putrification while drying (Howard, *et al.*)[3] The protein content of alfalfa is markedly reduced unless it is dried immediately after cutting. Solar fish drying is a particular challenge because higher temperatures are required than for drying other foods.

Biomass: Biomass includes forest products and manure, and is considered to be solar energy because these products are indirectly produced by the sun. Biomass has been the main source of energy in many areas of the world. Utilization of biomass has resulted in deforestation, investment of a disproportionate amount of effort in wood gathering, and loss of a valuable potential fertilizer source.

Cooking: Wood, and/or dung cooking, though traditional, causes such serious deforestation, fertilizer loss, and sanitation problems, that alternative solutions, though unfamiliar, may become more acceptable.

Water Heating: Those unaccustomed to using hot, or even warm water, are likely to give water heating a low priority, certainly below easily, available potable cold water. Nevertheless, the lack of hot water contributes to sanitation problems. The need is for simple, reliable, inexpensive hot water systems which cope, if necessary, with freezing.

Wind Power: Although wind power was traditionally used for irrigation and general power, it has generally been replaced by electric power for economic reasons. With the rising cost of petroleum and of electric power and the inaccessibility of such power sources for remote communities, the interest in wind as a power source is growing. Windmill generators of about 10 kW, constructed of local materials are needed.

Heating/Cooling of Structures: Solar heating/cooling has been categorized into active systems (those using a working fluid such as air or water, pumps, storage tanks, distribution systems and automatic controls), and passive, or natural systems (in which there is no mechanical control system). Every solar heating or cooling system has three components: collection, storage, and distribution. Only natural systems are considered here because of their low incremental cost, reliability, and simplicity. The collection relates to orientation of the structure, its color, presence of glass, trees, and wind velocity in the area. Storage may be in stone, earth, adobe, or water (in bottles or drums). Distribution involves natural heat transfer means: radiation, conduction and convection. Problems arise in filling these three functions (collection, storage and distribution), optimally using locally available materials. Solar heating and cooling applications include homes, shops, greenhouses, or chicken breeders. The objective of a solar heating/cooling system is to smooth out the annual and diurnal temperature swings to permit greater comfort, or food production.

Electricity Generation: In remote sites where electricity is not otherwise available, solar powered electricity generation using photo cells offers a viable alternative to small petroleum powered generators or batteries, wind or hydro power systems. Depending on the local situation, the cost of solar electric generation may be less than the cost of a petroleum powered system.

Refrigeration: Food preservation, health, and commerce would all benefit from low cost refrigeration methods which would reduce temperatures below what is possible with evaporation cooling.

III. PROPOSED SOLUTIONS TO ARID LAND DEVELOPMENT PROBLEMS

As this paper has indicated, small scale farms, businesses and housing efforts in arid areas face numerous difficult development problems, and increasingly problem solvers are turning to appropriate technologies for solutions to these problems. A recent U.S. House of Representatives Report highlights the importance of regionally adaptive low cost intermediate technologies as follows:

"The experience of more than a quarter century of development assistance programs overseas has clearly demonstrated that much of the technology used in the United States and other industrialized countries is not well suited to the economies of developing countries. It is too big and expensive, and it does not create the jobs needed to absorb rapidly expanding labor forces in countries which already have an abundance of labor. It is not appropriate for use on the very small farms and in the very small business enterprises that make up so much of the economic activity in the developing world."

Proposed appropriate technology solutions should be carefully evaluated to assure that they comply with the following technical merit factors (which are expanded from factors listed in reference 3):

a. Capital/labor ratio. Is the ratio low, in balance with local resources?

b. Affordability. Is the cost of purchased items very low? Moderate and high cost solutions are often available, but cost prohibitive. Furthermore, any commercial product output must also be low cost.

c. Efficiencies/economies of scale. Would the solution permit an efficiency such that less time is required, and/or increased production results which more than compensate for the extra time required to pay for the capital? The challenge is to cope with the above for small scale applications.

d. Simplicity. Is the solution easily operated, maintained, and repaired? It must be to be adopted and used continuously.

e. Replicability. Would it be easily adopted in other areas?

f. Intersectional linkages. Would the solution contribute toward community wealth generation by enhancing market linkages with other communities?

g. Cost of technological development. Does the potential benefit of the solution far exceed development costs?

h. Intensive use of abundant production factors. Would the solution strengthen local production capacity?

i. Social, cultural acceptability. Would the solution fit into local social and cultural patterns?

j. Potential long term impacts. What long term impacts are predicted if the intervention is successful, unsuccessful?

Having identified solar energy problems in the previous section, this paper will focus on the status of the development of solutions to these problems. Before looking specifically at contemporary approaches to these problems, it may be illustrative to look at the technologies developed by the ancients which were appropriate for their times to support their populations. A Canadian group is studying Inca ways for possible lost technologies that may be used now.
J. Godard[4] noted that Sahara natives did not adopt European growing techniques, rather that successful Europeans adopted native techniques. The New Mexico ancients, the Anasazis, made wise use of natural forces in house design. They used cliff overhang to protect from summer sun and child winter winds. In winter the overhang permits the sun to heat the stone walls. The walls store the heat collected and radiate it inward at night. Both comtemporary Pueblo and Altiplano Indians use adobe construction which ameliorates the heat of the summer and the cold of the winter. In a contemporary Navajo Indian hogan, a large amount of earth provides protection against temperature extremes.

With the experience of the ancients as a background, let us now look more specifically at the status of solutions to the solar energy problems posed in the previous section.

FOOD DRYING

Solar driers without moving parts using natural forces have been successfully used in New Mexico for drying fruit and vegetables. Such driers have flat plate collectors, angled for the sun at the desired season, and a path for the heated air to flow over trays and out the top. The driers provide rain and some insect and dust protection. Aside from using glass or plastic for the collector, a drier may be built of local materials. Since the solar drying process is faster than the traditional approach, care is necessary for quality control. The higher temperatures preferred for fish and meat drying may be provided by a parabolic focusing collector. Such a collector requires a reflecting surface such as aluminium foil, and a support structure such as wood or earth. Las Gaviotas is experimenting with such a system in Colombia.

Solar driers for bulk crops such as alfalfa, offer a bigger challenge because of the bulk, the heat per kilogram required, and the effort and time required for collection, movement into and out from the drier, and into storage. It appears that it will be feasible to use solar driers for only small volumes of crops in the forseeable future.

BIOMASS

Methane and fertilizer production from livestock wastes and community sewage systems are cost effective in some places and contribute significantly to community health, but have the disadvantage of requiring a relatively high initial capital investment.

Brown and Howe[5] calculated that dung from two cows (20 kg/day) would provide about 1 m^3/day of methane, enough for cooking two meals. One cow would be sufficient for lighting needs, and five cows for pottery firing every four days (thereby preserving wood and charcoal). They calculate the capital amortization cost for cooking three meals and lighting using gas in Tanzania to be 1 1/4 ¢/kWh, or about 22 ¢/day. This is the same cost as an engine-generator system there. Unfortunately, many families cannot even afford the livestock required. A practical one cow enclosure (expandable) plus human latrine methane/fertilizer system at an amortized annual cost of about \$20 would be very useful. As an incidental advantage, the enclosure should provide theft protection for the livestock.

COOKING

Whether or not methane biogas were available for cooking one meal per day, parabolic spherical or cylindrical reflecting cookers might be used most days of the year for mid-day cooking. It is easy, with a wood frame and aluminium foil, to develop the heat required. It is much more difficult to build a solar cooker which is easy to use, which requires little attention, which is rugged, and cheap. Increases in petroleum costs, and decreases in readily available wood for cooking mean that there will probably be substantial innovation in this area during the next two decades.

WATER HEATING

Thermal siphon water heaters have been used successfully in New Mexico for several years. They require a minimum of piping, a storage tank (and a second tank if the system is drained at night into storage to prevent freezing and damage to the system), and a flat plate collector. Inventors presently are seeking solutions to the freezing problem which would eliminate a need for draining. Although a recycled 55 gallon oil drum may be too expensive for some, the thermal siphon system provides an attractive alternative to petroleum or electric water heaters in many cases.

WIND POWER

Where wind speeds exceed about 5 m/sec. for about half a day, it may be feasible to use windmills for direct power applications such as irrigation, and threshing, as well as for electric power. Sail mills, such as long used on Crete, were recently built for water pumping in Ethiopia for \$ 375 each.

HEATING AND COOLING OF STRUCTURES

There has been a great deal of innovation recently in New Mexico with respect to solar heating of structures, especially using passive, or natural means. The structure should be painted a dark color, have a maximum southern exposure, deciduous trees and roof overhang to keep out summer sun but admit winter sun. The cold north side should be well insulated. The southern side of the house acts as the collector, utilizing many windows (direct gain), a trombe wall, or a greenhouse for this purpose. The heat coming through the windows is stored either in a stone floor (direct gain), for radiation into the structure at night, or in the surface blackened trombe brick wall for conduction into the structure at night, or in the stone floor or water filled recycled 55 gallon drums in a greenhouse for conduction into the structure at night. There are limitless variations possible. Careful engineering and architecture is required to balance cost objectives with temeprature range required. This is particularly important for applications such as chicken breeders. Since the relative humidity is low in arid regions, water evaporation may be used to provide additional cooling beyond that accomplished by natural means (light, color, shading, chimney effect to create air flow, ventilation posts, double ceilings, for example). Again, the designs are limitless.

ELECTRICITY GENERATION

There are three electricity generating systems worth noting for small applications: wind, hydro, and photovoltaic.

A 10 kW sail mill generating system, with batteries, operating about 10 hours daily, would produce electricity at about 18¢/kWh compared to about 2.8¢/kWh for a 6 kW diesel generator system in Tanzania (Brown and Howe).[5] Where sufficient water flow and head is available, hydroturbine generating systems are particularly attractive. The system costs in Tanzania (Brown and Howe)[5] were estimated to range between 5¢/kWh and 12¢/kWh. Unfortunately, arid lands rarely have sufficient hydro-power potential.

Photovoltaic generating systems, like windmill systems, require batteries to provide a continuous supply of electricity since the sun does not always shine, and the wind does not always blow. Photovoltaic systems are currently about ten times as expensive as competing systems, but, with predicted reductions in photovoltaic costs, would be cost competitive within a decade. The reliability and convenience of photovoltaics are so significant that planning such systems may be justified in about five years.

REFRIGERATION

Where refrigeration would permit a community to store food products long enough to market them, and economic considerations indicate the investment is warranted, ammonia based solar systems may now be comparable in cost to electric power systems. Packaged systems need to be developed for these applications.

All of the preceding "solutions" involve decentralization. This carries with it operating, maintenance and repair requirements. There needs to be sufficient effective and supportive infrastructure for them to function satisfactorily.

IV. SUMMARY

Technological problems confronting small scale businesses, farms and housing efforts are outlined since small scale enterprises comprise a sizeable segment of the productive sector of arid areas in the U. S. and Latin America. Of the technological problems outlined, particular attention is given to the following direct and indirect solar energy applications: food drying, biomass, cooking, water heating, wind power, electricity generation, heating and cooling of structures, and refrigeration. Under certain conditions to these problems are now economically competitive for village applications. However, there must be available capital, sufficient expertise, community involvement and support, and supportive infrastructure for them to function satisfactorily.

V. UNM VILLAGE TECHNOLOGY PROGRAM

The University of New Mexico College of Engineering has begun developing programs aimed at developing and analyzing devices and systems appropriate for small scale agriculture which may be used in New Mexico and comparable areas. This newly started program has focused upon low cost wind power systems, Savonius and sail mill.

The UNM program has joined with the northern New Mexico Ghost Ranch project in developing a three hectare test facility for use testing new devices. The program also cooperates with other educational efforts in the State. UNM Engineering faculty members welcome the opportunity to explore possibilities for joint programs with interested faculty members in other countries.

APPENDIX

Problems which must be resolved and technological applications to facilitate strengthening and expansion of small scale businesses, farming and housing efforts in arid areas. (See also references 6, 7 and 8).

A. Food Generation

1. Water management is of central importance, including water collection (rainwater and wells), storage, irrigation, recycling, prevention of water loss, and enhancement of water quality (through desalinization and purification).

a. Soil conditioning including desalinization or prevention of salinization.
b. Drip irrigation using low capital cost local materials.
c. Ditch lining using little or no concrete or expensive materials. The feasibility of pressed and/or stabilized blocks, could be investigated.
d. Water storage tanks and well liners, especially for non-gravel wells, utilizing locally available materials such as pressed and/or stabilized blocks.
e. Well digging efficiency in areas where there is rock or gravel.

2. Micro- and macro land management to correct undesireable climax vegetation, desertification, erosion, flooding, creation of gullies, and lowering of the water table as a result of overgrazing and poor cropping practices.

a. Cost effective substitutions for traditional plowing implements to increase productivity.
b. Low moisture micro-agriculture such as the Hopi Indians have traditionally used involving pre-sprouting, deep stick planting, sand cover, wind breaks, and dew collection.

3. Controlled environment agriculture, including greenhouses, to increase productivity when growing conditions are not optimal.

4. Driers for fish, fruit, vegetables and other agricultural products to avoid loss due to putrification, storage, shipment, rain and poor weather conditions, and to improve the qualify of naturally sun dried products.

5. Chicken breeder heating to maintain temperatures within an acceptable band.

6. Hydro and wind power systems for energy conversion of 25 kW or less, where such energy sources are available.

B. Food Preparation.

1. Storage to protect against losses due to deterioration and pests.

2. Small grain processors for threshing and shucking, possibly utilizing wind or pedal power.

3. Cooking utilizing solar energy or gas produced by a methane digester as alternatives to using wood or manure for cooking.

4. Water heating utilizing thermal siphon solar water heaters which don't require pumps to enhance sanitation.

C. Shelter

1. Construction utilizing locally available materials which resist rain and earthquakes. Methods of adobe brick stabilization and structure rigidification without using concrete or steel are of particular interest.

2. Alternative roof construction techniques which substitute locally available materials for wood ceiling beams and vigas.

3. Secure enclsoures to reduce theft of livestock.

4. Small electrical energy conversion systems to meet the demand for energy in remote areas as the cost of petroleum products increases.

5. Heating and cooling systems which utilize "passive" design and construction techniques rather than "active" mechanical systems.

6. Elimination of pests such as rats and termites.

7. Development and testing of solar collectors which can be locally fabricated using small size glass or plastic panes since transportation of large panes to remote areas is often infeasible.

D. Clothing

1. Small, economically feasible wood carding or spinning units, possibly utilizing pedal or wind power.

E. Transportation

1. Temporary bridges as a means of coping with flash floods and other emergencies.

2. Pedal powered systems which adapt the bicycle and utilize existing bicycle infrastructure.

F. Commerce

1. Intermediate management training to give small scale businessmen essential information about cost accounting, planning management, marketing, and other basic business practices.

2. Quality control of ceramic composition and firing to reduce breakage during firing.

3. Publication of a loose leaf notebook giving production specifications for appropriate technology devices in lanaguages understood in developing countries.

G. Impact of a major water reservoir on the area served. The results of an in-depth study of the impact of the Lake Powell reservoir on the U.S. Southwest may be useful to those considering development of a reservoir. The following potential problems were contributed by UNM Professor David Kidd who participated in the Lake Powell study: Numbers in parentheses refer to published Lake Powell report numbers.

1. Multiple pressures from agriculture and energy consumers for use of water.

2. Increased downstream users exert pressure on use and allotment of the impounded water (21).

3. Maintenance of potable water quality.

4. Concentration of heavy metals and organics either from cultural or natural sources in food species which live in the impounded water (34).

5. Conditions favor development of various parasitic diseases within and along the outlet river (37).

6. Major political-legal decisions relating to: a) developmental needs of the basin, b) legal constraints relating to water compacts, c) economic wisdom of storage project, d) possible conservation values of any resources to be disturbed by developing the water source, e) engineering feasibility and geologic effects (40).

7. Sovereignty of native peoples living along the river in interactions with non-natives: a) mechanisms for assimilation, b) taxing powers of native unit versus taxing power of the state (19).

8. Regional energy development may not raise the standard of living or cash flow (7).

9. Impact of changes in native residence patterns on fertility, contraceptive use, educational level, modes of communication regarding family planning (21).

10. Differences between populations living near a new resource and those living at remote distances: a) incomes, levels of employment and education are higher in impact areas, b) unemployment and welfare expenditures, lower in the impact area and c) household heads younger and family groups, smaller in the impact area (22).

11. Variation between mortality and fertility rates in response to economic variations (20).

12. New developments impact trade patterns and lead to expansion of energy-consuming activities (11).

13. If applicable, the long term surface water supply and under ground supplies in the new area should be analyzed (18).

ACKNOWLEDGEMENT

The author gratefully acknowledges information about problems receives from many people in New Mexico and Latin America. Among these, the following were particularly helpful: Phil Reno for his wide ranging familiarity with the southwest and its people, Gerald May and Job Ebenezer for their Village Technology efforts at UNM and for their photographs, Sam Keller for information about Anasazi practices, Peter Van Dresser for his creative concern for people, ecology and economics, and David Kidd for his extraction from the Lake Powell effort of reservoir- related questions which may be important to Egypt.

REFERENCES

1. ------, "The World's Economic Systems," Time, March 13, 1978, pp. 26-27.
2. ------, New Mexico Business, Jan. 1978.
3. E. Howard, et al, "Project Paper, Appropriate Rural Technology, Project No: 527-0162, U.S. State Department, Peru, 1978.
4. J. Godard, "A Plan for North Africa," Landscape Magazine, Winter, 1955,6, (from L'Oasis Modern, La Maison des Livres, Algiers, 1954).
5. N.L. Brown, and J.W. Howe,"Solar Energy for Village Development," Science, Feb. 10, 1978, pp 651-57.
6. ----------, More Water for Arid Lands, NAS, Wn. D.C., 1974.
7. M. Harper, Consultancy for Small Business, Partnership for Productivity, PO Box 170, Annandale Va., 22003, 254 pp. Explains management concepts required for small business and gives 49 training sessions for training consultants.
8. R.L. Bulfin, and H.L. Weaver, Appropriate Technology for Natural Resources Development: An Overview, Annoted Biblioorgraphy and a Guide to Sources of Information, Arid/Semi-Arid Natural Resources Program, University of Arizona, Tucson, 1977.

THE INTRODUCTION OF LARGE COAL-FIRED ELECTRICAL PLANTS IN THE SOUTHWEST DESERT OF THE UNITED STATES, AND ITS CONSEQUENCES

ORSON L. ANDERSON
Director, Institute of Geophysics and Planetary Physics,
University of California, Los Angeles, CA 90024 U.S.A.

ABSTRACT

For the past decade plans have been made to supply electricity to the large cities of coastal California from the proposed construction of large (3000-megawatt) coal-fired plants in the desert regions of Utah, Arizona, and New Mexico. Many of these plants were to be built near famous recreational areas and national parks. As a result, interests supporting and fighting these desert plants finally came into collision over the proposed Kaiparowits plant. Environmental interests defeated the industrial interests on this particular issue, and now future plans for desert projects are dominated by concerns about air quality.

INTRODUCTION

The unfavorable balance of foreign payments in the United States is generally attributed to oil imports. In order to counter this unfavorable fiscal situation, the United States has adopted a policy of favoring coal as an energy source to be substituted for imported oil.

The response of the nation's total market to this policy is sluggish, if one uses imported oil as a scale: the amount of oil imported into the United States is still rising, year after year. Nevertheless, the effect of this policy is pronounced in the coal-producing areas of the country. The activity of increased coal production has brought heightened prosperity to coal communities that were, only a decade ago, declining because of lost markets.

Although the United States has roughly one-quarter of the world's coal resources,[1] petroleum and natural gas captured the railroad and household energy markets in the 1940s and 1950s. The decline in coal production started to reverse in about 1960[2] and has been steadily increasing ever since. The increased coal production has been due primarily to the use of coal in the electrical utility industry. Coal is beginning to recapture the house-hold market lost many years ago, by recapturing the electrical generating market. The emphasis of national policy is to accelerate the growth of coal-based electrical power production, because the most immediate and economical use of coal is for burning to create electrical energy.

In response to these pressures, the western part of the United States is undergoing a very rapid growth in both coal production and coal-based electrical power demand. The southwestern United States (comprising the states of Utah, Colorado, New Mexico, Arizona, Nevada, and California) has an arid climate and therefore is in some respect analogous to Egypt. It is these states that are the focus of this paper, and the subject is the production of coal for the purpose of generating electricity. The most populous state--California--has virtually no coal resources. The coal resources in question are found in the geological province called the Colorado Plateau (Figure 1), an area which is the least populated zone of the four states: Utah, Colorado, New Mexico, and Arizona. The urban areas of these states are outside the energy-rich Colorado Plateau. This very arid land has little water resources and the surface water is largely transported out of the Colorado Plateau for use by the larger population centers of the Southwest (Figure 2). There are relatively large distances (500 to 1000 kilometers) from the coalfields to the major population centers which consume most of the electrical energy produced from the Colorado Plateau (Figure 3).

The impact of national policy for coal-produced electrical energy is large in the desert lands of the southwestern United States. It amounts to the industrialization of the desert, and the following list of issues arises because of this industrialization:

How will the demand for water needed for processing the electricity affect the expectations of agricultural interests in the southwestern United States?

To what degree will emissions from powerplants pollute the skies above recreational areas of the desert?

What will be the public reaction to this pollution?

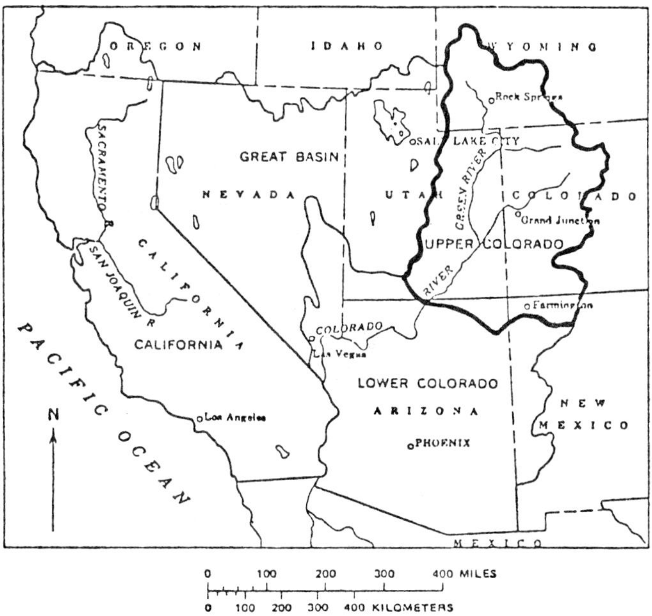

Figure 1. The Colorado Plateau--outlined with a heavy line--comprising part of five western states of the United States. This region contains the energy resources used in the Southwest--coal as well as uranium, oil, and gas.

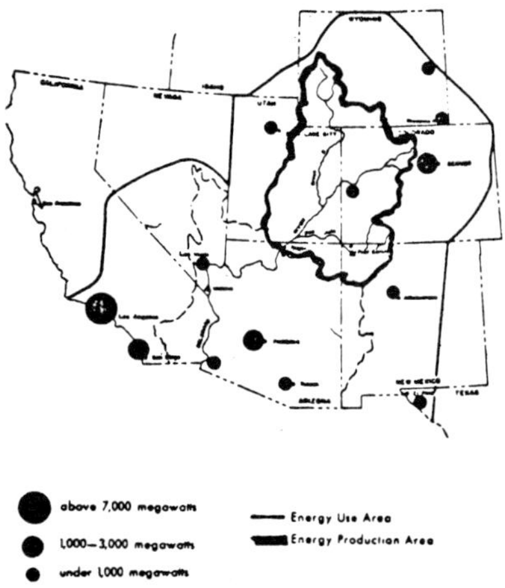

Figure 2. A diagram showing the load centers--consumers of energy--with regard to the Colorado Plateau. The sizes of the circles are proportional to the electrical energy consumption.

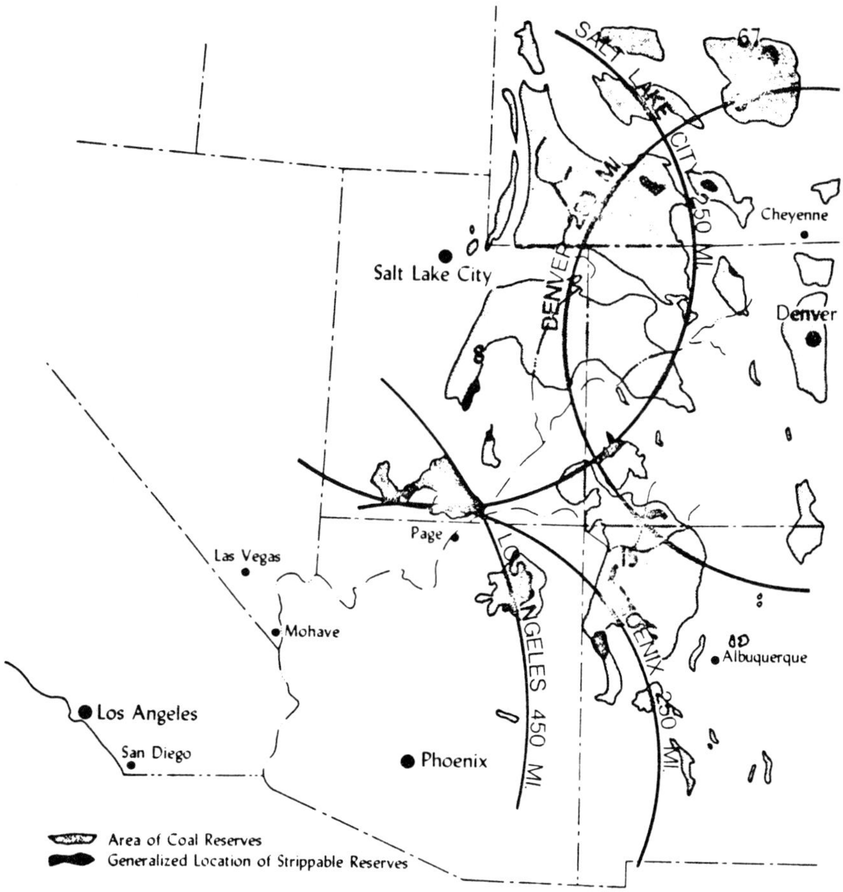

Figure 3. Location of the major coalfields of the Southwest with regard to distance from the coalfields. The Kaiparowits coalfield is in southern Utah, just above Page, Arizona.

How severe will be the socioeconomic impact of industrialization upon the present rural, sparsely populated communities in the area?

What implications will this industrialization have for the transportation network of the southwestern United States?

Only a decade ago, the creation of a large coal-mining and power-generating enterprise consisted of a few key steps: (1) obtain control of the basic resource; (2) obtain the necessary capital; (3) establish access to a growing market; and (4) assemble the requisite labor supply. In the western United States, much of the resource is on Federal land, but mining permits were readily obtained if certain standard procedures were followed.

Today the situation is quite different. There are far more complex requirements for the successful launching of a mining and power-generation enterprise in the western United States. The situation is exemplified by the case of the Kaiparowits Project, a $3-billion enterprise involving a 3000-megawatt coal-fired powerplant with its associated new coal mines. The project was recently cancelled after 13 years of expensive preparation for construction.

The Kaiparowits experience well illustrates the present situation in the southwestern United States with regard to the launching of new mining operations. From this story it is clear that if mining and powerplant operations of the future are to be successful, they must incorporate substantial considerations in political science, law, sociology, atmospheric physics, and biology, as well as the traditional technical and economic factors.

THREE MODELS FOR THE GENERATION AND DISTRIBUTION OF COAL-BASED ELECTRICAL ENERGY IN THE SOUTHWESTERN UNITED STATES

There are good economic reasons to site a coal-fired electrical generating plant at one of three typical sites: (1) at the coal mine; (2) at the source of water needed for boiler cooling; or (3) at the load center, close to the ultimate consumer. Political considerations may impose a fourth kind of site: a place where opposition focused on air pollution is minimized.

From a systems analysis point of view, there are three modes: (1) the plant sited at the mine; (2) the plant sited at the load center; and (3) the plant sited at some intermediate position.

In the 1960s extensive plans were made for the first mode. A large 3000-megawatt powerplant was successfully sited near Farmington, New Mexico; the plant used strip-mined coal from the neighboring area. This is the so-called Four Corners Plant, and an adjunct to it, the San Juan Plant, is presently being completed. The unsuccessful Kaiparowits plant (also to have been 3000 megawatts) was to have been sited at new mines in an area in southern Utah where up to now no commercial mining has been possible. In both these instances the plans were to transport the electrical energy by a.c. transmission lines to the load centers (the big Southwest cities of Los Angeles and San Diego, California; Phoenix, Arizona; and Albuquerque, New Mexico).

In the intermediate mode, there are two successful plants. Coal mined at Black Mesa, Arizona (on the tribal lands of the Hope and Navajo Indian Nations),is transported by slurry line over 350 kilometers to a site on the Colorado River, the so-called Mohave Plant (1800 megawatts) in Nevada. Part of the coal mined at Black Mesa is hauled by a closed railroad, constructed for that sole purpose, to a site on Lake Powell, the so-called Navajo Plant (2100 megawatts) in Arizona. The electrical energy is transported by a.c. transmission lines to the load centers. In these two plants the criteria of water availability dictated the choice of the sites.

The third mode, locating the plant near the large cities which actually consume the power, was not a serious consideration until recently. In the cities of the western United States, where urban values dominate, campaigns against air pollution have received a high priority, and the installation of large coal-fired powerplants near them has been unthinkable, at least until about one year ago.

In the 1960s and the early 1970s the pattern of the electrical industries was to plan on the installation of giant centralized coal-fired plants, sited far out into the desert wherever the water was available.

The Kaiparowits Project plant was planned in this pattern. The plant was to be sited in the rich coalfields of southern Utah which make up the Kaiparowits Plateau (a Piaute Indian name). Moreover, the coalfields were situated adjacent to Lake Powell, so the requisite water was available. Up until

1976 the Kaiparowits Project seemed assured to its proponents. It could count on all the classical ingredients to ensure the successful completion of a large coal-mining and power-generating enterprise in the western United States. The power companies involved in this enterprise had (1) control of the basic resources--coal and water; (2) the requisite capital (or so it seemed); (3) access to a growing energy market; (4) access to the requisite labor; (5) a favorable political climate in the producing states; (6) enthusiasm for the project in the rural region where the mines were to be located; and (7) cooperation of the Federal Land Managers.

Yet these favorable ingredients, always successful in the past, were not enough to ensure success of this particular project. The project was cancelled by its proponents in April 1976, because of environmental resistance and the escalating costs resulting from this resistance. This was a turning point in energy planning in the Southwest. The ingredients for success have to be greatly broadened for future power projects. A new principle has entered the arena: *The desert has a right to be protected from air pollution*. The desert gained a formidable champion as the environmental forces organized themselves to prevent the Secretary of the Department of the Interior from releasing to the power companies the final permit to proceed with construction.

The projected costs of the Kaiparowits Project escalated in large part due to the regulatory delays and additional required studies to the point that the project became economically unfeasible.[3]

Scenic values of the desert, especially those connected with designated recreational areas such as national parks, have now become an important ingredient in decision-making for new powerplant sites.

THE KAIPAROWITS PROJECT STORY

The Kaiparowits region of southern Utah is part of the canyonland complex traditionally associated with the Colorado River canyons. It is immediately adjacent to Glen Canyon. The first energy enterprise in this region was the construction and operation of Glen Canyon Dam, which created Lake Powell by flooding Glen Canyon. The dam was designed to produce 800 megawatts of hydroelectric power to be sold so that the cost of the dam could be retired. Hydroelectric energy production was a major goal in the planning of Lake Powell.

However, far greater significance is attached to other energy-production activities in the area arising indirectly from the existence of Lake Powell. The creation of Lake Powell resulted in the establishment of a network of roads into the region and in the establishment of a reservoir of water available for powerplant boiler cooling.

Before Lake Powell was created, the Kaiparowits region was one of the most inaccessible regions in the United States. It was explored and mapped in the era of explorer-scientists who did their work on horseback. It received early attention from these scientists who regarded it as part of the Colorado canyon complex, and it appeared, for example, in the John Powell map "Lands of the Arid Region" and in other monographs of the 1880-1890 period.[4] In 1931 Herbert Gregory issued a monograph on the Kaiparowits written in the grand style, encompassing the history, ethnography, economic geography, and geology of the region.[5] Gregory called attention to the vast coal reserves of the region.

Nevertheless the region received no economic attention and was by-passed by railroads and by roads until the advent of Lake Powell. The small farming communities nestled at the base of the Kaiparowits Plateau were substantially unchanged 100 years after they had been colonized as part of the Mormon hinterlands. These communities (Kanab, Escalante, Boulder, Tropic, Panguitch) tended to perceive that mining enterprises would result in economic benefit to themselves, as they were economically impoverished by modern standards due to their isolation.

At the same time, surrounding the Kaiparowits Plateau in all directions there had been established a number of national parks and recreation areas of great renown, all based upon the picturesque canyonland vistas. To the southwest, Grand Canyon; to the west, Zion; to the northwest, Bryce Canyon; to the north, Capitol Reef; to the northwest, Arches; and to the west, Glen Canyon National Recreation Area (Figure 4). Visitors by the millions come to these national parks and recreation areas each year seeking recreational values, but geographically central to them all the Kaiparowits Plateau remains, in the words of Durron (1880), "...the burning plains... glowing with bright color...the extreme of desolation...the blankest solitude...a superlative desert..." It remained largely unvisited until 1963, when a great surge of exploration was undertaken.

Los Angeles, San Diego (both in California), and Phoenix (Arizona) have grown beyond any land developer's dreams, and

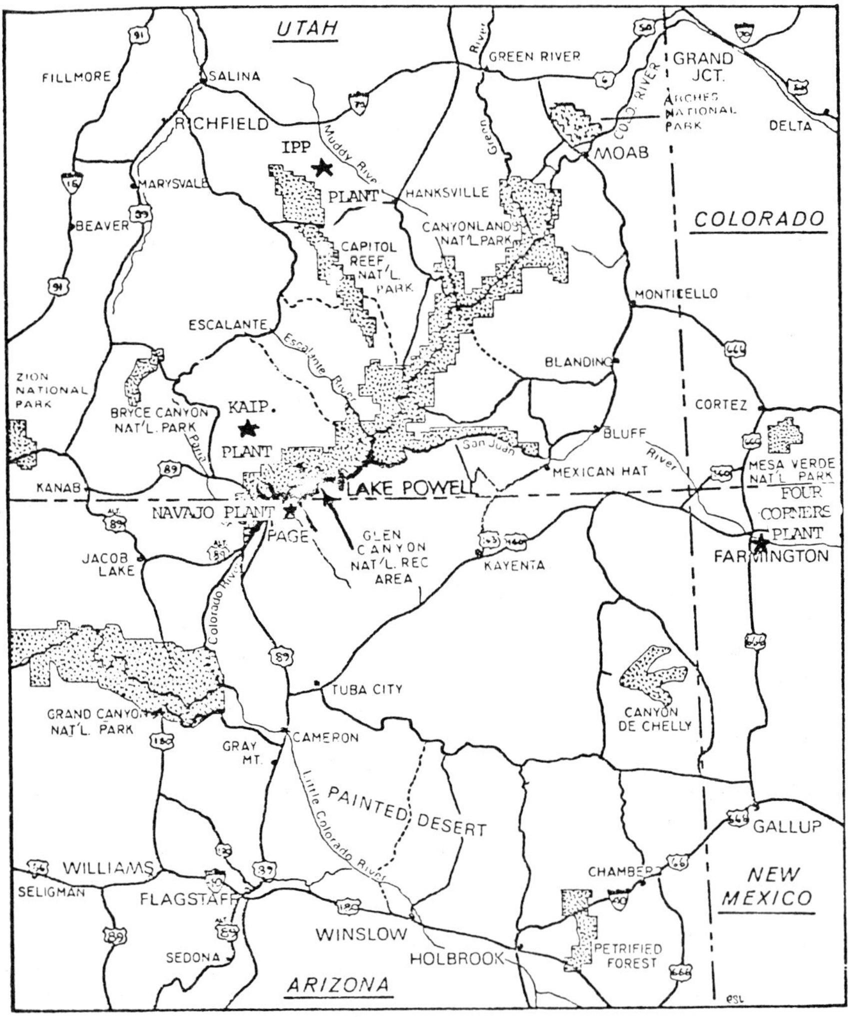

Figure 4. A map showing the location of national parks in the vicinity of the Kaiparowits plant. The proposed location of the Intermountain Power Plan, next to Capitol Reef National Park, is also shown.

all have escalating appetites for energy. Glen Canyon Dam was completed in 1963 and began forming a large lake (Lake Powell) at the foot of the Kaiparowits Plateau. Simultaneously, while the dam was under construction, the electrical industry achieved new capabilities in long-distance transmission and claimed the achievement of extra-high-voltage transmission which doubled the distance over which electricity could be transmitted competitively with rail transport of coal. A consortium of power companies formed the Kaiparowits Power Project and began leasing coalfields in 1962. In 1963, the United States Geological Survey began systematic coal investigations of the Kaiparowits region; these investigations continued for the next decade, and the industrial reports on possible plant sites began to appear in 1964. As the Kaiparowits Project gained momentum and success appeared certain to many, other large industrial organizations began to obtain leases to the coalfield so that a large-scale industrialization seemed likely. The Kaiparowits Plateau was about to be transformed from the most primitive region in the United States to an industrial complex likened by some to a "Ruhr Valley."

Most of the energy produced by the Kaiparowits plant would have gone to Southern California, one of the most urbanized areas of the United States. Producing energy for Southern California by burning coal in Utah has many advantages to Southern California. In a region where smog is a daily concern, there is little desire among either residents or officials to raise the levels of air pollution by establishing coal-fired powerplants in the urban areas.

Many forces in Utah were mobilized to build coal mines and to construct powerplants in southern Utah, and the Kaiparowits Project was welcomed by the Establishment. The few groups which opposed this industrialization were weak and scattered. The mines and powerplants were welcomed by the local residents as a way to ease a number of social and economic problems. The state political apparatus of Utah welcomed the tax revenues that would accrue, especially from the powerplant, and actively encouraged the project.

The creation of large powerplants and mines in southern Utah adjacent to Lake Powell for the production of power to be used in California would appear to solve major problems in both Utah and California. Indeed, the establishment of the proposed project was a trade pattern between the two states that has many historical parallels.

What was not anticipated was the extent and depth of the reaction by those concerned with environmental values. The opposition was regionwide and nationwide, transcending the issues of the principal parties of both states concerned with the project. While the pro-project forces concentrated on identifying the powerplant site as a desert region with little redeeming value, the anti-project forces concentrated on the effects of air pollution arising from powerplant operations on one or more of the six national parks surrounding the Kaiparowits Plateau.

The environmental forces of the nation became organized for what is now a classic confrontation between power industry interests and preservation value interests. Through their efforts, the struggle was transferred in part to the halls of the U.S. Congress, and some of the subsequent national legislation on air quality regulations represented the vector sum of the pro-Kaiparowits and anti-Kaiparowits forces acting in the Congress.

Simply stated, the basic issue was: Would the plume from the smokestack of this powerplant desecrate (and by how much) the air quality above the surrounding national parks? The air quality of the region had been carefully measured and the air was reported to be among the cleanest and quietest in the United States.[6]

To quantify in any rigorous way the amount of air quality degradation was beyond the capabilities of engineers and scientists. How much was too much? The magnitude of the impact of a plume on air quality depends on atmospheric variables, plant characteristics, and the properties of the coal. One side pointed to the awful example of polluted skies above certain notorious plants (the Four Corners Plant in New Mexico, for example). The other side focused on the great efforts by technology to ensure that the pollutants from new plants would be at a far lower level than those produced by the Four Corners Plant.

While the efforts of the design engineers to control emissions were substantial, the amounts of pollutants would have been large because of the enormous amount of coal to be burned. One fact stood out: the 3000-megawatt plant operating at full capacity would consume about 33,000 tons of coal per day and would inject into the atmosphere 2440 tons of fly ash per day.[7] This would result in a reduction of visibility and a false coloration of the sky by some amount in the affected national parks.

The issue was resolved when the project sponsors withdrew their support. Apparently the price became too high for a continued struggle to obtain the required permits from the Secretary of the Interior.[8]

There were repeated delays by the Department of the Interior. The project sponsors were required to find a second site, thereby undertaking all the costs that go with the environmental surveys. Scarce capital was tied up, with consequent loss of interest on that capital, while the issue lingered unresolved.

At the time of cancellation, project officials indicated that the Kaiparowits plant was originally conceived in 1963 as a 5000-megawatt plant costing an estimated $700 million. They stated that the current (1976) costs of the 3000-megawatt plant itself would be $5.7 billion, with plant environmental protection equipment estimated to be an additional $600 million. Costs of a new community and the mines would be $800 million.[8] Thus the cost estimate for the project increased by a factor of nearly 7 in 13 years. Some of this increase was technical, arising from new and projected air quality controls. Some was due to inflation. But much was due to the prolonged process of gaining approval on permits which involved environmental concerns. These are transactional costs, and they defeated the Kaiparowits Project.

In the environmental struggle over the Kaiparowits Project, the Department of the Interior lost its posture of a monolithic organization speaking with one voice. The National Park Service, responsible for managing the national parks, prepared and issued a report critical of the Kaiparowits Project because such a plant would have produced aesthetic intrusions into the park areas.[9] When the Park Service issued this report, it became apparent there was a public split within the Department of the Interior over this project, with one bureau (the Bureau of Land Management) preparing the documents aiding industry and another bureau (the National Park Service) actively compaigning against the project.

When the Kaiparowits Project sponsors withdrew their application, the Secretary of the Interior was released from the difficult decision of whether to sign the permit or to publically deny the permit. The Secretary was left, however, with the prospect that one of his bureaus, the National Park Service, had publically adopted a hardened view of the preservation of air quality in national parks.

The environmental forces were greatly heartened because they had found a successful formula: Bring about delay in the decision of one of the permit applications. Delay, and the costs mount. Delay the decision, and the local issue will evolve into a national issue where new allies can be found. Transactional costs became a lever to the environment movement.

The withdrawal of the project forced a re-evaluation of other power projects in the vicinity of Lake Powell. The Kaiparowits Plateau is still the desolate arid region it always has been, and the prospects for any major industrialization are slim. The opening of new coal mines in this region and the transportation of the coal elsewhere are still possible, if the financing can be obtained for building a very expensive rail line into the area.[10] (However, environmental opposition is being organized to resist the building of railroads and mines in this region.)

The power of the environmental forces gained in the Kaiparowits struggle was tested again in another proposed powerplant project one year later. To the north, and on the edge of Capitol Reef National Park, the Intermountain Power Project (IPP) plant was to be built (at the Salt Wash site). Essentially, the same pattern of forces became involved again. Again the Secretary of the Interior delayed making the crucial decision, and again the National Park Service publically opposed the plant. The issue was responsible for some of the legislation on air quality passed by the Congress in 1977. The Governor of Utah inserted the state apparatus and, largely through his efforts, another site in another desert region was proposed, far from the national park zones.[11] When this second site had been determined by the Governor's committee, the Secretary of the Interior declined the Salt Wash site. The electrical industry at first resisted moving to another location because of the transactional costs but finally abandoned their first preferred site. The Secretary of the Interior was spared making a hard decision when the Governor's committee found an acceptable alternate site. The state apparatus of Utah is aiding the industry in obtaining the necessary permits to build the IPP plant in Lyndale, Utah.

AFTERMATH OF THE KAIPAROWITS PROJECT DEMISE

There is no doubt that the direction of energy development in the southwestern United States took an abrupt turn in

April 1976 with the cancellation of plans to build a 3000-megawatt coal-fired electrical powerplant and its associated mines and transmission corridors.

Not long after this date, an agency of the State of California undertook a study of the feasibility of constructing coal-fired powerplants within California's boundaries, to be fired with coal shipped from the Colorado Plateau mines by railroad or slurry lines. The result of this study showed that it was economically and politically possible to build coal-fired plants in California.[12] Thus, the second mode (a plant sited at the load center) appears to be the most likely solution for future powerplants in the southwestern United States. Already there are plans being actively pursued by the State of California for a coal-fired plant in Southern California and by a private utility, the Pacific Gas & Electric Company, to build a coal-fired plant sited in Northern California. Coupled with this is a movement against the further construction of nuclear plants; this movement is aided by the Governor of California.[13]

In the academic study of siting coal-fired plants,[12] attention was given to a large variety of scenarios. This resulted in the following general economic conclusions that may be of interest:

> In no case are the total project costs cheaper when the energy is shipped by electrical power lines over long distances (say, 800 kilometers or more). Long distance a.c. transmission is by far the most expensive transportation alternative.
>
> Air quality controls sufficient to meet U.S. and California standards are comparable in cost to the plant capital costs.
>
> Transportation costs over long distances (say, 800 kilometers or more) are comparable to the cost of the coal at the mine.
>
> Transportation by railroad is always cheaper than by water slurry, if a railroad line already exists in the area, e.g., railroad access can be obtained by constructing spurs.[14] Transportation by a new slurry line is cheaper than by a new railroad line, providing the amount of coal shipped is sufficiently large (about 3 million tons or more annually).

Air quality considerations control the location of the plant site in the State of California. Water and coal are shipped from their sources to the site.

Some political factors are beginning to be felt which reinforce the economic factors above.

Some successful coal-slurry lines exist in the West. But the demand for new water sources in the Southwest is in excess of supply. It is not politically likely that any region would agree to exporting water (as it is conceived happens with a slurry) from its own region to another region. States in the Colorado Plateau region where the coal is produced are the most arid states in the West. For this reason, political approval of large-scale transportation of coal by slurry line is quite unlikely. Furthermore, the railroad network in the Southwest is adequate to serve the coalfields (Figure 5) and the load centers.

Thus, there is a convergence towards the concept of mining the coal for shipment by rail to the plant site which is close to the load center. Many in the environmental movement support this concept because it places the burden of improving the technology of air pollution squarely on the user of the electricity: the urban center. The urban center has the economic and the intellectual resources to solve this difficult problem. If the air pollution is far removed from the city--out in the desert--then the motivation for the urbanite to solve this problem is lacking.

Shortly after the Kaiparowits Project was cancelled, it became apparent that the power of decision-making regarding electrical plants had shifted away from the industry and toward the Governors of the states.

In the 1960s the planning of a major coal-fired electrical plant was controlled by the utility industry. A power company supplying electricity to its customers in one state could (and did) negotiate with the officials of another state for permits to produce energy and distribute energy. In this way three large coal-fired plants were constructed in the Southwest desert (the Four Corners Plant in New Mexico, the Navajo Plant in Arizona, and the Mohave Plant in Nevada). By mid-1970 many new factors had entered the decision-making process; these factors are symbolized by the Environmental Impact Statement (EIS) that is now required before a project can proceed.

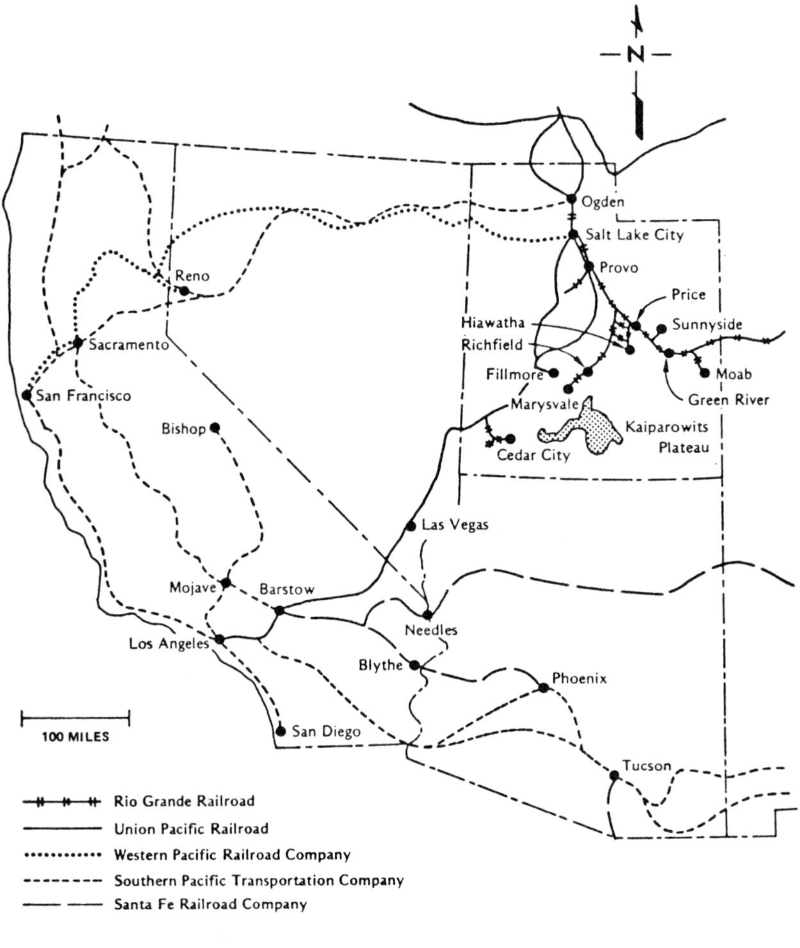

Figure 5. The railroad network of the Southwest United States which serves the coalfields and the load centers. The Kaiparowits Plateau is isolated from the railroad network.

It has been argued[15] that the EIS provides an opportunity to interested participants (the Governors, environmental groups, government agencies, local citizens) to serve their own ends, all of which tends to deny the industry its accustomed position of dominance in the decision-making process. The utilities are losing their ability to make private interstate arrangements for their own power projects in their traditional capitalistic mode.

The states of the Southwest (California, Colorado, Arizona, New Mexico, Utah, and Nevada) have begun to proclaim new individual state policies for themselves, and moreover these policies are beginning to stick. The Governors of these states seem to be emerging in a new power role, and the various state legislatures are sympathetic to the protection of their state's rights (versus national or regional accommodation).

The national policy, as focused in the Secretary of the Department of the Interior, seems to waiver as various interests compete in the United States capital. The procrastination of the Secretary of the Interior in the decision on the Kaiparowits Project permit is an example.

Since the Kaiparowits controversy, decisions about large powerplants in the southwestern U.S. desert will be made in an arena comprising an indefinite and vascillating national policy, coupled with emerging strong individualistic state energy policies and a receding ability of the industry to obtain permits in their traditional operating mode.

ACKNOWLEDGMENT

The support of contract X88-95-484 from the Los Alamos Scientific Laboratory of the University of California is gratefully acknowledged.

This is Publication No. 1854, Institute of Geophysics and Planetary Physics, University of California, Los Angeles, CA 90024 (U.S.A.).

FOOTNOTES

1. Paul Averitt (1975). Coal resources of the United States, January 1, 1974. U.S. Geol. Survey Bull. 1412:91, Table 9.
2. Ibid., Fig. 10, p. 93.
3. Gladwin Hill. Huge power dam stopped after a long fight in Utah. New York Times, April 5, 1976.
4. C.E. Dutton (1880). Report on the geology of the high plateaus of Utah. U.S. Geographical and Geological Survey of the Rocky Mountain Region. U.S. Government Printing Office.

   G.M. Wheeler (1889). Report upon geographic and geological surveys west of the 100th meridian. Vol. I, Geographical Report. U.S. Govt Printing Office.
5. H.E. Gregory and R.C. Moore (1931). The Kaiparowits region. U.S. Geological Survey Professional Paper.

6. E.G. Walther, M.D. Williams, R. Cudney and W. Malm (1974). Air quality in the Lake Powell region. Lake Powell Research Project Bulletin No. 3. Institute of Geophysics and Planetary Physics, Univ. of California, Los Angeles (August).

M.D. Williams and E.G. Walther (1975). Theoretical analysis of air quality: impacts on the Lake Powell region. Lake Powell Research Bulletin No. 8. Institute of Geophysics and Planetary Physics, Univ. of California, Los Angeles (May).

7. U.S. Dept. of the Interior, Bureau of Land Management (1975). Kaiparowits Environmental Impact Statement, Chap. III, p. 207.

8. Editorial. Washington Post, Washington, D.C., May 7, 1976.

9. U.S. Department of the Interior, National Park Service, Analysis of Kaiparowits: Powerplant impacts on national recreation resources, 1976, p. 47.

10. Michael B. Rogozen (1977). Costs of transporting coal from the Kaiparowits Plateau to Southern California. Lake Powell Research Project Bulletin No. 38. Institute of Geophysics and Planetary Physics, University of California, Los Angeles (March).

11. Ray Gibson, Plans for power plant in Lyndale area tentative, Salt Lake Tribune, Salt Lake City, Utah, March 7, 1978.

12. O. L. Anderson, D. L. Carey, R. Purtich and M. Rogozen, Colorado River Basin coal for electrical power generation in Southern California. Lake Powell Research Project Bulletin No. 58. Institute of Geophysics and Planetary Physics, University of California, Los Angeles (September).

13. The Sundesert Issue: Nuclear Safety or Brown Politics? California Journal, March 1978, p. 82.

14. O. L. Anderson and M. B. Rogozen (1977). The role of transportation of energy in the development of the Southwest. Lake Powell Research Project Bulletin No. 50. Institute of Geophysics and Planetary Physics, University of California, Los Angeles (August).

15. Helen M. Ingram and Scott J. Ullery (1977). Toward a framework for evaluating the impact of procedural change upon policy: the case of the Kaiparowits Environmental Impact Statement. Lake Powell Research Project Bulletin No. 53. Institute of Geophysics and Planetary Physics, University of California, Los Angeles.

HUMAN SETTLEMENTS IN NEWLY RECLAIMED LAND

HASSAN FATHY
Architect, Cairo

ABSTRACT

In planning for human settlements in newly reclaimed lands, the ecological balance between mineral resources, plant life, animal life, and human life must be studied in order to determine the population that a given geographical area can sustain at an acceptable standard of living. Local building problems include the use of local materials, such as mud-brick and stone, and developing traditional techniques for vaulted roofing using local materials.

Future emigrants to reclaimed lands should be chosen from already over-populated villages, as interrelated groups who will form neighborhoods of their own choosing and thus facilitate cooperative building in an owner/builder system. In-service training may be carried out in old villages as the future emigrants construct needed public service buildings for these villages and learn skills to use in the new lands.

House design and town planning for desert sites must utilize natural means for cooling, such as wind-catches in houses and provision of airy and shady public places. Town planning also serves to reinforce social relationships. These principles will be discussed in relation to the settlements planned for Abu Tartour phosphate mines in Kharga Oasis.

The theme of the Workshop, "The applications of science and technology for desert development," is one of vital importance at this moment in view of the discovery of huge amounts of underground water in both the Western and Eastern deserts and the forthcoming projects of land reclamation covering several millions of acres.

It is to the credit of AUC to have invited these leaders in all disciplines related to the subject to take part in what we hope will yield a working model, not only for the desert, but also for the rehabilitation of the existing rural settlements in the Nile Valley, since every acre added to the existing arable land will help in relieving congestion in the overpopulated villages, towns and cities, and in redressing the ecological balance in the country as a whole.

In our project, we are aiming at showing scientifically how to create a human settlement together with its hinterland to serve as a model for proposed projects for the desert.

In what concerns the hinterland, we have to consider and apply a whole complex of physical sciences: geology, hydraulics, irrigation, agriculture, etc., whose domains are well defined. But when we come to planning for the settlers and their dwellings we shall find the task much more complicated. We shall have to add a gathering of other sciences whose provinces are much less well defined: the sciences that concern man, society and environment.

These sciences -- demography, anthropology, sociology, social psychology, history, economics, climatology, theory of architecture, aesthetics, etc., are no less important than the physical sciences, for they are directly concerned with man and the ones to which the planner has to add his own creativity and artistic sensitivity.

As a matter of fact, man is the focus of architecture and the touchstone in evaluating any project for human settlements lies in the answer to the question: is it for man, or for something else? If it is for man, we can discuss and argue, but if it is for something else, such as politics, economy, etc., then anything could be done and there will be no discussion.

The problem of human settlement is vast and complex, dealing simultaneously with the analytical and the synthetic, the scientific and the human and the mechanical at the same time. This is not an easy task, but there is no reason to try to facilitate the task by skipping any one factor, especially in what concerns the human aspect.

Over many centuries people in each part of the earth have learned by trial and error how to deal with their housing and environment (which in our case is the desert). They have adapted the rhythm of their life to it, and their habitat and clothes, and they have built houses and villages which are more or less satisfactory. Their solutions to the problem of housing were not the result of scientific thinking. They grew out of the experience of the generations of builders who kept what worked and rejected what did not, and these solutions were handed on in the form of tradition.

The planner, if he ignores tradition, denies himself whatever required more than one life span to mature and crystallize, and thus denies himself of much that could improve and enrich his solutions.

HISTORY

The environmental condittions in the desert have not changed today from what they were yesterday, and we have historical examples in the Western Desert from all periods, the Pharaonic, the Ptolemaic, the Early Christian, the Islamic and the modern, from which most valuable information could be drawn in architecture, structures, planning, climatology, etc.

There is an outstanding example in Bagawat, in Kharga Oasis, which is of great value for us in our project. Bagawat is a desert settlement which goes back to the 4th century A.D., with more than 200 buildings: churches, dwellings, mausoleums, etc. It is a settlement of immigrants from the valley who fled from Roman persecution -- not from over-population, as we do today, though this is just as hostile! They had no materials other than the stone and silt which they had under their feet, and no means of transportation, no industrialized materials and no contractors, but they had an inherited tradition of techniques of building which goes back to the Old Kingdom, using mud-brick for walls and roofs. They used parabolic vaults and spherical domes for roofing, thus eliminating the bending and tensile stresses that mud-brick cannot take, and limiting the stresses to the compressions which mud-brick can safely resist.

These vaulted structures were built in an ingenious way, right over space with no supports or centerings.

In places, it may be seen that the builders used these domes and vaults in combination to cover the same spaces which challenge the modern architect to produce interesting solutions to the problem of roofing large spaces without raising the ceiling height.

From the socio-economic point of view, we have another interesting example in the new village of Gennah in Kharga Oasis. The old village was submerged by moving sand in the 1950's and the inhabitants built the new village cooperatively by themselves, with no aid from the government, with no architect or contractor, and with no equipment from outside the locality. This was possible because they build cooperatively, using the materials they had at hand, namely, mud-brick for walls and the palm trunks and frond stems from their old houses for roofing. The technique of vaulting had been forgotten long before because of the ease of using the palm trunks and frond stems which existed in abundance in the oasis relative to the limited number of houses to be roofed. But when we come to future land reclamation projects, in places where there are no palm trees, we will have to revive the technique of the vaulted roofs, as will be explained later. If Gennah does not give any new information about structure, it does give us a lesson on cooperative building.

Another historical example of modern architecture, with architects, in the desert is the New Village of Bariz, also at Kharga Oasis. The structures of this village are built of mud-brick and stone quarried from the nearby locality, rather than from the site itself, with vaulted roofs right through. The system of vaulting has been developed by engineering and architecturally to be universally used for covering large spans in all sorts of buildings: schools, mosques, markets and houses, large and small.

The project we are considering at this Workshop was conceived as a pilot project and center for research on desert living, including housing. To universalize the use of mud-brick architect-re, a trio of experts is required: the soil mechanics engineer, the structural engineer specialized in shell structures and the architect specialized in mud-brick architecture and vaulting. The design of Bariz was carried out by such a trio, composed of Dr. Mohamed Said Youssef, Professor Soil Mechanics at Cairo University, Dr. Ibrahim Gaafar, Professor of Structures at the same university, and the author. We collaborated closely in testing the soil for mud-brick making, calculating the forces and the stresses to which all the architectural forms that suit this kind of construction, thus subjecting technology to the economies of the poor and their governments.

Unfortunately, the project had to be stopped for lack of funds after it had just begun to yield most encouraging results. The buildings erected have stood there nobly for more than ten years without developing a single crack, waiting patiently for evaluation, recognition, and imitation of the results.

The applied research in this project included provision for thermal comfort, and the subjective results felt by the body would have to be assessed the the usual scientific instruments so as to be conclusive. Fortunately, the governorate of the New Valley has decided to conduct scientific research on climate and architecture in the oasis which will include an evaluation of the design concepts for natural cooling applied in Bariz, and we hope that the information obtained will be useful for our project and for desert architecture in general.

CLIMATE

Climate in the desert is severe with great changes in air temperature. In summer, air temperature in the shade reaches 48°C in Kharga Oasis, and great care has to be taken in the design for thermal comfort using economical, natural means.

Solar radiation and the reflectivity of the white sand are the major cause of physical discomfort in the desert, and shading is the major solution; the streets in the old oasis villages are narrow and covered in parts. Adopting this principle would mean that the houses be contiguous and introverted, opening onto inner courtyards, and shut to the hostile desert at ground level.

This configuration raises certain problems which may have been beyond the capacity of the village builders to solve, amongst which are the problems of hygiene and modern means of communications, as for ambulance and fire-fighting cars.

Here, the modern planner, guided by his scientific knowledge of heat and aerodynamics, and his creativity, is expected to give the appropriate solution to the problem of shading without detriment to hygiene or any other factor that has to be respected in the design, and which would have defeated the village mason.

Another concept is that of the vegetable-garden village which was tried in the plan for the remodelling of the town of Sohar in Oman. The idea consists of having row houses with each two rows having a meandering narrow street which is covered in parts, exclusively for pedestrians, and between each two of these double rows of houses, a large vegetable garden. This area has paths for cattle, agricultural machines and cars.

The vegetable garden has the function of the courtyard or private back garden in giving sun, light and air to the house. It also helps in giving better ventilation and air movement through the houses by convection, with the hot air rising in the large garden to be replaced by the cool air coming from the shaded street. At the same time, this concept answers the problem of segregating car traffic and cows from pedestrians.

NATURAL COOLING

During day time, air at the higher levels is cooler than near the ground. In many traditional houses in Egypt, Iran, Iraq, India, and the countries of the Gulf, this phenomenon is put to use for air cooling and for creating drafts indoors, by using "wind-catches," or "badgeers." The wind-catch is a shaft rising above the house with a large opening facing the prevailing cool wind, to catch it from the higher levels and carry it down into the room. The use of the wind-catch goes back to antiquity, as can be seen in the frescoes of the tomb of Neb-Amun in Gourna dating from the 19th Dynasty. The wind-catch, being small in size, solves the vexing problem of blocks of buildings screening one another from the wind. Cooling by the wind-catch can be increased by hanging wetted straw mats or panels of animal charcoal held in between two sheets of chicken wire.

A device was designed for accelerating air movement and increasing the cooling effect in the design for the good stores in the project for Bariz. This consisted of having another shaft opposite the inlet shaft to work as an outlet at the lee side. It has a sort of metal chimney pot painted black, with blades leaning downwards to the outside, as can be seen in some places in Italy. The effect can be markedly felt by the body even without this pot or the wetted panels. Evaluation by the hygrometer, the velometer and the thermometer gives results numerically.

THE DEMOGRAPHIC AND SOCIAL ASPECT

A naturally grown community normally consists of a certain number of members of different ages and sex, and a variety of families with different sizes or number of members, with different professions, habits and culture, who have adjusted themselves through time to their environment in a more or less balanced demographic and occupational composition.

In our project, these communities are to be created by conscious planning, in stages in place and time that match the rate of land reclamation projects. What poses the most vexing problem in planning is the choice and recruiting of the future inhabitants of the new villages.

Some proposals have suggested nomadic laborers and prisoners as settlers, but aprt from the insufficient number of both, it is not advisable to create communities out of psychologically abnormal and socially unsettled persons.

We know that there is over-population in the existing villages in the valley which need thinning out by emigration, but we know too that man, psychologically and socially, normally has attachments to his native place and to his society. So the best system for organizing migration is to choose the immigrants for every new village from a corresponding over-populated one, in interrelated family groups, and in a well balanced demographic composition with regard to sex, age, and profession, as in a naturally grown community. In this way, even if the immigrants are to lose their attachment to their place, they would not be deprived of their attachment to the community, and will still benefit from the traditional system of cooperation in satisfying those needs which demand collective effort. They will thus take roots in the new place more easily.

A human settlement made of farmers only does not make an organic community. There should be a diversity of occupational groups to ensure the provision of all the services required to maintain an acceptable standard of living for all. For instance, if there were no plumbers, there would be no sanitation.

In a survey of twenty agricultural counties in England, it was found that the working population engaged in farming was 41%, the remaining 59% being distributed among the various trades, professions and services. However, in the region of Hilla in Iraq, the working population engaged in farming was 92%, with only 8% in the other trades to serve the entire region, according to a survey carried out in the 1950's.

As to planning with regard to the problem of occupational balance, the I.L.O. formulated the "International Standard Classification of Occupations" in 1958. It gives a full list and specifications of the different trades and professions which can be used as a reference and check-list.

In planning, these professions can be decided upon by kind first, then by number and distribution over the geographic area, among the different villages and towns.

BUILDING MATERIALS AND TECHNIQUES OF CONSTRUCTION

From the economic point of view we are no better off in our projects for the desert than the emigrants of Bagawat and it is implicit that we have recourse to the same cooperative system, the same materials, mainly stone and mud-brick, and the same methods of construction for roofing by vaults and domes. If in the Nile Valley we have mud, in the desert we have silt which is found deposited in the water courses and in layers at the surface or not very deep under the surface in most parts. Fortunately, mud-brick had been engineered by the modern science of soil mechanics, allowing us to use it with the same security as any other recognized building material like reinforced concrete or steel.

As to building, in the past every village had its own skilled laborers who were integrated into the social and economic web of the community. In our projects the villages are yet to be built and we have to have a system of training these laborers in their old home villages in sufficient numbers before they emigrate to the new ones.

The best way of training the villagers in building construction is to train them while they are working or an actual project in the course of construction, usually called in-service training. Such a training course has been devised and actually tried in the Gourna Village project. By this system, one master mason trained 46 boys, from among those who were engaged as helpers, in only three months, to master all the operations that go into the construction of every element of a house, walls, arches, vaults, domes, stairs etc. By now, there are 80 of these master masons, seeing the benefits of the craft they taught one another in their village themselves.

As said before, for every new village to help in rehabilitating a corresponding old one in the valley, this training can take place in a new public service building which is lacking in the old village and which is to be built as part of the projects for the new land. That way we shall have doubled the gain. The newly trained master masons will be employed in the construction of the public service buildings in their new village for which they will be paid, and at the same time they will build their own family houses cooperatively on their own, with the aid of the government in providing them with the necessary equipment and tools.

COOPERATIVE BUILDING

The snag in cooperation is that one man cannot build a house but a hundred men can build a hundred houses quite easily. But for cooperative building, families have to know each other. Studies have shown that beyond a certain number, anonymity is introduced, leading to abstraction of the individual and hindering cooperation. Experience shows that beyond thirty families, there is confusion. So if the village community can be broken up into neighborhoods of 20 families each, it will be possible for the neighborhood to build its houses easily. If the neighborhood of 20 families (about a hundred members) can raise a group of workers of 24 young men and girls, i.e., four teams of masons, the neighborhood can be built in one year. An incentive for cooperative building is that the family will have its own house besides the farm, and will have the chance of having a mason among its members. A mason's pay is L.E.5 per day, which is quite a respectable source of income.

THE AUTO-CONSTRUCTION CENTER

If a village is to be built by its own future inhabitants, it is not enough to give them skills; we have to add to that the necessary tools and equipment for quarrying the raw materials and for building construction. A striking proof of the practicality of this system appeared in the building of the school at a village called Fares. This is an isolated village on the Western bank of the Nile opposite Kom Ombo, with no easy communications, in fact an oasis like Bagawat. The contract for building this school had been offered to contractors for three successive years, yet no one submitted a tender. The School Building Organization gave me the project for research and I proposed to have it built by the local masons who still had mastery of the technique of vaulting, using it in their village for roofing. We bought L.E.200 worth of equipment: Scaffoldings, trowels, adzes, ropes, etc. and lent them to these local builders against 10% of their pay as rent. The school cost only L.E.6000, which was one third as much as normal ones in more accessible places in the region. This school has 10 classrooms, a specially designed library and a large multi-purpose room backing an open-air stage for theatricals. It was built in mud-brick with vaulted roofs. The building construction including the roofing cost L.E.4000 so the School Building Organization got back its loan plus L.E.200, and kept the equipment which it used in building another school. It was this experiment that brought to me the idea of the auto-construction center as a new amenity to be included in every new and even every old village.

THE AESTHETIC FACTOR: UNITY IN VARIETY, NOT IN UNIFORMITY

The architecture of a house, its location in the village and its size are all expressions of the personality of the owner and the social status of his family. In nature, no two men are alike. Even if they are twins and physically identical, they differ in their dreams. The architecture of the house emerges from the dream; this is why in villages built by their inhabitants, we will find no two houses identical.

This variety grew naturally as men designed and built their thousands of dwellings through the millenia. But when the architect is faced with the job of designing a thousand houses at one time, rather than dream for the thousand whom he must shelter, he designs one house and puts three zeroes to its right, denying creativity to himself and humanity to man. But the architect has at his command the prosaic stuff of dreams. He can consider the family size, the wealth, the social status, the profession, the climate and last but not least, the hopes and aspirations of those he shall house.

As he cannot hold a thousand individuals in his mind at one time, let him begin with the comprehensible, with a handful of people or a natural group of families which will bring the design within his power. Once he is dealing with a manageable group of say 20 to 30 families, then the desired variety will naturally and logically follow in the housing.

Compartmentalizing the village community into such neighborhoods will bring it to the human scale. At the same time this allows the creation of closer relationships and breeds cooperation and concern within the group. This is a prerequisite to the workability of the cooperative or owner/builder system which is the only realistic solution to the problem of housing great numbers of people in newly reclaimed lands.

POPULATION PRESSURE AND DESERT DEVELOPMENT IN EGYPT: A SOCIOLOGICAL OVERVIEW

SAAD EDDIN IBRAHIM
Professor of Sociology, American University in Cairo

ABSTRACT

Egypt doubles its population every twenty-five years. Major Egyptian cities double theirs every fifteen years. By any socioeconomic indicator, Egypt is over populated and over urbanized. The pressure is mounting and is undermining the viability of the societal and eco-system.

One imperative solution is to match Egypt's surpluses -- population and deserts. Creating desert settlements is an embodiment of such a matching process.

A desert settlement, however, is not an easy operation. It is a multifaced task in which human beings, resources, environments and social organization must be synchronized. The vision and leadership committment are no less important factors in such an undertaking.

This paper attempts to open up some of these considerations both theoretically and operationally.

SUMMARY

This paper focuses on Egypt's "surpluses:" population, urbanization, and deserts. Egypt is said to be "overpopulated," "over-urbanized," and its landscape dominated by desert. So far, each of the three "surpluses" has impeded Egypt's development, and stunted any meaningful self-sustained economic growth.

We submit that Egypt's development is contingent on its ability to evolve a new symbiosis among its population, urbanization, and deserts. Matching and synchronizing the two powerful forces of population and urbanization with that of Egypt's deserts is not an easy task. Yet the implications of demographic and urban dynamics make it unavoidable. The question, therefore, is not whether or not we should expand into and settle in the desert. The question is rather, how, where, and with what speed. The premises underlying our argument in this paper may be summed up as follows:

1. Whereas population and urbanization growth operated as stimulants to socio-economic development or followed immediately as a consequence of such a development in the First and Second Worlds, this is hardly the case in Egypt. In fact, with the turn of the century the dynamics of Egypt's population and urbanization have stunted Egypt's development.
2. Population and urbanization growth in Egypt are integral parts of a vicious syndrome of imbalances and disequilibria - set in motion by the early incorporation of Egypt in a world-system dominated by Western industrial capitalist powers.[1] Contacts with and domination by such powers undermined Egypt's traditional structures, including its demographic and urban equilibria, without allowing the evolvement of modern functional equivalents to replace them. The net result of this "false start" has been a distorted development.
3. There are one thousand and one faces of such distorted development and disequilibria. They include imbalances in (a) human and material resources; (b) growth rates of rural and urban areas; (c) quantity and quality of services among Egypt's governorates; (d) growth rates among urban areas themselves; (e) spatial distribution of the population among Egypt's major regions; (f) distribution of high-level manpower between rural and urban areas, and among urban areas themselves.
4. Developmental planning efforts in Egypt are bound to fail unless there is a commitment to a doctrine of population and urban dispersal. The successful operationalization of such a doctrine may not be sufficient for ensuring a self-propelling development. But it is a necessary condition for a successful "take-off."
5. There is only one substantial direction for channeling the population and urban surplus: Egypt's deserts. The alternatives of large scale migration to the sparsely populated Arab countries and/or dramatic reduction in birth rates are not likely options in the near future--for a multitude of political and socio-

cultural reasons. Egypt's deserts will remain the only realistic option for "habituating" most of the thirty million population increment in the next twenty years.

6. Former and current attempts to settle Egypt's deserts have been captive to conventional concepts and traditional modes of settlement typical of the Nile Valley and Delta. Imageries of flat, green cultivation have dominated such attempts, made them very costly, and with meager pay-offs socio-economically. Therefore, new efforts to disperse Egypt's surplus population in Egypt's surplus desert must be freed from, or at least not be limited to, such conventional concepts.
7. The maximum potential of land reclamation for agriculture in Egypt, under the most optimistic conditions, is two million acres by the year 2000.[2] At an average of five acres per family, such potential, if materialized, would accommodate 400,000 families, i.e., two million people out of the projected 30 million which will be added to Egypt's population by the year 2000. In other words, even though a must, agricultural expansion into the desert is hardly sufficient to accommodate these many millions. Other types of expansion and settlements into the desert must be pursued.
8. At least five other types of desert-settlements are feasible in Egypt (beside those revolving around agriculture). Each type is economically and socially based on a major function, the demand on which is both growing and mostly "inelastic." These are settlements around (a) natural resources; (b) higher education and research centers; (c) low water-consuming industries; (d) tourism and health resorts; and (e) military and defense sites.
9. Settling Egypt's deserts is for Egyptians and by Egyptians. Yet such an undertaking will be challenging to most of their values and behavioral patterns. There are ample reasons, however, to believe that given the proper physical and human planning and a built-in incentive system, Egyptians will rise to the challenge and respond positively. The important point to be emphasized here is that settlements in the desert must be envisioned and implemented as full-fledged communities--not as mere housing projects, or,

worse still, as adventurous land speculations. Furthermore, they should project the best of what the rest of Egypt aspires to be. To express it dramatically, planning and building the new communities would be an exercise in *"miniature nation-building."*[3]

10. The most likely population groups, capable of ensuring and enhancing desert settlement, at least in the initial stage, are youth and high-level manpower (i.e., highly educated and highly skilled population). For reasons elaborated in the paper, these should be the target groups spearheading the "long march" into the Egyptian deserts.

The above contentions have been substantiated by this author in several papers over the last ten years.[4] In the remainder of this paper, we give an overview of Egypt's "overpopulation" and "over-urbanization," followed by a proposed doctrine for population and urban dispersal.

I. THE QUESTION OF OVERPOPULATION

The concept of "overpopulation" is a relative term. Whether a country has too many or too few people must be in reference to some agreed-upon standards. Such standards may include density, GNP or income per capita, quantity and quality of services, rate of economic growth, and the like. Sheer population increase, no matter how high, does not by itself mean overpopulation. Libya and Saudi Arabia, for example, are now increasing at the rate of 3.2 per cent annually -- one of the world's highest. But they are still considered underpopulated, as indicated by their heavy importation of manpower. Egypt, on the other hand, with an annual rate of increase of 2.4 is considered overpopulated, as evidenced by the exportation of nearly 1.5 million workers to neighboring countries.

In other words, a country is said to be overpopulated when it is (a) unable or unwilling to absorb its incremental manpower in gainfully productive employment at home; and (b) unable to continuously raise the standard of living for the bulk of its population. The two conditions of overpopulation are, obviously, interrelated; and they summarize the import of many other indicators.

Growth Trends in Egypt

Much has been written on Egypt's population problem for the last half century. Most of the facts are well-known. At the time of Napoleon's expedition (1798), Egypt's population was estimated to be around five million. A hundred years later, the first complete census put the figure at 9.7 million in 1897. Fifty years later, the population doubled again, to reach 19 million in 1947. Less than thirty years later, in 1976, Egypt's population more than doubled, to 38.2 million. Egypt's population had remained "stationary," regulated by a traditional equilibrium, which balanced births and deaths, for hundreds of years. But at the beginning of the nineteenth century, this equilibrium was shattered, and Egypt's population grew steadily and moderately to double itself in one century. In the twentieth century the population growth has continued steadily, and much more rapidly. Thus, the second doubling took half as much time, i.e. fifty years, and the third doubling took only thirty years.

Table 1 summarizes the dramatic evolution of Egypt's population in the last eighty years (1897-1976). Underlying Egypt's rapid population growth is a constantly declining death rate and a persistently higher birth rate. Annual death rate has declined from 40 per thousand in the beginning of this century, to 26 per thousand in 1940, to 12 per thousand in 1975--i.e., 28 points, or more than two-thirds decline. Birth rate, on the other hand, decreased from about 45 per thousand to 38 per thousand in the same 75-year period--i.e., seven points or less than one-sixth decline.[5] At the present annual rate of natural increase of 2.4 per cent, Egypt's population will double itself in twenty-nine years.

Structure of Egypt's Population

Egypt's population is generally young, dependent, misemployed, and maldistributed.

Like most Third World countries, Egypt's population is dominated by young age groups. The proportion of those under 15 years of age stood around 40 per cent during the 1960's, compared to less than 30 in the Second World and less than 20 in the First World. Egypt's young age structure reflects on the socio-economic conditions of the society. Their high proportion implies mounting pressures on the scarce resources of the country, for individuals under 15 are, generally,

TABLE 1 Population Size at Census Counts and Annual Intercensal Growth Rates 1897-1976[1]

Census Year	Population Size in Thousands	Annual Growth Rate % (3)
1897	9,749	
1947	19,022	1.34
1960	26,085	2.38
1966	30,076	2.54
1976	38,228[2]	2.31

Source: Population and Family Planning Board (PFPB), Cairo, 1977.

[1] For brevity we did not include data for the 1907, 1927 and 1937 censuses.

[2] Preliminary results of the 1976 census. Includes Egyptians abroad on census date estimated at 1.4 million by the Central Agency for Public Mobilization and Statistics.

[3] Assuming exponential growth between censues.

consumers, not procedures. Along with those over 65 years of age, they make up the "dependents" who are to be taken care of by the society's labor force (those between 15 and 65). The "dependency ratio" in Egypt was 70 and 86, in 1947 and 1960, respectively. The preliminary results of the 1976 census indicate a decline in Egypt's dependency ratio back to about 71.2, but the figure remains high by First and Second World standards.[6]

The labor force, i.e. those who are 15-64 years of age, stood around 58 per cent of total population in 1976. Almost half of them are women not "gainfully" employed. In other words, the real, not crude, dependency ratio is even more unfavorable. It implies a heavy burden placed on the relatively small labor force that is gainfully employed. In 1976 the economically active population was 11.5 million, or 31.5 per cent of the total population.[7]

The quality of population may also be indexed crudely by their level of literacy. Among those who are ten years of age or older, the percentage of illiterates was 57 per cent. The overall percentage of illiteracy has declined from 80 in 1947 to 65 in 1965, to 57 per cent at present.[8] The 23 percentage point reduction in thirty years is quite modest by any standard. The high illiteracy rate must, in a multitude of ways, reflect on the productivity of the Egyptian labor force.

The maldistribution of Egypt's population is evidenced by its density and spatial distribution. The most striking characteristic in this regard is the tremendous concentration of 99% of Egypt's population in less than 4.0% of its territory. Thus, while Egypt's one million square kilometers may give it a moderate density of 38 persons per square kilometer, this is highly misleading. The Nile Valley and the Delta are at present the habitat of 99% of the population. This habitat is no more than 35,000 square kilometers, which means that Egypt's real population density is over 1,092 persons per square kilometer -- one of the highest in the world. The figure was 278 around the turn of the century, and rose to 543 in 1947. The per capita cultivated land has sharply declined as a result of rapid population growth. In the last 75 years, as we have seen, the population doubled itself twice (from 9.7 to 38.2 million), or a percentage increase of nearly 300%. Cultivated land, meanwhile, increased by only one-sixth or less than 20 per cent. The per capita acreage declined from 0.5 to 0.2 between 1897 and 1976.[9]

In brief, Egypt's growing population has not been managed, deployed or employed well enough to ensure a sustained rise in its standard of living. For these reasons, we provisionally conclude that Egypt is an "over-populated" country. Its over-population is not an eternal state. Should the country redistribute its population, improve their quality, manage their utilization more efficiently, Egypt would cease to be "overpopulated," even with the same rate of annual growth and the same size of population.

II. THE QUESTION OF OVER-URBANIZATION

Like overpopulation, "over-urbanization" is a relative state of affairs. A country is "over" or "under" urbanized depending on some standard of reference. One such standard or criterion is the stage of socio-economic development. It has been observed, for example, that a minimum level of urbanization, say 10-15 per cent, is essential for initiating what Rostow calls the "take-off" stage.[10] Throughout the "take-off" stage itself, urbanization grows to about 30 per cent. In the following stage, self-sustained growth, the level of urbanization rises to 50 per cent or more of the society's total population.

Another criterion is the degree to which the growth of cities is organically linked to the development of the country side in the same society. This is usually indexed by income and service per capita differential between urban and rural areas. The wisdom of this criterion is to detect the presence or absence of "cancerous" or "parasitic" growth of "oversize" cities. Urbanization in this sense implies that cities pick up over-weight at the expense of the countryside, often reflecting a marked income differential caused by appropriation of surplus-value from the latter by the former.[11]

A third criterion is the presence or absence of an urban balance, or a smooth urban hierarchy as measured by the "Rank-Size-Rule" (RSR) of the "Four-City-Index" (FCI).[12]

A. Egypt's Urbanization Trends

We have already pointed out the tremendous concentration of Egypt's population in 3.5 per cent of its total area. There is even more glaring concentration in its two great metropolitan areas of Cairo and Alexandria.

First of all, we observe, as Table (2) shows, that Egypt's urban population has been growing steadily -- from 2.1 million in 1907 to 16.2 in 1976, or more than 700 per cent increase in 70 years. The proportion of urban population rose from 19 per cent to 44 per cent of Egypt's total population during the seventy-year period. The annual rate of population growth in urban areas seems to have always been more than twice that of general population growth. Thus, at the turn of the century, while the latter was 1.5 per cent, the former was 3.2 In 1947 the annual rate of total population growth was 2.4 per cent, and that of urban areas 5.9 At present, while population as a whole is growing at 2.4 per cent annually, that of urban areas is over 5 per cent. In

TABLE 2. Population of Urban Areas and the Cities of Cairo and Alexandria, and Ratio to Total Population -- 1907 - 1976

Census Year	Population in Thousands			Ratio to Total Population %		
	Urban Areas	Cairo*	Alexandria	Urban Areas	Cairo	Alexandria
1907	2,145	678	356	19	6.1	3.2
1947	6,277	2,091	919	13	11.1	4.9
1960	9,651	3,349	1,516	37	12.9	5.9
1966	10,434	4,220	1,801	40	14.2	6.1
1976	16,204	5,084	2,319	44[1]	13.9[1]	6.3[1]

Source: PFPB, Population Situation in Egypt, Present Conditions and Future Prospects, Cairo 1977, pp. 25 and 29.

[1]Preliminary results of the 1976 census. For population inside the borders of the country.

*Cairo Proper--Does not include urban Giza (1.2 million) on the west side of the Nile river, or Shubra El-Kheima (363,000) to the north, both of which are socio-economic extensions of Cairo. Greater Cairo, including these two extensions, was 6.7 million in the 1976 census.

other words, while Egypt as a whole doubled its population once (from 9.7 to 19.0 million) in half a century (between 1897 and 1947), Egyptian cities doubled their population twice (from 2.9 to 6.3 million). In the last thirty years (1947-1976), Egypt's population doubled once (from 19.0 to 38.2 million), but the city population increased nearly 300 per cent (from 6.3 to 16.2 million). At the present rate of growth, urban population would double itself every fourteen years, i.e., it would reach 32 million by 1990, and 64 million by the year 2004.

To appreciate how dramatic this rate of urban growith is, we may recall that most western societies took nearly a whole century to double their urban populations. In the last quarter of the 19th century, France, Sweden and Switzerland were successfully completing their economic take-off with only one-third the level of urbanization Egypt now has. The three countries, with no more than 15% of their population is urban areas, had, nevertheless, between 30 and 45% of their labor force employed in industry.[13] In contrast, Egypt with 45% urban population has no more than 20% of its labor force engaged in industry. In other words, in terms of development, Egypt's urbanization has outdistanced its socio-economic growth. In this sense, Egypt may be described as over-urbanized.

B. Egypt's Megalopolis

Another measure of "over-urbanization" in Egypt is the domination by two "primate cities" of its demographic and urban landscapes. The reference here, of course, is to Cairo and Alexandria. Cairo proper has grown from about half a million around the turn of the century to over five million in 1976 -- i.e., more than ten-fold increase. Thus if Egypt's population has grown in this century alone by 300% and its urban areas by 700%, Cairo has grown by over 1,000%.

Considering Greater Cairo, which includes Giza, across the Nile, as well as Shubra El-Kheima to the north, the population of this single urban sprawl tops the 6.7 million mark. It represents nearly 18 per cent of Egypt's total population. (In 1900, Cairo represented only 5 per cent of Egypt's total population.)

Alexandria, the second largest city, has equally grown from about 300,000 at the turn of the century, to over 2.3 million in 1976. This is over 770% increase in seventy-five years. As a result, Alexandria's share of Egypt's total population has risen from less than 3.0 per cent in 1900 to 6.3 per cent in 1976.

Together, Greater Cairo and Alexandria had a total of 9.0 million people in 1976, i.e., about 24 per cent of Egypt's population, and over 56 per cent of all urban areas.

According to the Rank-Size-Rule, Cairo should not exceed twice the size of the second largest city, i.e., Alexandria. The latter should not exceed twice the size of the third largest city (Suez), and the third should not exceed twice the size of the fourth largest (Port Said), and so on. Applying the rule, Cairo should not have exceeded 4.6 million (twice the population of Alexandria) in 1976. That means Greater Cairo is at least 2.1 million "over-weight" or "over-size." Alexandria itself should not have exceeded 800,000 (twice the size of Suez, the third largest with 396,000). In 1976, accordingly, Alexandria had at least 1.5 million too many people. Applying the Four-City-Index[14] yields a similar revelation about the overweight of Egypt's primate cities.

The growth of Cairo and Alexandria in the last half a century is due to two major sources: natural increase and internal migration. The first source, natural increase, ranged from between 2 and 3 per cent annually, i.e., similar to that of Egypt as a whole. Internal migration from the countryside, as well as from other small towns and middle-size cities, has been responsible for the rest of their annual growth--i.e., between 3 and 4 per cent. In other words, Greater Cairo and Alexandria have been growing by pulling as many migrants or more as they have been by natural increase (the difference between birth and death).

The evidence shows that two-thirds of the migrants to Cairo come from other urban areas in Egypt, and one-third come directly from rural areas.[15] Cities other than Cairo and Alexandria get most of their migrants from the countryside. Thus we have one-step as well as multi-step rural-urban migration along with an active inter-urban migration.

For Egypt's two primate cities to pull as many migrants from urban and rural sources testifies to the immense concentration of wealth and power and culture in Cairo and Alexandria. With only one-quarter of Egypt's total population, the two cities have between one half and two-thirds of Egypt's industry, banks, university seats, hospital beds, doctors, telephones, television sets, and theaters.[16] This disproportionate share by the two cities has been and can only be primarily at the expense of rural areas, and secondarily at that of small towns and middle-size cities. The overgrowth of a given entity at the expense of its surroundings is termed "parasitic" or "cancerous." It retards the development of the surrounding entities, i.e., Egypt's hinterland and smaller cities.

C. Consequences of Egypt's Over-Urbanization

Unlike their counterparts in the West, the rapidly growing cities of Egypt are neither a product of a highly efficient agriculture nor a function of an expanding modern industry. Instead, they are associated with conditions where rural productivity is low, manufacturing is only partially developed, and the levels of education, skills, health, and welfare are still grossly inadequate. As a result, Egyptian cities are not generating sufficient indigenous capital and revenue to meet the rising demands of their own population. Living off the countryside and/or allowing their services to stagnate or deteriorate become ready options, at least in the short run. The unfavorable consequences of Egypt's overurbanization may be summed up under three headings: diseconomies of scale, mounting pressure on services, and breakdowns of healthy social relations.

1. Diseconomies of Scale

At present Egypt must spend per capita more, not less, on Cairo and Alexandria than on other smaller cities and the countryside, to keep them from deteriorating (much less improving them). It is estimated that urban per capita income is nearly twice (L.E. 171.0 in 1975) that of rural areas (L.E. 91.0 in 1975). Yet per capita public spending in urban areas is nearly four times that of its rural counterpart.[18] In other words, urban dwellers are consuming proportionately twice as much public services and utilities as they are contributing to Egypt's GNP. Some of this lopsidedness in resource allocation is due to the urban bias of policy-makers. But it is also due to the ecological nature of Egyptian cities, which are old and "pre-industrial" in their physical structure and lay-outs. Most of them are neither fit for modern-age technology nor for the huge populations that are presently crammed into them.

The rate of Egypt's urban growth is estimated to be twice that of new job opportunities.[19] Yet the added population's mere physical presence in the city requires a minimum of overhead capital to be allocated for public services. The surplus labor force remains either unemployed, under-employed, or is crammed into government bureaucracies. If they remain unemployed, social and political unrest threatens the whole society. If they are forced into state jobs, they bog down an already fat and halfparalyzed bureaucracy. It is estimated that only one of every three newcomers from rural areas finds fully productive work in urban areas.[20] The inability of Egypt's big cities to absorb many of those newcomers into modern productive activities creates a situation of urbanization without "modernization." Had their flow been smaller in volume and slower in pace, rural migrants would have been more easily incorporated into the city's modern institutions and life-style.

In brief, the optimum size which would have enabled Egyptian cities to provide services and perform modernizing functions at an economic scale has been surpassed by at least Cairo and Alexandria. The two primate cities, with their present size, epitomize diseconomies rather than economies of scale.

2. Mounting Pressures on Services

Nowhere is the problem of over-urbanization felt more than in the area of housing. If our projections of future urbanization are accepted, then Egypt must build, before the year 2000, more housing units than it has built throughout the last several centuries, including those now in existence.[21] The magnitude of pressure on this one service in urban Egypt today is staggering.

The housing shortage in Greater Cairo alone at present (1978) is 1.5 million units.[22] This means that many families are either in dilapidated sub-standard housing or without housing altogether. It is worth recalling that according to the Rank-Size-Rule, Cairo was 2.1 million people "over-size" in 1976--which came very close to the volume of the housing shortage. At the present rate of governmental response, Cairo's housing problem will take at least 15 years to solve.[23]

The presssures on housing and governmental response to them represents one of the many vicious circles in Egypt's policies (or the lack of them) toward over-urbanization. The visible conditions compel the government to act. But no sooner does it act than the problem worsens--either because the government does not do enough after having raised expectations, or because more rural migrants flow in and occupy the recently vacated squatters and dilapidated housing. In a typical year, the government would spend probably twenty times as much on public housing in large cities as it would in small towns and rural areas.[24] The result is obvious-- the average citizen believes, rightly or wrongly, that his chances of getting public housing are better in larger than in smaller cities or rural areas.

What applies to the housing situation also applies to other services: transportation, communication, water and electric supply, health, and education. The increasing sprawl of Egypt's large cities, for example, has led to a grotesque, noisy, congested and polluting transportation system. Even the most impressionistic observation of Cairo cannot fail to detect the horrifying dimensions of the transportation problem.

The mounting pressure on service institutions has meant deteriorating quality and diminishing per capita quantity. The irony, however, is that such deterioration has not discouraged newcomers from migrating to the city. This can only be explained by the fact that deteriorating as urban services may be, they are still markedly "better" than what is available in the countryside or smaller towns. In this sense, the average migrant is a relativist, and, in a way, very "rational" in his decision to migrate and remain in the city.

3. Breakdown of Healthy Social Relations

The rapid growth of Egyptian cities has led to social disorganization. The steady loss of the sense of community is a natural result of "succession," "invasion," and rapid turnover of population groups and urban functions. Neighborhood cohesion, individual integration in the collectivity, and tight social control have been rapidly disappearing. The usual urban social problems, well known in the West, are surfacing in large Egyptian cities, e.g., crime, mental illness, prostitution, drug addiction, alcoholism, and violence.

Large-scale violence, for example, is almost totally new on the Egyptian scene. Rising incidents of individual, collective and mob violence have been recorded in recent years.[25] The rising frustration, deprivation, and congestion of the urban poor makes it quite easy to incite mass violence in Egyptian cities at any moderate provocation. The riots of January 1977 are an example. The intensity of violence was a function of city size, and within each city was a function of the district's congestion and poverty.[26]

In conclusion, the overall population problem of Egypt has created a negative chain-reaction: mounting pressure on agricultural land, overflow from the countryside in the form of rural-urban migration streams, rapid urbanization growth, primate cities, mounting pressure on urban services diseconomies of scale, and break-down of healthy social relations. Over-population and over-urbanization, therefore, are integral parts, causes and effects of Egypt's development problem.

Controlling Egypt's population and urbanization are necessary measures to halt the deterioration of its standard of living and quality of life. Both population and urbanization growth have been totally unguided; and until recently no vision or sound policy existed. We have suggested elsewhere curative and preventive measures to deal with over-population and over-urbanization.[27] But we indicated that such measures are still predicted on traditional settlement patterns. We now turn to a futuristic strategy that matches Egypt's surplus of population, urbanization, and deserts.

III. THE LONG MARCH INTO THE DESERT

The imperative of population growth in Egypt is a compelling fact. We may be able to slow it down; but even if we succeed, the absolute increase will still be phenomenal. The grand question, obviously, is where will the thirty million increment over the next twenty-two years go? What is the best we can imagine in the way of a solution?

The two questions are far more difficult than it may seem. To conceive of ideal living conditions for twice as many millions of people as we now have calls for a titanic exercise of imagination and intellect, and for encyclopedic knowledge. Fortunately, men and women of the requisite imagination and boldness do exist. And we must draw on the fruits of their work.[28]

The expanse of Egypt is tremendous; but only 4.0 per cent of it is now inhabited. The other 96 per cent is empty deserts. Agricultural arable land is limited and is already over-populated to the last inch. This reality is so glaring that the only direction we can go in the future is into our deserts. This is not an option or a choice, it is the only alternative. Despite the singlehood of that alternative, it seems to have taken us too long to turn to it; and even now there is timidness and hesitation.

Present Values and Perception of the Desert

Historically, there are few examples of successful adaptation to the desert, such as could be found in the achievement of the ancient Arabs in their original habitat.[29] They adjusted their needs and means to its conditions, and established themselves as masters of the desert.

Over the past few centuries, however, the bulk of the Arab population has progressively lost contact with the desert hinterland. It is no longer a friendly terrain. It is now described in their literature as "ugly" and "treacherous." A value system has developed in which green is beautiful and yellow is loathful. But even when some of us have recently begun to think timidly of the desert as a possible extension of our habitat, we are still conditioned by the same anti-desert value system.[30] We only entertain ideas and schemes of "conquering" the desert, making it submit to our rigid values. It must yield itself to green pastures and fields. We measure our success in terms of how many acres we managed to snatch from the mighty, inhospitable ocean of sand that surrounds us.

There is nothing, of course, objectionable to greening the desert and increasing our hard-pressed agricultural land. What is objectionable, however, is to straitjacket our thinking about the desert in this narrow fashion. What this limited perspective does is to make us conclude that if we cannot reclaim the desert agriculturally, then there is nothing we can do with it. The counter thesis is that there is much more that we can do with the desert. In view of Egypt's demographic and urban explosion, it would be an ill-afforded luxury not to take our deserts more seriously. Indeed, this may be our chance, by taking advantage of the population and urbanization thrust, to renew and strengthen our acquaintance with the desert as our ancestors once did: To explore and exploit whatever mineral resources it may have; to re-establish human bridges among the Arab countries whose populations are now separated from one another by hundreds of miles of emptiness. Furthermore, our very recent history (after 1967) should have taught us a lesson: *land unsettled is a land unclaimed*. If we do not settle our deserts, other hungry, over-crowded nations may very well do it by force of arms.

The Long March into the Desert

We are deliberately coining the phrase "Long March" to dramatize bold and revolutionary strategy necessary in thinking about the desert. Admittedly, the "March" will be neither supersonic nor easy. It will start by inching into the desert slowly and arduously. Early results will not be spectacular and frequent frustrations must be expected. But once the first several miles are covered, the march will gain a momentum all of its own. To ensure the success of our march into the desert, we must be aware of possible problems and increase our knowledge of the desert.

A prerequisite to the introduction of medium and large scale settlements in the desert is a scientific understanding of what it takes to live in the desert. Fresh water, generally scarce in arid areas, must be provided for. While the water needed will not be as much as that required for agriculture, it will be substantial, nevertheless. A good part of our effort, therefore, must be devoted to the technology and economics of desalinization and water transportation over long distances. Fortunately, we do not have to start from scratch. Promising research and applications in this area are under way.

There is an equally pressing need to understand the changing role and mechanisms of human settlements in arid regions. That is, technical and scientific developments in the fields of regional and urban planning must be related to the particularities of conditions prevailing in the desert. Again, we do not have to begin from a zero point. The American Southwest (Arizona and Nevada), which bears many similarities to our region, has gone a long way toward settling its desert. Arizona has been the fastest growing state in the U.S. We can learn from this experience. Closer to home, developments of the last hundred years around the Suez Canal and the Arab-Persian Gulf should be even more relevant.

We must acquaint ourselves with research done and undertake some of our own, regarding means of adaptation to conditions imposed by dry climates. This includes problems of internal physical structure of urban settlements (e.g., site planning, specialized building materials, construction techniques, desert architecture, etc.) as well as those relating to continuity of viable functions and survival. The pioneer works of Frank Lloyd Wright and Paolo Soleri[31] in the Arizona desert and that of Hassan Fathy in Upper Egypt must be digested and utilized. It goes without saying that blind importation of ideas and concepts should be avoided. Egypt has accumulated substantial experience in resettlement schemes. Many mistakes were committed, especially in Nubia, and these should be as valuable a learning source, by negative example, as successful efforts elsewhere.

Where Do We Settle in the Desert

The question of how and where to proceed in settling the desert is an important one, for it will determine our success or failure. One can't plant a city in the middle of nowhere just because it happens to conform to a vision. In developing and settling the desert we cannot afford to be indiscriminate. We must be selective on the basis of carefully tested criteria. Such criteria have been implicitly or explicitly stated throughout this paper. We envision desert settlements which can absorb some of the millions now crowding our densely populated areas as well as many of the millions of people who will be born in the next twenty-two years. But to persuade people to live in the desert there must be economic and social meaning to the venture, along with a sense of national mission. We can think of five types of objectives and functions which desert settlements can serve: development of natural resources; water extensive industrial projects; education and scientific research centers; health, recreation and tourism; and defense and strategic settlements.

Settlements around the development of *natural resources* should rank high on our priority list. Here there is obvious economic potential, e.g., oil, mining, and quarrying. The tangibility of these enterprises will make them less controversial in the early years of our desert march. Sites where minerals are being extracted include six along the gulf of Suez and three in the Egyptian eastern desert along the Red Sea and several in or near the oases of the western desert (now grouped in the New Valley Governorate). Unfortunately, these sites have remained as mere work camps. The living conditions at present are closer to a Spartan life than to a humanly integrated one. So much is this the case that employees usually leave their families behind in the Valley and Delta (where there is some civilization). This need not be the case, however. These desert sites can be turned into flourishing cities. Their economic activities provide for a sound "economic base" in which for every worker in primary industries (like oil) there is room for two or three workers in services (public utilities, government jobs, schools, hotels, etc.). Thus, if there are 150,000 workers in oil fields and mines, there is a potential for urban growth of new towns to accommodate about half a million. And as these towns grow into sizeable cities the ratio of "basic" (primary industry) to "non-basic" (service) workers changes in favor of the latter.[32] In Egypt, most of the oil fields and mining sites are closer to either the Mediterranean or the Red Sea and the Gulf of Suez, which offers a double opportunity for mining and resort towns, thus creating a whole network of mutually supportive communities. Indeed, three new demographic concentrations can feasibly be created: One in the area extending from Suez to El Kosseir; the second from Alexandria to the Libyan borders; and the third in the New Valley (from Bahriyya to Farafra Oases, and from Kharga to Dakhla Oases). A similar demographic pull center can be stimulated in Sinai. The four areas may very well accomodate ten million people, instead of the 300,000 who live in all these areas now.

Recent history in our area and elsewhere should convince us that building sizeable population centers around extractive industries is possible and feasible. The Arab-Persian Gulf area, where only a few thousand people lived until recently, is now supporting settlements totalling three million people. Similarly, the Suez Canal grew, in a matter of fifty years, from a virtual desert emptiness to a teeming population center with three sizeable cities and other smaller communities totalling nearly two million people in 1976.

The Chinese have built new towns of this type around newly discovered natural resources. A notable example is the model city of Ta Ching, built around new oil drilling and refining operations in the Sungari basin of Manchuria. The city is laid out around a small central town which provides the usual urban services, but is surrounded by a ring of satellite communities of about 6,000 each which are involved in both manufacturing and non-manufacturing activities.[33] Much could be learned from Ta Ching's social organization, since it embodies an attempt to eliminate the traditional distinction between mental and manual labor and between city and countryside.

A second type, settlements around educational and research centers, would also be minimally controversial and promise early success. Recently, an Egyptian scientist[34] suggested the establishment of a university in Sinai, not only to serve the educational needs of the people in that long-neglected region, but also to foster research and development of the desert. The warm reception of the idea by his countrymen makes ventures of this type highly commendable in other desert areas of Egypt.

During the past twenty-five years, the number of colleg and universities in Egypt has tripled. These institutions, however, are all located in major urban centers. The need f higher education could have been used to stimulate newly opened regions. It is not too late to rectify the situation More institutions of research and higher learning will be needed in the next two decades. They should be located in t desert. Not only will they attract thousands of people, but they can also serve as model communities for significant experiments in social change. The demand on higher educatio is so great that many Egyptian youngsters go as far as Beirut, Baghdad, and Khartoum to obtain admission in the undergraduate programs of these respective universities. Higher education in this sense is an "inelastic" commodity --i.e., distance and expense do not markedly diminish the demand for it. Thus locating new universities in new desert communities will ensure the attraction of a sizeable student population, along with their teaching faculty, administrativ and other support staff.

A third type of desert community may be centered around low water-consuming industries. The initial investments required for capital goods and social overhead may be enormous. But this is a problem that can be solved by allocatin infrastructure cost to the overall prices of developing dese regions, and by indirectly transferring costs that otherwise would have been incurred in existing urban centers. Another way of reducing cost here would be by locating such industri settlements near the first two types of desert communities. Factories for assembling imported bulk-goods (tractors, cars refrigerators, televisions, etc.) could be ideally located i desert routes, say between Alexandria and Cairo, along the Suez Canal, or between Ismaelia and Cairo. Around such factory sites towns and middle-size cities could be built up and out. The Tenth of Ramadan and Sadat Cities, now under way, are cases in point.

Several small communities do now exist in Egypt's north west coast, along the Mediterranean between Alexandria and Mersa Metrouh, such as Borg El-Arab, Alamain, El-Hamam, and Foukah. These small communities could serve as nuclei for larger towns with both summer and winter tourism as their economic base. There is already a small sweet-water pipelin stretching along the three-hundred kilometer distance betwee Alexandria and Mersa Metrouh. If its capacity is increased to supplement drilled underground wells, this area could support several millions in the next three decades. Tourism here could be coupled with health centers for respiratory diseases a few kilometers inland from the coast, where weather is pleasant and dry all year around.

A similar network of sizeable cities could be established by expanding the three towns of Ghardaqa, Safaga, and El-Koseir on the Red Sea. Among them, a complex economic base of tourism, health resorts, mining and import-export activities is potentially sufficient to provide a habitat for three millions. Its relative proximity to the Nile Valley on the one hand and the heavily populated west coast of Saudi Arabia on the other hand would naturally enhance the development in this area. Its only serious problem is a sweet-water shortage which could be overcome by a pipeline.

A fourth type of desert urban settlement could successfully be developed as health resorts and for the purpose of tourism. Egypt has about 1,500 kilometers of coastline area with climate that permits outdoor recreational activities most of the year. Sharm El-Sheikh may serve as a reminder of what could be done. And again, if we do not do it, somebody else will, to meet the needs of the vacation-hungry, affluent Europeans and Japanese, not to mention the growing Arab middle-class.

The fifth type of settlement is of strategic and defense nature. These would be frontier towns having military and paramilitary intensive presence as their "economic base." Naturally, what is meant here is for army personnel to take along their families and dependents which, as a matter of course, would necessitate infrastructures and a multitude of services to insure appropriate amenities. Such services, in turn, would be operated by people who are encouraged to take along their families. In many parts of the world, there are thriving communities whose existence revolves solely around military bases. It is unnecessary to dwell on the multi-functional advantages of this type of settlement in Egypt. The Army's Corps of Engineers, like that of the U.S. and China, should be utilized in building the physical base of such communities.

In practice, desert settlements will have greater chances of thriving and success if each can combine two or more of the above functions, in "multi-function settlements." For example, settlements located along the Cairo-Alexandria desert axis, such as the proposed Sadat City, could combine research and higher education, water-extensive industry, and health functions. The northwest settlements can combine functions 2, 3, and 4 along with agro-industries based on olives, figs and citrus. The New Valley oases settlements, likewise, cou combine functions 1, 2, 3 and 4. As a matter of fact, in al the proposed sites multi-functional settlements are not only possible but quite desirable from both economic and sociolog ical points of view. The multifunctionality reinforces a community's economic base. Since every function attracts certain types of human beings, multifunctionality would enhance the sociological viability of the community by diversi fying its human base. The quality and intensity of social interaction would be much enriched. Although the Tenth of Ramadan and Sadat Cities are already under way, it may not b too late to suggest that each should plan to have at a minim a university, a medical complex, a water-extensive industry, and a military base. These, along with their needed infrastructures, would ensure socio-economic viability and save them from turning into mere white elephant housing projects.

High Level Manpower and the New Strategy

High level manpower (HLM), here defined as population educated and trained beyond secondary schools, is a crucial factor in the population and urban crisis facing Egypt. The hold the key to instrumental success of the proposed strateg for population and urban dispersal. It is axiomatic that a more balanced distribution of HLM would trigger general proc esses of sectoral and structural balances in Egypt. Underlying this axiom is the following constellation of factors.

Every single HLM element requires, on the average, between eight and ten elements of middle and lower manpower as support staff. Thus, dispersing 1,000 doctors from the national capital to either the populated hinterland or the new proposed desert communities implies the dispersal of nearly 10,000 support personnel. If each element is accompanied by his family (averaging four or more persons), then we are tal ing about an aggregate of 44,000 people. If, for purposes o illustration, we consider these doctors and their support staff as manning as medical complex as the "economic base" o a new community, then there will be a need for nearly five times as many to service them, i.e., in "non-basic activitie such as transportation, education, commerce, entertainment, and other public utilities. This easily adds another 55,000 people. Thus we could be talking about a sizeable town of

100,000 - initially triggered by only 1,000 HLM elements.

HLM, naturally, include the most articulate groups in any society. They are capable of generating pressure for the establishment of essential infrastructures and quality services. Governments are usually more sensitive to their demands. Thus, mere dispersal of HLM away from national capitals and primate cities almost automatically insures a more equitable share of such services to the hinterland, long neglected, and to the new proposed communities.

The mere presence of sizeable HLM elements tends to have other beneficial spillovers and chain reactions. Among other things, HLM attracts economic activities and circumvents the flow of migrants from the hinterland to national capitals.

All the measures suggested in our futuristic desert strategy (the five types of desert settlements) are HLM-intensive, i.e., they require sizeable numbers of highly educated manpower, whether in new industries, universities and research centers, or health and tourism.

It is essential to remember that the strategy of dispersing urban population into the desert requires: (1) massive road-building, communication networks, intermediate service stations, water-well drilling and water pipelines (or desalinization plants if near the sea) to make these desert settlements not only viable but also attractive to an ever increasing number of people; and (2) initial incentives to the first waves of settlers, especially those who take their families along. The rush to the Gulf states, which are essentially new desert communities, came about because of financial rewards, and proves the willingness of Egyptians to leave familiar surroundings for economic opportunities.

The incentive system must build on progressive rewards, proportionate to settlers' involvement. To begin with, wages and salaries should be at least double what they are in metropolitan Egypt, with incremental increase as distance into the desert increases. Housing should be for nominal rent. As the settler brings in his family, he should have the option of buying and owning his house with easy credit, provided he does not sell it for twenty years. Other services, likewise, should be offered at reduced prices initially, and then made free to people who stay, say, beyond ten years. The underlying principle is to make the new desert communities attractive and expedient for potential settlers right from the beginning, and then build up more stakes and interests for them in the new communities over time. This way they become more attached and find it costly to desert such communities. The second generation, i.e. settlers' children, may also be granted a variety of privileges, so that roots get struck deeper and pride in belonging to the new community grows over time.

Youth and Desert Settlements

The other important human component in a successful settling of the desert is youth. Not only are young men and women more idealistic, adaptive, and adventurous, but the new communities would offer them a variety of ways out of their current problems in Egypt.

Many of them feel strained by the grip of custom and tradition. At present, one outlet is to travel to Europe during the summer months, and some of them stay longer. Abroad, conventional social controls are relaxed or suspended Many youngsters discover the joy of being independent from their family's suffocating protection. New desert communitie must be socially planned to enhance greater measures of freedom and self-reliance for young settlers. Recreation and entertainment, along with hard work, must be built into these communities, not as luxuries to be added last, but as integra features to be initiated at the very beginning.

Another problem experienced by Egyptian youth today is the tremendous difficulty of new family formation. At the core of this difficulty is their inability to save enough money to find housing, to furnish it, and to pay the bride's dowery. Many of them find it impossible to get married unles they go to one of the oil-rich countries, work there for several years, and save enough money to accomplish the above "feats." But of every four who try such a temporary sojourn, only one ever gets there; the rest wait in hope or frustratio New desert communities can provide so many Egyptian youngster with functional equivalents. They may not be able to match the salaries and wages given in Kuwait and Saudi Arabia, but they can provide free or minimal-cost housing, higher incomes than their counterparts in metropolitan Egypt, a real opportunity to shape their new communities and to evolve a new lifestyle.

The ultimate success of marching and settling into the desert depends on how deeply the leadership is committed to this national objective. Slogans and rhetoric aside, the bes the leadership can do is to act out its commitment, i.e., the or their children should be the pioneers.

FOOTNOTES

1. For a detailed discussion, see Charles Issawi (1963), Egypt in Revolution, An Economic Analysis. Oxford University Press, London. pp. 18-32.

2. This projection is made by Egypt's agricultural and land reclamation authorities. See Saad Eddin Hanafy (1976), "The evolution of Egypt's agricultural sector until the year 2000," submitted to the Joint International Symposium for Long Range Planning and Regional Integration, Cairo, January 14-21.

3. Kurt Janson (1971), "The planning of new settlements," Workshop on Human Settlements in New Lands, Cairo, September 19-30; Wafik Hassouna (1971), "Means of Implementing the design: organization administration," Workshop on Human Settlements in New Lands.

4. Saad Eddin Ibrahim (1971), "Over-urbanization and under-urbanization, the case of the Arab world," submitted to Arab League Conference on the Development of Manpower in Industry, Bloudan, Syria, July 5-12. A shortened version is published in Int. J. of Middle Eastern Studies, V(1) (January 1975).

   Saad Eddin Ibrahim (1974), "Urbanization in the Arab world, the need for an urban strategy," submitted to First Regional Population Conference of ECWA, Beirut, February 18 - March 1. Published in Population Bulletin of the United Nations Economic Commission for Western Asia, No. 7 (July, 1974).

   Saad Eddin Ibrahim (ibid.), "A doctrine of urban dispersal, an imperative for developmental planning."

5. Nader Fergany (1978), "Population growth and manpower development in Egypt," in Saad Eddin Ibrahim, Ed., Egypt: Social Change and Development 1952-1977. Arab Development Institute, Beirut (Arabic).

6. Sweden, for example, has 66% of its population in the 15-64 age group, and only 22% in the under-15 age group. Its dependency ratio in 1967 was 48.0 (compared to Egypt's 71). See Saad Eddin Ibrahim and N. Hopkins, Eds., Arab Society in Transition. American University in Cairo Press, Cairo, 1977, pp. 310-316.

7. Figures derived from Table 18 in N. Fergany, op. cit., "Population growth and manpower development in Egypt."

8. Ibid., Table 8.

9. Helmi Tadros (1978), "Rural resettlement in Egypt." The Cairo Papers Series, Monograph 4, Cairo. pp. 1-2.

10. For a discussion of his theory, see Walt W. Rostow, Stage of Economic Growth, a Non-Communist Manifesto. Cambridge University Press, London, 1959.

11. Saad Eddin Ibrahim (1975), "Over-urbanization and under-urbanism," op. cit.

12. The "Rank-Size Rule" was first recognized by Felix Awerba while studying German cities. It was restated and elabor ted by Alfred Lotka in 1925 in his book, Elements of Phys cal Biology (Williams and Wilkins, Baltimore, 1925, pp. 306-307). The Four-City Index was elaborated in Kingsley Davis (1969, 1972), World Urbanization 1950-1970, Vols. I and II, University of California Press, Berkeley.

13. Saad Eddin Ibrahim, in S. Ibrahim and N. Hopkins, Eds., op. cit., "Urbanization in the Arab world."

14. The Four-City Index (FCI) is a measure of urban balance. It divides the population of the biggest city by the combined population of the next three. Mathematically, the minimum value of FCI must exceed 0.33, but there is no necessary maximum value. In practice, the FCI seldom fal below 0.50 and seldom rises above 8.00. In 1950, Egypt's FCI was 1.81. It rose to 1.91 in 1960, 2.08 in 1970, and 2.17 in 1976. This means that Egypt's urban imbalance is being accentuated rather than being corrected over time.

15. The Demographic Situation in Egypt, Present Conditions ar Future Prospects (1977). Population and Family Planning Board (PEPB), Cairo. p. 38.

16. Wafaa H. El-Zomor (1978), "Demographic paths and socio-economic development, a case study of the population of egypt," in Population Studies 5(45):32-33.

17. Karima Kourayyem (1978), "Income redistribution between Egypt's rural and urban areas," in S.E. Ibrahim, Ed., Egy Social Change and Development 1952-1977; see especially Table 1.

18. S.E. Ibrahim, "Urbanization in the Arab world," in S.E. Ibrahim and N. Hopkins, Eds., op. cit., pp. 372-373.

19. Ibid., p. 374.

20. Ibid.

21. Ibid., p. 375.

22. Quoted from an interview with Ahmed Talaat, Egyptian Minister Housing, Misr Newspaper, August 22, 1977.

23. The Egyptian Government plans to build about 100,000 units annually, hoping that the private and cooperative sectors will supplement its effort. Ibid., p. 3.

24. Ibid. The Ministry of Local Government announced the allocation of LE 4 million for housing and public utilities in rural areas in the fiscal year 1978-1979. On the same page of Misr Newspaper is announced the LE 80 million allocation for urban public housing.

25. See "Report on violence in the Egyptian society," National Center for Sociological and Criminological Research, Cairo, 1977 (Arabic).

26. Mark Cooper, "Coorelation between the January 1977 violence and voting results in Egypt," unpublished paper, 1977. The author found significant correlation between violence (as indicated by the number of casualties and arrests), poverty, and voting for the opposition in the election held a few months earlier. In Cairo, for example, the districts of Shubra, Bab El-Shariyya and Imbaba, and Sayeda Zeinab were the hardest hit. All of them are overcrowded and contain sizeable slum areas. Bab El-Shariyya, for example, has the highest density in Cairo, 100,225 persons per square kilometer, according to the 1976 census. The district of El-Gomrok in Alexandria, also hard hit by the January riots, has the highest density in Alexandria, 133,464 persons per square kilometer.

27. See footnote 4.

28. Examples of Middle Eastern scholars who have given this issue great attention include Professors Thomas Seaty (The Compact City, Freeman, San Francisco, 1973), Nohad Toulan, Salah El-Shakhs, and Z. Zahlan.

29. Professor Nohad Toulan has also suggested a similar approach to urbanization in the Arab world. See his "Urbanization in the Arab states: a potential for regional settlement, growth and development," presented at the Industrial Development Center of the Arab States Conference (IDCAS) Conference, Bloudan, Syria, July 5-10, 1971. We have incorporated many of the ideas suggested by Professor Toulan in this paper.

30. An illustration of this hesitation is well dramatized by the following two news items appearing in the same day in the same issue of an Egyptian daily: "Dr. Ibrahim H. Abdul Rahman, Minister of Planning, calls for the creatio of desert settlements to absorb 20 million Egyptians in the next 25 years." "An official in Greater Cairo Water Authority announces the transfer of 13 employees from Cairo to desert and remote governorates as a punishment for their neglect and lack of orderliness." (Al Ahram, November 30, 1975.)

31. See an article on Paolo Soleri's innovative ideas in Horizon, Autumn, 1970, pp. 32-39.

32. John Alexander (1954), "The basic - non-basic concept of urban economic functions," in Economic Geography, 30: 246-261.

33. Rhoads Murphy (1976), "Chinese urbanization under Mao," i Brian J.L. Berry, Ed., Urbanization and Counter-urbanization, Vol. II of Urban Affairs Annual Reviews, Sage Publications, Berkeley. pp. 324-325.

34. Osama El-Kholy, "Sinai University," reprinted from Al-Ahra (Cairo daily newspaper) in El-Shabab Al-Arabi (January 28, 1974).

THE BIOSALINE CONCEPT

JAMES C. ALLER
OSKAR R. ZABORSKY
National Science Foundation, Washington, D.C.

ABSTRACT

The biosaline concept envisions the harmonious interplay of solar energy, a high temperature environment, saline water, and biological organisms as the foundation for abundant and renewable resources. The implications and potentials of this concept, proposed in 1976, were assessed in a preliminary manner at the International Workshop on Biosaline Research. The five major research areas examined were: (1) Food and feed from desert environments; (2) Food and feed from marine environments; (3) Biological waste treatment at elevated temperatures and salinities; (4) Salt tolerance in microbes and plants, and (5) Enzyme technology. These research areas and various other aspects of this concept will be discussed and recent developments in research and program development will be emphasized.

The "biosaline concept" envisions the harmonious interplay of *bio*logical systems with *saline* environments for the ultimate benefit to man. In particular, the desire is to provide alternative options to critical material needs from renewable resources in an environmentally acceptable manner. The essential elements of the concept are biological systems and a saline or marine environment. Usually, the environment of primary interest is at a high solar flux and elevated temperatures. Biological systems represent whole animals, plants or microorganisms or their individual cellular constituents, especially the critical enzymes. Thus, the boundary conditions for "biosaline research" are biological organisms, or their essential constituents, living in or being able to adapt to and tolerate saline or marine environments. Arid lands become an intrinsic part of the biosaline concept because they comprise a resource that is currently underutilized.

The biosaline concept is both simple and complex. It is simple in that most of man's activities to convert the enormous resources offered by desert areas to productive uses are based on modifying the environment to be suitable for plants and animals that are already understood and used. One exampl of such an environmental change is the removal of salt and minerals from brackish water and the use of the desalinated water on plants which accept only fresh water. The simple part of the biosaline concept is that nature does not levy this fresh water requirement. In fact, throughout the world -- in the seas, brackish waters, and the salt ponds and lakes of the deserts -- many plants and organisms exist which not only tolerate the presence of minerals, including salts, but actually relish them.

We can follow two separate paths in reaching a goal of useful biosaline-derived products. We can start with a plant such as barley or rice that has a known useful product. Then we can capitalize on genetic diversity, or perhaps a known salt tolerance as in the case of barley, and use modern plant science techniques to obtain a plant that has both useful pro ducts and can utilize salty water. This approach is shown on the left side of Figure 1 and characterizes the work of Epstein[1] and Akbar[2] and co-workers.

There is another, perhaps more powerful, method available. Modern technology provides new tools for converting various biomass sources into products that can be used in a variety of ways. So far these techniques, which can be loose ly termed biotechnology, have been applied primarily to plant and microorganisms using fresh water. The techniques are not limited except by the state of knowledge. We show this concept as an alternate path on Figure 1. One should note that now we start with sources which, over the ages, have been selected by nature because they are naturally adapted to saline waters. The heritage is precious and is not yet fully identified. While we may plan to use the desert for the fina utilization step, we should expect to concentrate intitially on water-rich environments for our genetic sources. Figure 1 also shows that the products may not be just the cereals of present agriculture but can well be chemicals or other essential products of the various industries. Both paths shown on Figure 1 are in principle straightforward.

The complex part of the concept is that we lack the know ledge on how to capitalize on what nature has already given us. The International Workshop on Biosaline Research,[3] held a year ago, brought together a number of diverse scientists and research administrators who showed that there is a base of knowledge on which to build a program. However, much more needs to be known, for we can do little without research results.

We can speculate that we should anticipate the unexpected if biosaline research leads to new ways of employing what are now underutilized land areas. For example, let us consider the following possibility. We know from experience that most products need to be consumed near their source for efficient use. U.S. agriculture is energy intensive partially because of long transportation links, large-scale storage and warehouse facilities, and the use of refrigeration to protect perishable products. As the results of biosaline research reach the product stage, they probably will be initially on a small scale. One can easily visualize a saline water-based microbial factory using a small area to produce a specialty product, such as hormones, pharmaceuticals or feed additives.

However, what do we do if our objective is to use our presently unproductive arid lands to support large numbers of people? For example, let us imagine planning for a dense population, similar to that found on rich farmlands, placed in what is now an unpopulated desert. If we were to build a fresh-water irrigation system, we have some reasonable idea of the requirements for the needed town-city complex. Engineers can carry out preliminary plans for schools, roads, water systems, etc. Although there are many difficulties to be encountered, we do know a great deal about how to solve these problems. Yet, we know virtually nothing on how to develop a community which capitalizes on the use of sea water irrigation to produce products most of which would then be used to support that community. We can speculate on some items. For example, mangroves could produce wood products used for structural materials, fiber needs and energy. Other salt-tolerant plants could provide for other human needs directly or indirectly as animal feed. Thus we might imagine a spectrum of possible practical resources. However, to capitalize on these individual research results requires complete system experiments if a biosaline-based community is to result.

In time, these experiments would have to include changes in living patterns. Mann[4] has summarized some of the experience in the arid western United States by pointing out that early settlers were influenced by life styles which had been developed for a humid setting. The shift was one from humid to arid environments. Perhaps we should expect similar attitudes if we proceed to a biosaline society. We should be candid and state that we are dealing with a complex system which will eventually require equally complex utilization experiments.

To convert promising research into stable, viable societies may lead to different types of research facilities. As Devey[3] phrased this at the Kiawah conference, "An arid lands village program as a living laboratory to demonstrate results in real time settings. Costly and ambitious?" We should note, however, that in some ways the interaction of complex sociological and engineering problems in capitalizing on biosaline are even more imprecise than some of the disciplinary areas. We can say only that research is critically needed and that the prize for success can be enormous. We can outline some research that can be tackled with existing laboratories and scientists. Table 1 shows a general list of possible objectives for biosaline research.

With regard to the research areas currently comprising the biosaline research portfolio, the identification process started with the generation of the concept. Initially, enzyme oriented research objectives and projects were emphasized. An unsolicited proposal submitted by San Pietro identified many topics for possible inclusion -- from biophotolysis to salt-tolerant crops. However, in order to arrive at a coherent and broad-based scientific view, a steering committee was established to focus on a limited number of high-priority research topics consistent with the embryonic concept. A host of topics was considered but ultimately the number was reduced to the five general research areas identified in Table II. A number of topics were to be excluded from open discussion because of the time constraints of the conference as well as the need to focus on new topics. For example, energy and climatology were not included in the research list because a number of conferences, workshops and articles had been devoted already to those topics. Of course, this omission does not imply that they are not important or do not form integral elements of any biosaline research plan. Also, the emphasis of the research initially envisioned was to focus on nonconventional agricultural systems.

At the Kiaweh Island meeting, the five research areas listed in Table II were examined in the following manner. A position paper was prepared by an acknowledged expert who also delivered a summary of his paper and held general discussions with participants to arrive at the best possible listing of research needs, priorities and associated research costs.

Finally, a research document was written for all five areas and a Summary Report was issued by San Pietro.[3]

A diverse array of research topics was identified and enumerated at the Kiawah Island conference and in the various documents generated during the further evolution of the concept. The topics span from basic to applied research to demonstrations or field projects. Many disciplines and fields of science and engineering are involved -- more than enough for everyone's interest and opportunity. The spectrum of some of the particular research topics identified is presented in Table III.

The next step in the process of developing a coherent research plan is to arrive at a realistic research portfolio consistent with the objectives of a particular program. Not all of the topics mentioned may be germane to all countries or regions of the world.

We welcome your suggestions for modifying the topics listed to be included in any biosaline research program. Execution of this program will convert the promise of the biosaline concept into reality.

REFERENCES

1. E. Epstein (1977). In, Plant Adaptation to Mineral Stress in Problem Soils, M.J. Wright, Ed. Cornell University Agricultural Experiment Station, New York, pp. 73-82.

2. M. Akbar, A. Shakoor and M.S. Sajjad (1977). In, Genetic Diversity in Plants, M. Muhammed, R. Aksel and R.C. von Borstel, Eds. Plenum Press, New York.

3. Summary Report on International Workshop on Biosaline Research (1977). A. San Pietro, Ed. Indiana University, Bloomington.

4. D.E. Mann (1963). In, Aridity and Man, C. Hodge, Ed. American Association for the Advancement of Science, Washington, D.C.

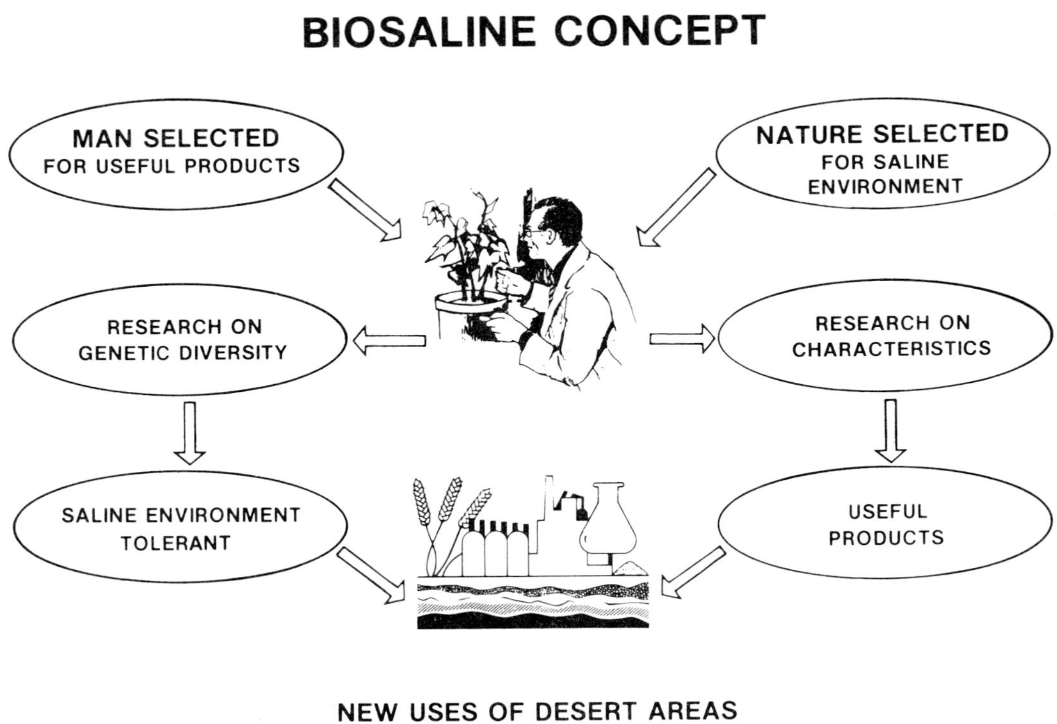

Figure 1. Schematic diagram of the information flow of biosaline research to useful products.

TABLE I. BIOSALINE CONCEPT: GOALS AND OBJECTIVES

GOALS

*Provide alternative options to material needs from renewable resources

*Provide environmentally compatible approaches to an enhanced standard of living.

*Apply modern biological sciences and technologies to derive essential materials in an environmentally harmonious manner, especially from marine and arid land biomass resources

OBJECTIVES

*Material Needs
- Food (protein, carbohydrates)
- Feed (supplements)
- Energy
- Fuels
- Chemicals
- Fertilizers
- Structural materials
- Fibers
- Medicinal compounds

*Environmental Compatibility
- Pollution control and conversion

TABLE II. BIOSALINE RESEARCH AREAS

* Land plants and controlled environment agriculture
* Photosynthetic marine organisms
* Salt-tolerant plants and microorganisms
* Biological waste treatment
* Enzyme technology

TABLE III. SPECIFIC BIOSALINE RESEARCH AREAS

LAND PLANTS AND CONTROLLED ENVIRONMENT AGRICULTURE

* Identify and characterize arid land plants for useful constituents
* Establish effective cultivation and harvesting practices
* Examine biochemical pathways and physiological control points
* Devise more effective and inexpensive controlled environment agriculture (CEA) systems
* Improve plants for CEA compatibility

PHOTOSYNTHETIC MARINE ORGANISMS

* Identify and characterize photosynthetic marine microorganisms and higher plants
* Establish effective large-scale production practices for promising candidates
* Investigate relevant biochemical, physiological, genetic, and environmental factors
* Examine general toxicity and nutritional characteristics of marine organisms
* Establish integrated use of marine resources (by-product development)

SALT-TOLERANT PLANTS AND MICROORGANISMS

* Determine the structure/function relationship of membranes from salt-tolerant organisms
* Select salt-tolerant cereals, legumes and forage crops for agricultural production
* Transfer salt-tolerant genetic traits to salt-sensitive lines
* Select halo-tolerant mycorrhizae for enhanced mineral nutrition of plants
* Enhance or introduce N fixing capability in saline species

cont.

BIOLOGICAL WASTE TREATMENT

* Determine the influence of salinity and temperature on algal growth and bacterial oxidation of wastes
* Devise new procedures utilizing local materials and easily harvested algae
* Develop methods which optimize solar energy utilization, reclaim nitrogen and other resources and minimize loss of water
* Investigate the processing of waste-grown algae
* Develop effective detection and control systems for infectious agents and refractory residuals

ENZYME TECHNOLOGY

* Identify and characterize thermo- and halo-tolerant enzymes
* Develop effective procedures for the stabilization and immobilization of biosaline-derived enzymes and intact cells
* Examine the utility of biosaline microbes for processing advantages (sterilization, energy effectiveness)
* Assess the feasibility of using microbes and/or enzymes to produce industrial chemicals (hydrogen, alcohols) and other material needs
* Explore the use of enzymes to upgrade low quality biosaline-derived foods and feeds

INTERNATIONAL ACTIVITIES IN BIOSALINE RESEARCH: SEVERAL EXAMPLES

ANTHONY SAN PIETRO
Department of Biology, Indiana University, U.S.A.

ABSTRACT

Scientists in many nations are attempting to develop to the maximum the food, energy and chemical potential of (salt-tolerant or halophytic) microorganisms and plants. Clearly, availability of fresh water is a serious limitation to crop productivity in large areas of the world. Several examples of this approach, for example, growth of the halophytic green algae, Dunaliella, seaweed and algae culture in the ocean, higher plant growth under saline irrigation, etc., will be emphasized as prototype investigations in biosaline research.

In 1975, there was convened in the United States an "International Conference on Crop Productivity" which was planned with the recognition that an exploding world population and an increasingly affluent society impose a more acute challenge to agriculture than ever before. At that conference, the internationally acclaimed nutritionist Jean Mayer brought this challenge into clear focus with a simple -- but to me -- rather profound and sobering statement.[1] Based on a projected doubling of the world's population by the turn of the century, he stated, "This means that we have to find in the next twenty five years, food for as many people again as we have been able to develop in the whole history of man 'til now." I believe we are here at this Workshop because we are all painfully aware that this challenge[2] still remains before us essentially undiminished in its enormity. Surely, inadequate food, as well as energy, supplies are continuing global concerns.

The theme for my presentation is, perhaps, best exemplified by the statement *"Man Inhabits This Planet as a Guest of the Plant Kingdom."* Since his beginning, man has depended almost entirely on plant productivity, through the process of photosynthesis, for an abundance of energy in the form of food, fuel and fiber. The reserves of fossil fuels (coal, oil, and natural gas) upon which his wellbeing so critically depends, are the result of photosynthetic conversion of solar energy over geologic time.

Many, if not all, of you know better than I that enhancement of crop productivity is inseparable from the resource base. As we have heard several times during this Workshop, fossil fuel energy, together with arable land and fresh water, are important components of this resource base. Unfortunately all three -- arable land, fresh water and fossil fuel energy - are finite resources. Although in the United States more land can perhaps be put into production than is now being cultivated, much of that land is thought to be marginal and may require substantial inputs of water and fossil fuel energy to be productive. There must therefore be a continuing search for new ideas that could provide alternative feedstocks for materials now petrochemically derived. The "*Biosaline Concept*" as described at this Workshop by Drs. J. Aller and O. Zaborsky, and elsewhere,[3] is an important first step in this direction and should be more fully exploited in the future.

I will not pursue in depth the possibility of higher plant growth under saline conditions since this topic will be considered more adequately by subsequent speakers in this session. I would but recall the successful growth of barley under irrigation with undiluted seawater as reported last year by Drs. E. Epstein and J.D. Norlyn of the University of California at Davis. Also, we have just seen a short film on this work during Dr. G.B. Devey's presentation. They reported[4] that the seawater culture produced grain of satisfactory feed quality and the seed from the undiluted seawater treatment was viable, as demonstrated by germination tests. The three cultivars included in their experiments had a mean yield corresponding to 833 kg/ha. To put this number in perspective, the mean world yield of barley for 1975 was estimated at 1710 kg/has and for the United States at 2370 kg/ha.

Let me now turn to two biosaline studies of lower plants, namely algae. A rather visionary variant of photosynthesis is found in the schemes to redirect the reductive capacity of the photosynthetic process to yield molecular hydrogen--rather than reduced carbon compounds--to serve as a non-polluting, storable and transportable fuel. The absorption of light quanta by the photosynthetic apparatus of chlorophyll-containing plants results in the formation of assimilatory power, i.e., the chemical free energy necessary for the reductive assimilation of carbon dioxide to the level of carbohydrate (Fig. 1). It is possible under certain conditions to modify the end result of the photosynthetic process such that molecular hydrogen is evolved in lieu of carbon dioxide assimilation. Under the stimulation of light, the photosynthetic process can provide electrons released from water at a reducing potential more negative than the hydrogen electrode. It is possible, therefore, to couple this reducing potential with a suitable hydrogenase -- hydrogen activating enzyme -- to effect the reduction of protons to form hdyrogen gas.

The first published report of photo-hydrogen evolution appeared about five years ago when Kaplan and Kamen and their colleagues reported hydrogen evolution with a reconstituted *in vitro* system.[5]

In the absence of an oxygen trapping system, the rate of hydrogen evolution was extremely low due to the: (a) autooxidizability of the reduced ferredoxin (FD in Fig. 1); and (b) inactivation of the hydrogenase enzyme. Although the rate of hydrogen evolution was increased about sevenfold by inclusion of the oxygen trapping system, the duration of hydrogen evolution was still only about 30 minutes due to inactivation of the chloroplast system. Subsequent studies by investigators in the United States and in Europe have extended the duration to about 3 hours and increased the rate of the reaction to about 10% of the capacity of the photosynthetic electron transport pathway.[6] While these studies indicate biophotolysis under laboratory conditions, there may be severe limitations to application on a large scale in light of: (a) the requirement for fresh water as the source of hydrogen; (b) the instability in light of the *in vitro* chloroplast system; and (c) the necessity to continually remove oxygen by means of a costly oxygen-trapping system.

Dr. A. Mitsui of the University of Miami has taken the alternative approach. He has isolated from the natural environment a number of marine algae and photosynthetic bacteria which exhibit hydrogen evolution in the light. Of particular interest is a blue-green alga which appears capable of sustained photohydrogen evolution at a rather high rate for about 4 days.[7]

This blue green alga evolves significant quantities of hydrogen in the light when maintained under anaerobic conditions and in the absence of a nitrogen source (Table I).[8] This is a very exciting result and the system should be intensly studied in the future. While extremely promising, it

Table I - Rates of Hydrogen Evolution

Author	Source	Rate[1]
Rao et al (6)	Higher Plants (Chloroplasts)	50[2]
Mittsui (7)	Marine Blue Green Alga (Whole Cells)	230[2] (4.2[3])
Weissman & Benemann (8)	*Anabaena Cylindrica* (Whole Cells)	16[3] (120[2])

[1]Values in parentheses are calculated values and provided only for purposes of comparison.
[2] µmoles hydrogen evolved/mg Chl.hr.
[3] µmoles hydrogen evolved/mg dry wt.hr.

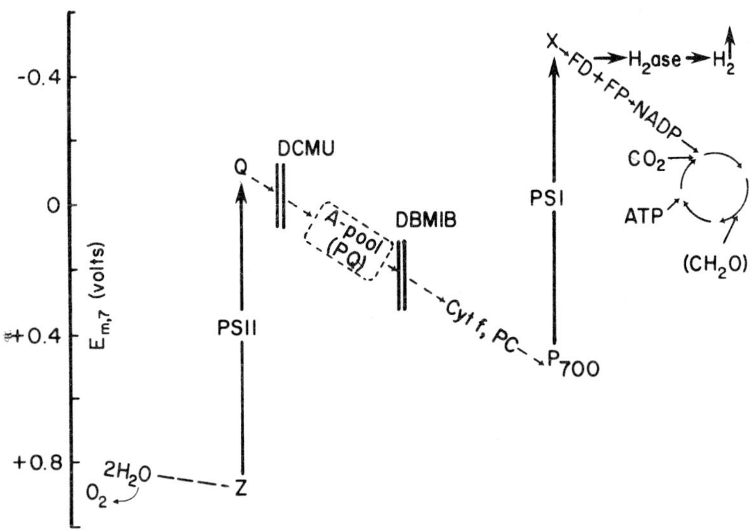

Figure 1: Schematic Representation of the Photosynthetic Electron Transport System.

must be determined if:(a) the system will produce hydrogen and oxygen in the proper stoichiometric ratio; and (b) the system will operate continually for periods beyond 4 days and with high efficiency.

I would like now to mention briefly a biological process for glycerol production under serious consideration by Drs. A. Ben-Amotz and M. Avron at the Weizmann Institute in Israel.[9] The organism is the unicellular, halophylic green alga *Dunaliella*, which lacks a rigid cell wall so that the cell is enclosed by a cytoplasmic membrane only. *Dunaliella* species are marine organisms and the organism used in these studies was isolated from the Dead Sea.

One of the most remarkable characteristics of *Dunaliella* is its ability to survive and withstand changes of the osmotic concentration of sea water. To do so, the alga produces photosynthetically from carbon dioxide and stores large quantities of glycerol. Further, the intracellular concentration of glycerol is iso-osmotic with the medium concentration of NaCl and no leakage of glycerol occurs at NaCl concentrations greater than 0.6M NaCl. Under natural growth conditions, the cell membrane is not permeable to glycerol or other small molecules.

Maximal growth of *Dunaliella* under optimal laboratory conditions is 5×10^6 cells/ml (5×10^9 cells/liter). The alga may accumulate up to 40% glycerol (w/v) per cell volume which translates to 0.4 kg glycerol per liter of cell volume. With a doubling time every 24 hours and daily harvest of the culture, one cubic meter produces about 0.5 kg glycerol per day. A pond some 20 cm in depth may provide potentially about 50 gm of glycerol per square meter per day.

An elementary analysis of the economics of large-scale culture indicates that the cost of bio-production of glycerol may be about 50% of the present cost of glycerol. Further, a major by-product of the proposed process is dry algae of high nutritive value. The algae powder remaining after glycerol extraction is about 60% protein before pigment extraction and about 70% protein following pigment extraction.

I believe this is the direction we should pursue in future with regard to large scale culture of algae. Profitability from biomass production of algae in large scale outdoor facilities for a high protein product will be improved when the algal material is first used for production of high value natural products.

Before closing, let me return to the desert environment and a point made by Prof. D. H. Baepler in his presentation, namely growth of plants for petroleum products. Prof. M. Calvin has for the past few years championed the idea of "petroleum plantations" and growing "green factories" for the production of hydrocarbonlike materials.[10] The plant under study is *Euphorbia lathyris*, an annual that grows to some 4 feet in height in 7 months. By proper seed selection, a yield of about 20 barrels of oil per acre could be realized. Furthermore, these hydrocarbon producing plants can be grown on land that is today non-productive. There is a great deal of land of this type in the U.S. and other areas of the world.

To quote Prof. Calvin: "This type of exploration gives two practical approaches to renewable resources. First, we can use the hydrocarbon as it comes from the plant itself as a crude oil, refine it, rescue the sterols which it contains, crack the rest of the compounds to ethylene, propylene, etc. and then reconstruct the desired chemicals from these products. This particular approach can be developed immediately. Later, we may learn how the molecular weight of the hydrocarbon is controlled and manipulate the plant to construct materials of the desired molecular weight for whatever purpose is desired. This approach will be longer and more complex and will use the plant as the collecting and constructing vehicle."

In summary, we have only begun to explore the applicability of the *"Biosaline Concept"* and the potential long-term benefits which may derive therefrom. As Dr. G. B. Devey indicated earlier, a two year study program is under consideration presently to define more completely a future biosaline research and development program.

REFERENCES

1. J. Mayer, in Crop Productivity: Research Imperatives, A.W.A. Brown, T.C. Byerly, M. Gibbs, A. San Pietro, Eds. Waverly Press, 1975, pp. 97-108.

2. In a comparative sense, Dr. Geni Graetz at this Workshop puts this challenge in mathematical terms when he defined a Standard of Living Index (SLI) as the ratio of the summation of resources divided by the summation of the product of population times demand. Granted that resources may not increase dramatically in the future, the SLI will decrease as population and demand increases. To maintain the SLI requires that population and/or demand decrease or that resources increase.

3. J.C. Aller, L.G. Mayfield and O. Zaborsky, as reported by G.B. Devey, in Summary Report, International Workshop on Biosaline Research, A. San Pietro, Ed., 1977, pp. 13-14; also, The Biosaline Concept: An Approach to an Underexploited Resource, A. Hollaender, Ed., Plenum Press (In press).

4. E. Epstein and J.D. Norlyn, Science 197:249-251.

5. J.R. Benemann, J.A. Berenson, N.O. Kaplan and M. Kamen, Proc. Natl. Acad. Sci., U.S. 70:2317-2320 (1973).

6. K.K. Rao, I.N. Gogotov and D.O. Hall, Biochimie 60:291-296 (1978).

7. A. Mitsui and S. Kumazawa, in Biological Solar Energy Conversion, A. Mitsui, S. Miyachi, A. San Pietro and S. Tamura, Eds., Academic Press, 1977, pp. 23-51.

8. J.C. Weissman and J.R. Benemann, Applied and Environmental Microbiology 33:123-131.

9. A. Ben-Amotz and M. Avron, Personal communication.

10. M. Calvin, in Bioenergy: Energy from Living Systems, Gottlieb Duttweiler Institute, Zurich, 1978, pp. 89-117.

SALINE AGRICULTURE UNDER DESERT CONDITIONS

RAFIQ AHMAD
Professor of Botany, University of Karachi; Principal Investigator, Brackish Water Irrigation Research Project, Pakistan Science Foundation

ZAIBUN NISA ABDULLAH
Research Officer, Brackish Water Irrigation Research Project, Pakistan Science Foundation

ABSTRACT

Plants were raised from seeds on desert sand and subjected to the "Survival Test" by irrigating them with different dilutions of chemically amended sea water. Another lot of seedlings was pre-conditioned for salt tolerance by irrigating with gradually increasing concentrations of chemically amended sea water reaching up to the maximum permissible limit determined through the above mentioned experiments. These seedlings were later transplanted in drum pots filled with sand having a basal outlet for drainage. They were irrigated with the same concentration of amended sea water which they were receiving before transplantation. Plants such as wheat, taramera, maize, cotton, sugarcane and sugar beet were grown to maturity on sandy soil using various concentrations of sea water for irrigation. Plants showed luxuriant growth up to certain levels of salinity, while at higher concentrations reproductive growth was more adversely affected than vegetative growth. This technique was found quite useful for saline agriculture where vegetative parts are used as fodder or for human consumption. However, the reduction in reproductive growth may still be within compromisable limits in some other plants.

INTRODUCTION

Deserts constitute a greater part of arid zones of the world. They have been considered useless for cultivation, but modern plant physiology approaches have revealed their agricultural potentialities. However, conventional practices are being replaced by the special technology developed for this purpose. Recent developments in hydrology show the presence of ground water resources in the deserts which could be exploited for irrigation. Though such subsoil water often contains salts beyond the maximum permissible limit recommended for plant growth, the knowhow of chemical amendments and growth through partial root contact techniques have made it possible to raise salt tolerant crops on sandy deserts. Similarly, if underground water is exploited at a suitable distance from the sea coast or sandy beach, the sandy strata is expected to filter out some of the salts and the water could be utilized for irrigation. In the triangle of plant, soil and water relationship, if an area containing coarse textured sand is selected for growing plants, saline irrigation water with chemical amendments becomes a special kind of culture solution and the whole plantation gives an idea of plant growth in an artificial medium.

Low and medium salinity water (EC 0-250 and 250-750 mmhos/cm) can be used for irrigation on most of the soils where a certain amount of leaching occurs. High salinity water, EC 750-2250 mmhos/cm, can be used for irrigation on the soils with adequate drainage. Water having EC 2250-5000 mmhos/cm is classified as very highly saline and is not suitable for irrigation under ordinary conditions.[24,29] However, it may be used on most permeable soils for growing salt tolerant plants. Sandy deserts provide an excellent strata for quick percolation of water and chemical amendments further improve the conditions for saline agriculture.

The critical level of salts in a soil solution which result in reduction in growth varies at different places depending upon external environment, quality of saline water and porosity of the soil substrate.[29] Durand[10] has rightly proposed a revised classification of irrigation water based on soil type with reference to salt tolerant plants. The addition of proper chemical amendments antagonistic to high sodium would further increase the maximum permissible limit under the above mentioned conditions.[6,9] Irrigation with saline water having NaCl concentration 1-1.5 gr/l did not cause any difficulty in growing cotton and maize on well drained soils of Beni Amir, Morocco.[5] Boyko,[6] author of the "partial root contact theory," has used extremely saline water for growing salt tolerant plants under desert conditions. His group has used salt concentrations up to 22,000 mg/l in sea water experiments. One can always find a maximum permissible concentration of salts in irrigation water which is capable of giving a compromised yield in sandy deserts.[3] Some very informative books have been published through FAO/UNESCO[24,27,28] covering world wide information on this subject. The present paper deals with some of the work undertaken for growing plants on coastal sands of Pakistan using various dilutions of chemically amended water from the Indian Ocean for irrigation.

CHEMICAL AMENDMENTS

A formula was worked out for the amendments to provide the maximum possible balance in ionic ratio, to bring down the pH and to supply essential minerals in various dilutions of sea water used for irrigation Na^+ ions antagonize other cations of amended salts and thus reduce sodium toxicity for plant growth. Salts capable of furnishing K^+ and Ca^{++} ions in the solutions have beneficial effects. Appropriate amounts of $Ca(NO_3)_2 4H_2O$, KH_2PO_4, H_3PO_4, Fe-EDTA and micronutrients were added in various dilutions of sea water to improve the quality for irrigation.[3] Concentrations of major ions present in sea water decreased, with the factor of its dilutions, whereas the concentrations of amendments in each case remained the same. (Note: "Sea water" henceforward in the text means "chemically amended sea water.")

Ionic Composition of Major Ions in Sea Water and the Amendment			
Composition of major ions in sea water (ppm)[25]			
Proportion of Cations		Proportion of Anions	
Na^+	12,000	Cl^-	20,000
K^+	42	PO_4^-	nil
Ca^{++}	59	NO_3^-	nil
Mg^{++}	3000	SO_4^{--}	3000
Composition of major ions in amendments (ppm)			
Proportion of Cations		Proportion of Anions	
Na^+	nil	Cl^-	nil
K^+	200	PO_4^-	200
Ca^{++}	600	NO_3^-	400
Mg^{++}	nil	SO_4^{--}	nil

CULTURAL PRACTICES

Growth studies on wheat, taramera, maize, cotton, sugarcane and sugar beet are included in the present paper. Seeds were germinated in plastic bags containing about 500 gr coastal sand and irrigated with half strength Hoagland[13] solution. Perforations were made at the bottom of the bags to allow rapid drainage of the irrigation medium. Only four seedlings were left in each bag. Further treatment started when they reached the three leaf stage. Sets of sugarcane were planted in plastic bags and further treatment was started when the sprouts were about four inches in height.

1. Survival Test

This test was designed to find out the maximum limit of sea water concentration in irrigation for the survival of seedlings. Plastic bags containing seedlings were irrigated by ten different dilutions of sea water starting from 10% up to full oceanic strength. Three replicates were kept for each treatment. Effects of different dilutions of sea water irrigation were recorded on the growth after every 24 hours for a week.

2. Seedling Treatment

Pre-conditioning for salt tolerance in seedlings was started at the three leaf stage. At first they were irrigated with 2% sea water but subsequent irrigations were given after every two days with gradually increasing concentrations of sea water depending upon the nature of experiment. Seedlings reaching various concentrations of sea water through gradual increments (see Growth Studies, below) were transplanted in drum pots where they were maintained by the same concentration which they were receiving at the time of transplantation.

3. Drum Pot Culture

This setup was based on the technique of Boyko.[6] Drum pots were filled with 300 kg of coastal sand. They were fitted with a basal outlet to ensure rapid drainage of saline water. The amount of water retained at field capacity was found to be 44 liters. Over-irrigation was practiced to avoid salt accumulation in the root zone. Seedlings pre-conditioned for salt tolerance and being irrigated with various concentrations of sea water were transplanted in drum pots and irrigated with the same concentrations which they were receiving at the time of transplantation. Plants were raised up to maturity and various physical and biochemical parameters of the growth were studied.

4. Growth Studies

The concentration of sea water for irrigation of wheat and taramera (*Eruca sativa*) was increased gradually and the drum pots were kept aside when they reached 20, 40, 50 and 60% sea water respectively. Since the upper limit of this range was found to be unproductive it was lowered to 3, 9, 17 and 25% for cotton, maize and sugarcane. The upper limit of sea water in irrigation for sugar beet was increased up to 50%, as it is reported to be a salt tolerant plant.[26] There was scanty rain fall during the course of this investigation which amounts to the naturally occurring supplemental irrigation beneficial for leaching out salts accumulated in the substratum.[16]

Shoots were harvested at their exponential phase of growth and fresh weights were taken. They were further oven dried for determination of dry weights. Some plants were harvested from time to time for various analyses.

Dry powders of roots and shoots were prepared separately and used for acid digest tests. Analysis of sodium and potassium was performed by flame photometer.[26]

OBSERVATION AND RESULT

Exposure of seeds to the saline during germination appears to be most critical in the life of the plant. Our earlier experiments showed that germination was totally inhibited when wheat seeds were irrigated with 30%, and taramera with 40% amended sea water.[3] The percentage of germination when seeds were irrigated with 20% amended sea water was about 50% in both cases. The effect of salinity in lowering germination is reported in the literature.[1] Hence it was considered preferable to germinate the seeds under normal conditions and transplant the seedlings before proceeding with saline irrigation.

Survival tests proved useful for indicating the range of salt tolerance at the seedling stage. Salinity of the growth medium no doubt has different effects at different stages in the life of the plant but its effect at the seedling stage reveals the probability of establishing vegetative growth under saline environment. Irrigation with 10% amended sea water did not result in any adverse effects under our experimental conditions, but 20% resulted in inhibition of growth

(Fig. 1). Seedlings irrigated with 30% amended sea water showed severe inhibition after four days and later died. Seedlings of most of the plants mentioned in these experiments behaved in the same way in this respect with the exception of sugar beet, where irrigation with even 40% sea water did not show any deleterious effects (Fig. 2). Seedlings of the latter irrigated with 50 and 60% amended sea water showed slight inhibition in growth, but higher concentrations proved lethal.

Seedlings of various crops were pre-conditioned for salt tolerance, transplanted in drum pots and then irrigated with various concentrations of sea water. Their growth responses are given below.

Wheat

China crocks of 15 kg capacity sand and a basal outlet for drainage were used in these experiments instead of drum pots. The rest of the procedure was as described above. The data presented in Table 1 gives an overall idea of the effect of irrigation with various dilutions of sea water on the vegetative as well as reproductive phases of growth in wheat. Plants irrigated with 20% sea water showed improvement in growth in all parameters except in height. In the rest of the irrigation levels, there appears a decrease in various growth parameters along with increased salinity of the irrigation medium, though the values are still greater than those of the controls for certain parameters. It is interesting to note that the increase in chlorophyll corresponds to the increase in sugar content even up to the plants irrigated with 40% dilution of sea water. Increase in the number of tillers at lower salinities is probably due to the imbalance in growth hormones.[20] At higher levels, however, the salinity is decisive for all parameters of growth.

Taramera *(Eruca sativa)*

This plant is capable of growing in the moderately saline soils of Pakistan. The seeds yield a poor quality cooking oil. The same experimental conditions and cultural practices were carried out as with wheat. Plants irrigated with 20% sea water showed better growth than controls on all parameters (Table 2), beyond which there appears a reduction in vegetative and reproductive yield. Sugar content of the leaves remained higher than controls in all treatments though the amount of chlorophyll started decreasing in higher salinities. Reduction in size and number of seeds beyond 20% sea water irrigation level was similar to that observed in wheat except

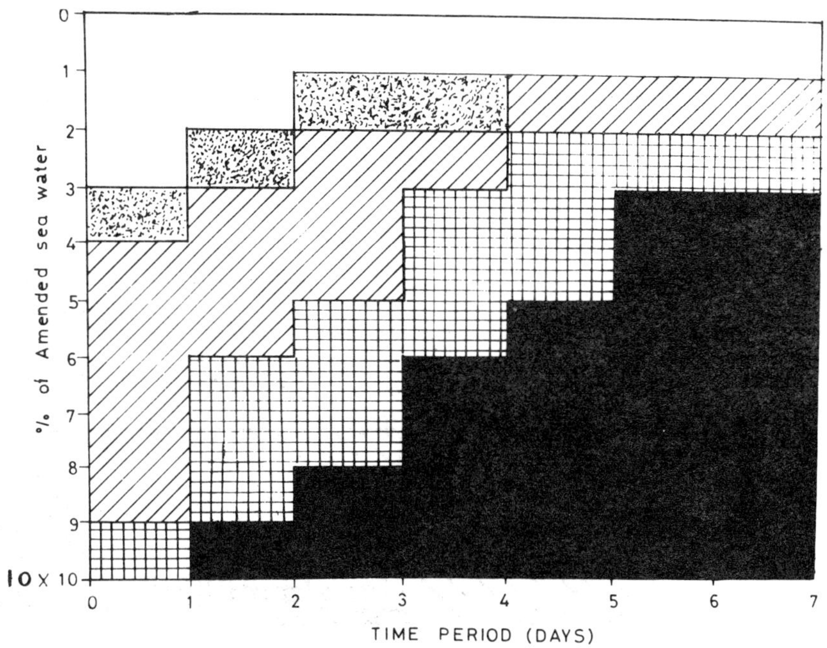

Fig. 1: Survival test of seedling irrigated with various dilutions of amended sea water.

Note: Seedlings of wheat, taramera, maize and cotton were used in this experiment seedlings remained.

as food as control.

Leaves loose their turgidity and margins dry up.

Leaves become pale, plants show wilting.

Plants show permanent wilting and bleaching of chlorophyll.

Plants die due to complete dehydration and disintegration of cellular integritry.

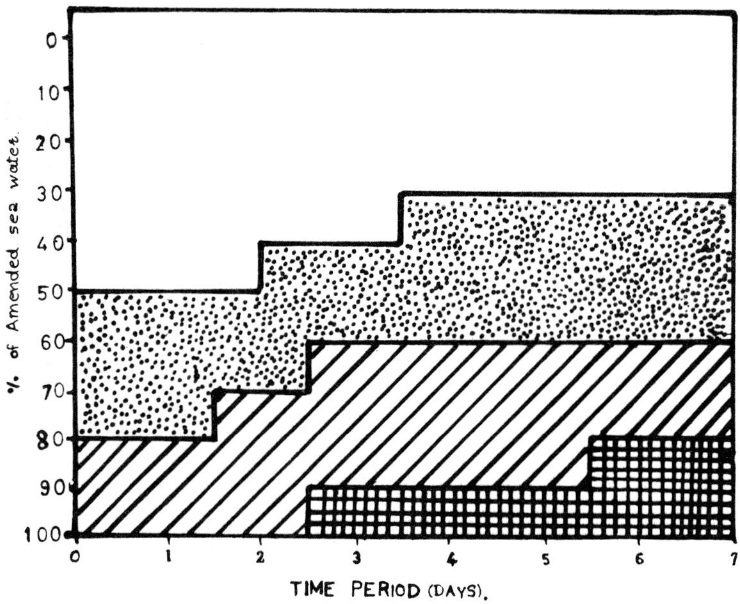

Fig. 2: Survival test of sugar beet seedlings irrigated with various dilutions of amended sea water.

Note:

Plants as good as control.

Leaves loose their turgidity and margins dry up.

Leaves become pale, plants show wilting.

Plants show permanent wilting and bleaching of chlorophyll.

TABLE I. GROWTH RESPONSE OF WHEAT TO SEA WATER IRRIGATION

Parameters	Control	Dilutions of Amended Sea Water Applied in Gradual Increments				
		20%	30%	40%	50%	60%
Height (cm)	72.0	68.5	59.7	56.3	42.8	41.8
Fresh weight of shoot (g)	73.3	75.1	47.2	11.4	7.5	3.0
Chlorophyll content in leaves (mg/g dry wt)	38.5	58.2	50.1	49.3	32.9	28.8
Sugar content in leaves (mg/m dry wt)	3.6	10.2	7.6	4.2	4.0	3.8
Number of tillers/plant	7	16	14	11	8	3
Number of ears/plant	280	400	240	24	0	0
Weight of seeds/plant (g)	12.6	16.8	9.84	0.672	0	0

TABLE II. GROWTH RESPONSE OF TARAMERA (ERUCA SATIVA) TO SEA WATER IRRIGATION

		Dilutions of Amended Sea Water Applied in Gradual Increments				
Paramters	Control	20%	30%	40%	50%	60%
Height (cm)	74.5	78.2	61.7	52.5	40.3	28.3
Fresh weight of shoot (g)	55.9	65.9	47.3	33.6	28.4	15.5
Chlorophyll content in leaves (mg/g dry wt)	90.4	95.6	80.3	70.9	38.5	31.5
Sugar content in leaves (mg/g dry wt)	3.12	12.9	9.6	5.4	4.9	4.08
Number of seeds/ plant	411	506	230	160	70	0
Weight of seeds/ plant (g)	7.316	8.804	3.91	2.688	1.05	0

that taramera managed to produce some seeds at 50% sea water irrigation. The deleterious effect of higher salinities in crossing the barrier from vegetative to the reproductive phase as noted in wheat was evident also in taramera.

Note: Since the higher dilutions of sea water were non-productive in the above experiments, the maximum salinity in irrigation was reduced to only 25% in some other plants.

Maize

Plants irrigated with 17% sea water had 9.8% more fresh weight of the shoot than that of controls (Table 3), which implies prospects of growing maize by saline agriculture as a fodder crop. There was an increase in the amount of total chlorophyll of the leaves corresponding to the increase of photosynthates (sugars) up to the maximum salinity level of these experiments. The harmful effect of salinity at the reproductive stage is evident in this crop as well, in the reduction of the number of cobs per plant and in the number and weight of seeds per cob. It is interesting to note that the weight and size of individual seeds increased by 15.4% at higher salinity levels (Fig. 3). However, due to the reduction in size of cob and number of seeds, the overall of the plant was not increased. Reduction in reproductive yield is proportional to the salinity level of irrigation water, whereas there is some inhibition in the vegetative growth only at the highest salinity level (Fig. 4).

Cotton

The vegetative growth of cotton decreased at all salinity levels except in those plants irrigated with 3% dilution of sea water (Table 4). Increase in the chlorophyll content of the leaves, coupled with increase in photosynthates, is in accordance with the increase of sea water in the irrigation medium. Cotton showed the same trends as maize in this respect. Reproductive growth, measured in terms of bolls per plant, number of seeds per boll, weight of lint and delinted seeds per boll, shows a decrease corresponding to the increase of salinity of irrigation medium. It is interesting to note that total weight of lint per boll shows a slight increase over controls in the plants irrigated with 25% sea water, although reduction in number of bolls per plant reduces the overall yield on a per-plant basis. The reproductive growth appears to be more susceptible to saline irrigation than vegetative growth (Fig. 5).

TABLE III. GROWTH REPONSE OF MAIZE TO SEA WATER IRRIGATION

Parameters	Control	Dilutions of Amended Sea Water Applied in Gradual Increments			
		3%	9%	17%	25%
Height (cm)	119.25	133.50	120.75	119.25	115.25
Fresh weight of shoot (g)	227.00	282.05	276.02	249.04	184.02
Chlorophyll content in leaves (mg/g dry wt)	69.04	73.05	78.02	82.06	84.01
Sugar content in leaves (mg/g dry wt)	9.14	10.23	12.31	13.73	15.31
Number of cobs/ plant	8	7	7	6	5
Number of seeds/cob	321	347	304	240	173
Weight of 100 seeds,g	26.6	24.7	24.3	26.0	30.7
Weight of seeds(g)/ plant	682.09	599.90	517.09	374.40	356.55

Fig. 3: Effect of amended sea water irrigation on the size of maize cob.

1. Control
2. Irrigated with 3% amended sea water.
3. Irrigated with 9% amended sea water.
4. Irrigated with 17% amended sea water.
5. Irrigated with 25% amended sea water.

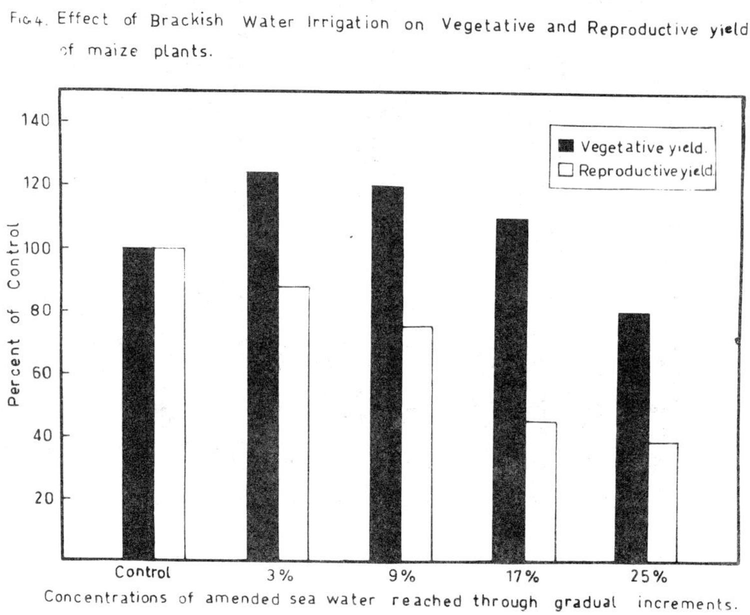

Fig. 4: Effect of Brackish Water Irrigation on Vegetative and Reproductive yield of maize plants.

TABLE IV. GROWTH RESPONSE OF COTTON TO SEA WATER IRRIGATION

Parameters	Control	Dilutions of Amended Sea Water Applied Gradual Increments			
		3%	9%	17%	25%
Height (cm)	71.75	71.75	63.25	60.75	57.00
Fresh weight of shoot (g)	144.00	154.10	135.70	100.70	82.00
Chlorophyll content in leaves (mg/g dry wt)	83.30	88.70	91.50	94.30	93.50
Sugar content in leaves (mg/g dry wt)	10.71	9.89	10.23	11.32	12.91
Number bolls/plant	28	24	21	19	16
Number seeds/boll	28	28	23	23	22
Weight of lint/boll,g	1.04	0.51	0.65	0.72	1.06
Weight of seeds/boll,g	2.33	2.02	1.63	1.25	0.97
Weight of lint/plant,g	29.15	12.31	13.67	13.59	16.96
Weight of seeds/plant, g	65.15	48.45	34.34	23.73	15.38

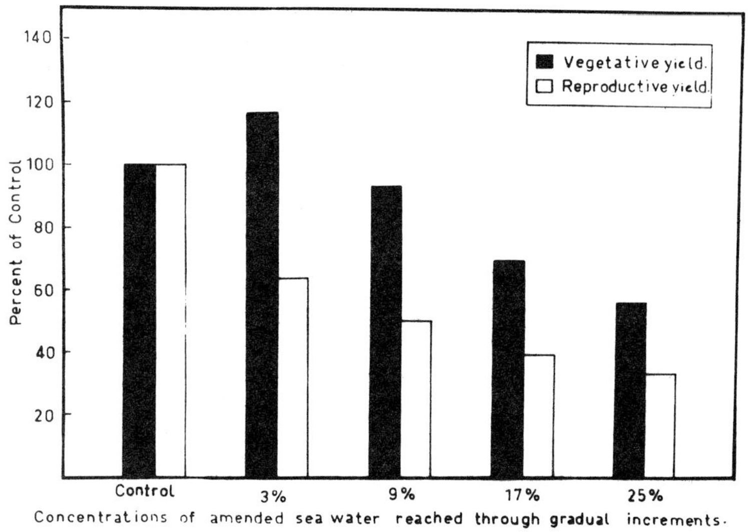

Fig. 5: Effect of Brackish Water Irrigation on Vegetative and Reproductive yield of cotton plants.

Sugarcane

Though increasing concentrations of sea water irrigation proved to be deleterious for the height and diameter of sugarcane (Table 5), the biomass of the shoot was more than that of the controls up to 17% sea water. The amount of chlorophyll in the leaves increased constantly up to 17% sea water but then decreased though the values remained more than in the controls. Sugar stored in the stem decreased with increased sea water percentage.

Sugar beet

As sugar beet is reported to be salt tolerant,[26] sea water irrigation was increased up to 50% (Table 6). The amount of foliage per plant was greater than that of the control even at the maximum concentration of sea water. The fresh weight of the shoot, per plant, in the first three concentrations of sea water was increased by 141.1, 92.5, and 74.4%, respectively, with a corresponding increase in the fresh weight of the root tuber by 33.7, 14.1 and 12%, respectively. Increased amounts of chlorophyll up to the 40% level of sea water corresponds with the increase in sugar content of the root tuber.

Intracellular Na^+ and K^+ accumulation

Data presented in Table 7 shows that plants differ in their response to Na^+ accumulation. They may be classified as Na^+-accumulating and Na^+-excluding groups.[19] It is interesting to note that K^+ absorption continues even in the presence of very high Na^+ in the irrigation medium, though the amount of its accumulation differs in various plants. The chemical amendments described earlier helped to improve this situation, which is an encouragment for inducing salt tolerance.

DISCUSSION

Chemical amendments in various dilutions of sea water proved to be useful in balancing the ionic ratio, lowering pH and providing essential minerals for plant growth. Cutler and Rains[8] have shown that osmotic adjustments created during the seedling stage in cotton plants by exposing them to various water stressed conditions have been quite helpful for the establishment of plants in arid environments. Our data show that pre-conditioning seedlings by irrigating them with

TABLE V. GROWTH RESPONSE OF SUGARCANE TO SEA WATER IRRIGATION

Parameters	Control	Dilutions of Amended Sea Water Applied in Gradual Increments			
		3%	9%	17%	25%
Height (cm)	279.0	257.75	239.0	230.25	145.25
Fresh weight of shoot (g)	227.0	282.5	276.2	249.4	184.2
Diameter at the base (cm)	8.5	10.3	7.2	6.5	6.4
Chlorophyll content in leaves (mg/g dry wt)	70.8	77.6	80.1	86.4	79.0
Sugar content in stem (mg/g dry wt)	459.0	354.0	378.0	323.0	350.0

TABLE VI. GROWTH RESPONSE OF SUGAR BEET TO SEA WATER IRRIGATION

Parameters	Control	Dilutions of Amended Sea Water Applied in Gradual Increments				
		10%	20%	30%	40%	50%
Number of leaves/ plant	12	20	18	16	16	14
Fresh weight of shoot (g)/plant	61.3	147.79	118.03	106.91	100.47	86.23
Fresh weight of root tuber (g)/ plant	66.25	88.60	75.59	74.17	57.26	48.78
Total chlorophyll (mg/g dry wt)	87.41	108.03	101.72	95.23	90.11	86.23
Total sugar content of root tuber (mg/g dry wt)	323.42	478.32	450.17	391.62	347.16	275.11

TABLE VII. SEA WATER IRRIGATION INDUCED CHANGES IN INTRACELLULAR IONIC COMPOSITION OF CROP PLANTS (mg/g dry weight)

TEST PLANT	CONTROL		20%		30%		40%		50%		60%	
WHEAT	Na^+	K^+	Na^+	K^+	Na^+	K^+	Na^+	K^+	Na^+	K^+	Na^+	K^+
Root	4.1	11.2	7.40	9.4	10.60	10.70	14.70	10.20	15.30	11.90	18.1	10.5
Shoot	2.3	22.3	3.00	21.6	5.60	23.40	7.90	20.80	8.60	21.70	9.3	22.2
TARAMERA												
Root	5.0	12.7	7.30	14.2	10.00	11.60	11.30	13.70	12.60	12.90	14.7	13.9
Shoot	3.3	20.3	4.70	19.6	5.40	21.30	6.70	20.60	9.30	20.10	8.1	21.0
MAIZE			3%		9%		17%		25%			
Root	3.6	14.7	4.30	13.9	5.80	13.30	7.20	12.60	8.78	11.10	-	-
Shoot	9.8	28.9	10.90	29.1	12.60	27.80	13.30	26.90	16.71	25.70	-	-
COTTON												
Root	5.2	16.3	6.90	10.50	10.00	8.00	13.30	10.40	15.50	8.00	-	-
Shoot	12.4	31.4	28.30	21.90	25.10	18.50	42.60	17.40	37.40	22.40	-	-
SUGARCANE												
Root	6.4	13.7	7.40	12.90	7.99	11.78	9.12	13.11	11.23	10.70	-	-
Shoot	10.7	35.9	11.32	36.10	12.73	34.70	13.81	33.70	15.23	30.70	-	-
SUGARBEET			10%		20%		30%		40%		50%	
Root	9.9	19.7	17.40	29.40	29.80	38.26	32.16	47.30	36.70	63.20	41.8	72.8
Shoot	27.3	39.5	37.30	49.70	50.92	59.90	57.15	63.50	76.50	81.40	110.2	110.2

gradually increasing concentrations of sea water has been beneficial for growth up to a certain extent in saline conditions (Tables 1-6). Our earlier experiments of "shock treatment," in which plants were irrigated directly with higher concentrations of sea water show that this method made survival impossible, whereas if some of these levels of salinity were reached through gradual increments, the plants were able to flourish.[4] Further increments of sea water irrigation however were fatal, even if introduced gradually.

Plants once established by irrigating with gradually increased concentrations of sea water show that vegetative growth is less effected than reproductive growth (Figs. 4,5). In certain cases,[14] within a suitable range of stress, the vegetative growth was even better than reproductive growth. In the present investigations, maize and sugar beet showed better vegetative growth than the controls (Tables 3, 6). It therefore appears that only these plants can be profitably grown with brackish water irrigation, where the vegetative portions are used either as fodder for animals or for human consumption.

The plants investigated in the present work are very clearly differentiated into two groups on the basis of the data obtained for Na/K ratio (Figs. 6,7). Plants in which this ratio is higher in the root than the shoot (taramera, wheat, maize and sugarcane) may be considered sodium excluding plants, while where the ratio is higher in the shoot than the root (cotton, sugar beet), the plants may be considered sodium accumulating, a finding reported by other workers.[17] It is also noticed that the uptake of K up to a certain extent is counter to its own absorption mechanism[11] which is later partly inhibited by the presence of high sodium in the substrate. This confirms the occurrence of a dual absorption mechanism[22] during mineral uptake.

Increased chlorophyll content of the leaves at lower salinities of substrate is reported in the literature,[7] while higher salinity levels cause serious damage to the photosynthetic apparatus.[15] However, gradual increases of sea water in the irrigation medium extended the toxic limit of salinity. It is evident from the data (Tables 1-6) that production of photosynthates (sugars) is in good accordance with the increase of chlorophyll content. Both of these factors are operative for the benefit of plant growth[12] within certain ranges of higher salinities. The energy-yielding primary products thus produced supply additional energy for controlling ionic fluxes during mineral uptake under saline environments.[7]

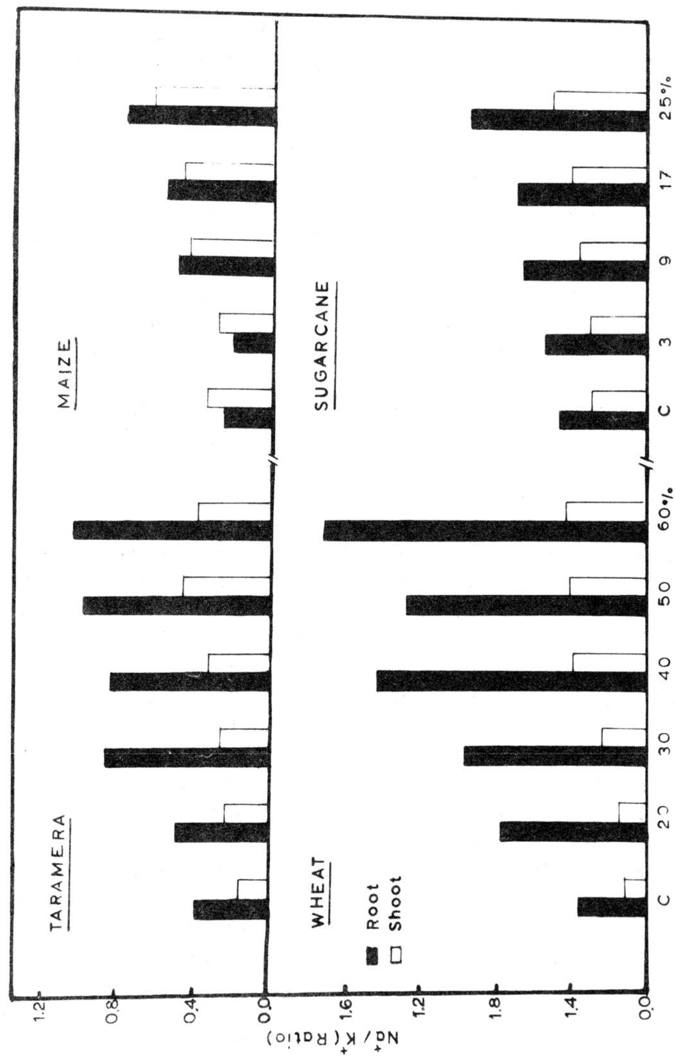

Fig. 6: Concentrations of Amended Sea Water

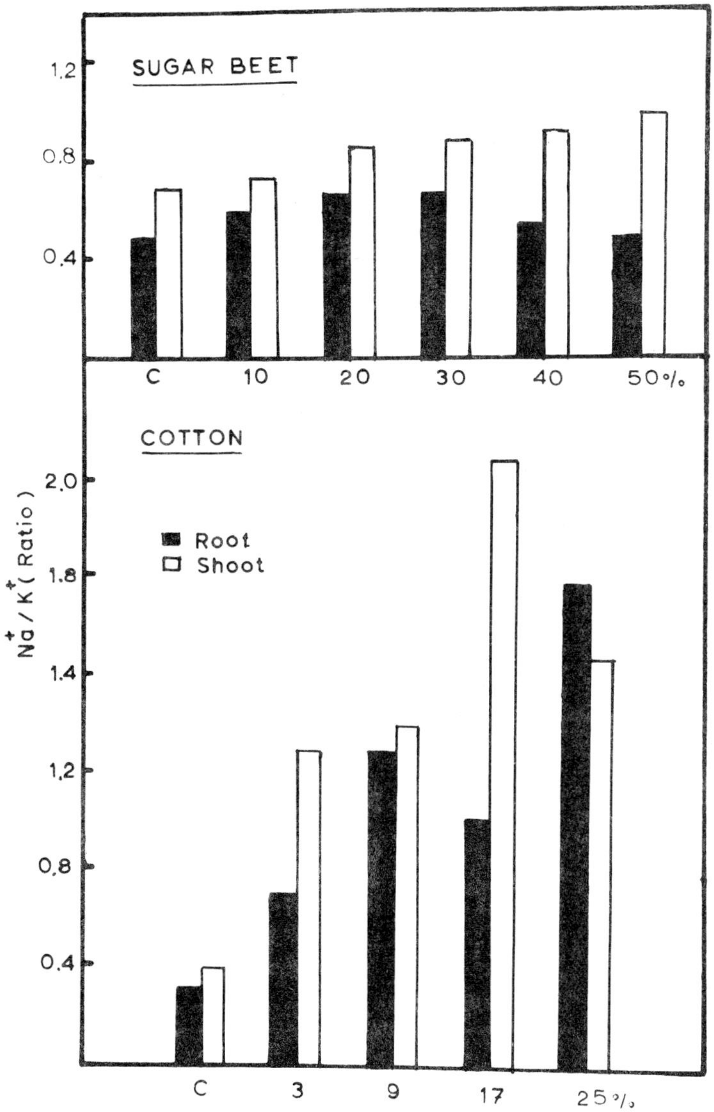

Fig. 7: Concentrations of Amended Sea Water

Data presented in Tables 1-6 further shows that reproductive growth is more severely affected than vegetative growth in response to saline water irrigation. The adverse effects of salinity on reproductive phases are due to reduction in pollen viability,[1] germination failure and fertilization inability. The barrier between the two growth phases under saline conditions could be crossed only by incorporating salt tolerance in the genetic makeup of the plant. The experiments of Rush and Epstein[22] in which salt tolerant, poor yielding wild species of barley and tomato with salt susceptible, good-yielding cultivated varieties, should be mentioned in this regard.

Our approach, based on pre-conditioning seedlings for salt tolerance, balancing the ionic ratio of sea water for irrigation, and devising special cultivation techniques, is useful in saline agriculture where vegetative parts of plants are used as fodder or for human consumption. Crops such as maize and sugar beet are quite promising in this respect. Reduction in reproductive yield may still be within compromisable limits in some other plants.

ACKNOWLEDGMENT

The authors are grateful to Pakistan Science Foundation for providing necessary funds to undertake this project. The facilities extended by the University of Karachi are also highly acknowledged.

REFERENCES

1. Z. Abdullah, R. Ahmad and J. Ahmad (1978). Pl. & Cell Physiol. (Tokyo) 19(1):99-106.

2. R. Ahmad and Z. Abdullah (1978). Irrigation Science (Germany; in press).

3. R. Ahmad and Z. Abdullah (1976). Utilization of brackish water for growing plants on sandy belts of Pakistan. Ist. Ann. Res. Rept. Pak. Sci. Found. SKU-Bio(13).

4. R. Ahmad and Z. Abdullah (1977). 2nd Ann. Res. Rept. Pak. Sci. Found. SKU-Bio(13).

5. G. Bryssine (1961). Experiments in irrigation with brackish water carried out in Moroko. Proc. Tehran Symp. Salinity Problem in Arid Zones. UNESCO. 245-249.

6. H. Boyko (1966). Salinity and Aridity. Dr. W. Jung, Publishers, Netherlands.

7. P.E. Chimiklis and E.P. Karlander (1973). Plant Phys. 51:48-56.

8. J.M. Cutler and D.W. Rains (1977). Crop Science 17:329-335.

9. J.G. Davis, M.N. Jones, C.Z. Lunney and G.M. Clark (1974). Pl. & Cell Physiol. 15:577-581.

10. J.H. Durand (1956). Sols Africains IV(3):53-88.

11. E. Epstein (1963). Desalination Res. Conf. Proc. N.A.S., N.R.C. Publ. 942:284-298.

12. J.H. Gale, C. Kohl and R.M. Hagen (1967). Physiologia Plantarum 20:408-420.

13. D.R. Hoagland and D.I. Arnon (1950). Calif. Agric. Expt. Stn. Circ. 347:1-32.

14. T.C. Hsiao, E. Fereres, E. Acevedo and D.W. Henderson (1976). Water and Plant Life, Springer-Verlag.

15. L.P. Lapina and B.A. Popov (1970). Soviet Pl. Physiol. 17:477-481.

16. J. Lunin, M.H. Gallatin and A.R. Batchelder (1964). Soil Sci. Soc. Amer. Proc. 28(4):551-554.

17. S. Maclachlam and S. Zalik (1963). Can. J. Bot., Part II, 41:1053.

18. N. Nelson (1944). J. Biol. Chem. 153:375-380.

19. J.H. Nola and H.N. Barber (1972). Search 3(7):259-360.

20. J.W. O'Leary (1971). Food, Fiber and the Arid Lands. University of Arizona Press, Tuscon.

21. M.G. Pitman (1970). Plant Physiol. 45:787-790.

22. D.W. Rains (1972). Ann. Rev. Plant Physiol. 23:367-88.

23. D.W. Rush, J.D. Norlyn and E. Epstein (1976). Crops and Soils Magazine 7-9.

24. B.P. Strogonov and V.V. Kabanov (1973). Plants in relation to waterlogging and salinity. Ann. Int. Source Book. FAO/UNESCO. Hutchinson & Co.

25. U.H. Sverdrup, W.J. Fleming and H.R. Fleming (1961). The Oceans, Their Physics, Chemistry and General Biology. Prentice Hall, N.J.

26. U.S. Salinity Laboratory Staff (1954). Diagnosis and improvement of saline and alkali soils. Agri. Handbook 60.

27. UNESCO (1960). Proceedings of Paris Symposium on the Problem of Arid Zones.

28. UNESCO (1961). Proceedings of the Tehran Symposium on Salinity Problems in the Arid Zones.

29. B.E. Yaron and Y. Vaadia (1973). Arid Zone Irrigation. Chapman & Hall Ltd., London.